W9-CRL-637

BIOPROCESS ENGINEERING

BIOPROCESS ENGINEERING

Kinetics, Mass Transport, Reactors and Gene Expression

WOLF R. VIETH

Department of Chemical and Biochemical Engineering
Rutgers University
Piscataway, New Jersey

A Wiley-Interscience Publication

JOHN WILEY & SONS, INC.

New York • Chichester • Brisbane • Toronto • Singapore

.596572X

CHEMISTRY

This text is printed on acid-free paper.

Copyright © 1994 by John Wiley & Sons, Inc.

All rights reserved. Published simultaneously in Canada.

Reproduction or translation of any part of this work beyond that permitted by Section 107 or 108 of the 1976 United States Copyright Act without the permission of the copyright owner is unlawful. Request for permission or further information should be addressed to the Permissions Department, John Wiley & Sons, Inc., 605 Third Avenue, New York, NY 10158-0012.

Library of Congress Cataloging in Publication Data:

Vieth, W. R. (Wolf R.)

Bioprocess engineering: kinetics, mass transport, reactors and gene expression / Wolf R. Vieth.

p. cm.

"A Wiley-Interscience publication."

Includes bibliographical references and index.

ISBN 0-471-03534-3 (cloth)

1. Biochemical engineering. 2. Biotechnology. 3. Genetic engineering. 4. Bioreactors. I. Title

TP248.3.V54 1994 93-44655

660'.63--dc20

Printed in the United States of America

10 9 8 7 6 5 4 3 2 1

TP 248
.3
V54
1994
CHEM

The author dedicates this book
to his faculty colleagues at Rutgers University
and to his research colleagues of the
Engineering Foundation Conference on Biochemical Engineering

FOREWORD

"In the practice of Chemical Engineering, the premium is on the man who can best apply a theoretical background and research aptitude to the formulation and solution of real problems." Much has changed since the time this statement first appeared.* At this stage, the author would like to offer an amended version, as follows:

"In the practice of Chemical and Biochemical Engineering, the premium is on the hybrid individual who can best systematize and apply a synthesis of chemical and engineering fundamentals, principles of biosciences, and a research aptitude, to the formulation and solution of real problems."

K. Venkatasubramanian in his Ph.D. thesis called attention to a quotation from Goethe, "Ein Gedanke kann nicht erwachen ohne andere zu wecken (A thought cannot arise without awakening others)." That is precisely the type of environment we have all been striving to create at Rutgers and to carry over into the Engineering Foundation Conference on Biochemical Engineering.

The appearance of this book is timed to coincide with the 25th anniversary of the biochemical engineering research program in the Rutgers Department of Chemical and Biochemical Engineering, as well as the gathering of research colleagues for Biochemical Engineering VII, inaugurated some fifteen years ago. Members of the Rutgers group chaired the first three biennial conferences and have maintained a lively interest and commitment ever since.

One of my students said a very kind thing recently, that reading my book, he could hear me speaking. With ongoing book projects, I have not had much time for attendance of technical meetings, other than the Biochemical Engineering Conference, for awhile. I hope my colleagues, especially those who read these books, will agree with my student, and understand my preoccupations.

*C. Selvidge and W. R. Vieth, Graduate Study at the School of Chemical Engineering Practice. MIT Publication Office (August, 1967).

PREFACE

As a microcosm, the evolution of the Rutgers program in Biochemical Engineering has parallelled the accelerated dynamics of the field globally.

In 1970, to focus our efforts, we embarked upon a long range research program in collagen technology for developing generalized biocatalyst/bioreactor systems. In the late 1970s we diversified and committed ourselves along the lines of alternate carriers (e.g., alginates) and reactors (e.g., fluidized beds) and, most importantly, to the use of recombinant cells.

We also made a commitment then to the Engineering Foundation biennial biochemical engineering conferences as a vehicle and as a testing ground for introducing our work along with that from many newly emerging laboratories. From these combined activities, I have enjoyed a rare opportunity to share and receive the best insights from my professional colleagues in participating in the development of the field.

The present work draws naturally on the results of research in our Rutgers laboratories, but even more so on the published results of seven Engineering Foundation conferences offered since 1978 and on many professional journal sources. In addition, it incorporates a state-of-the-art review and synthesis of a large body of the mushrooming current literature. I've tried to provide a lucid treatment of biokinetics and biocatalysis, and to show their integration with reactor concepts in bioprocesses, thereby tracing the rapid, recent evolution of biotechnology.

My goal is to provide the reader with a coherent synthesis of biochemical process systems. The book begins with simple enzyme and cell-based process kinetic models and then moves on to stress the kinetics of gene expression and product formation, with a unifying emphasis on operon concepts.

The book is intended for advanced undergraduate and graduate level students and professionals engaged in the field of biotechnology. Although written from an engineering standpoint, it should appeal to biochemists, microbiologists and food scientists as well.

In conjunction with my first two volumes* on membrane systems and on diffusion in and through polymers, this work completes and interrelates the trilogy. Increasingly we are reminded of the need to stimulate original thinking and dialogue across disciplinary lines. So most especially I have worked hard to ease the line separating large numbers of current chemical engineering practitioners from biochemical engineers.

Wolf Vieth

Piscataway, New Jersey
September, 1992

*W. R. Vieth, "Membrane Systems: Analysis and Design," John Wiley and Sons: New York, 360 pp. (1988).
W. R. Vieth, "Diffusion In and Through Polymers," Hanser Publishers: Munich; Dist. in U. S. by Oxford University Press: New York, 322 pp. (1991).

ACKNOWLEDGEMENTS

Publication of this manuscript marks completion of the fourth joint project by the author and his partner and wife, Peggy. (Actually, the eighth, counting our four children.) The first of these projects was the manuscript of the author's doctoral thesis, completed in 1961. She set a standard of cooperative effort at that time that has been unequaled in any work outside our partnership. In other words, I became a bit spoiled. But we enjoyed the activity so much that we resolved to resume it after our children had grown up, and did so in the form of these books.

Beginning about 1985, Peggy began mastering the uses of the computer and getting after me to begin, especially on a book dealing with bioengineering. This is no easy task, however, and I felt I had to work up to it a little more gradually, starting closer to my own technical foundations (the first two books). Meanwhile, Peggy began to wonder if this book would ever appear and began referring to it as "The Phantom of the Operon." At that point I had no choice but to proceed.

I am grateful to Peter Prescott and Dr. Ed Immergut, Hanser Publishers, who originated the project, for their assistance in the preparation of this book and to Professors Henrik Pedersen and Takishi Matsuura for their collaboration in research activities, which had an impact on the subject matter.

I wish to acknowledge the permission of the authors and publishers of the following journal and book to reproduce the portions of text specified: Ch. 10, pp. 303-318 and App. B. pp. 348-376, excerpts from pp. 245-298, J. M. Howell and W. R. Vieth, *J. Mol. Catal.*, Elsevier Sequoia S. A; and Ch. 2, pp. 24-29, excerpts from pp. 235-266, "Annual Reports of Fermentation Processes," Academic Press Inc.

CONTENTS

BIOPROCESS ENGINEERING

1

CLASSICAL BIOTECHNOLOGY

1.0 HISTORICAL PERSPECTIVE

Biotechnology had its origins when mankind first began to domesticate microorganisms by fermenting them in batch culture. The Congress of the United States, in 1984, defined Biotechnology as follows: "Commercial techniques that use living organisms, or substances from those organisms, to make or modify a product, including techniques used for the improvement of the characteristics of economically important plants and animals, and for the development of microorganisms to act on the environment." As Lee (1992) correctly points out, the area cannot be considered new. Indeed, fermentation processes have been practiced since early civilization and have been continually perfected with time, largely on an empirical basis. Historically, the technology evolved to the production of increasingly more valuable products such as antibiotics. More recently, the advent of recombinant DNA technology has dictated the development of novel processes for the production of totally synthetic products, such as monoclonal antibodies, as well as the scaling up of these processes, to such an extent that the incorporation of modern engineering concepts and methods in the development, design, and control of fermentation operations is becoming universal. Thus, biotechnological process developments are now parelleling and even fusing with developments in chemical technology (e.g., membrane separations in downstream processing).

"Fermentation" is no longer used in its original classic sense, but rather it is taken to mean "chemical reactions catalyzed by enzyme systems, which in turn are produced during the growth of microorganisms" (Gaden, 1955). The growth of a microorganism is a complex process and the transformation of a nutrient into a metabolic end product usually involves a large number of individual chemical reactions. This complexity of biological systems has, until the recent knowledge explosion brought about by systematic studies in molecular biology, contributed to delays in development in the areas of process modeling and optimization of fermentation operations (Rai, 1973).

As an outgrowth of earlier developments, the overall spectrum of activity which now exists can be conveniently represented in a bioconversion network, as shown in Fig. 1.1, which rather strikingly demonstrates the evolution of the field as it pushes outward from fermentation technology.

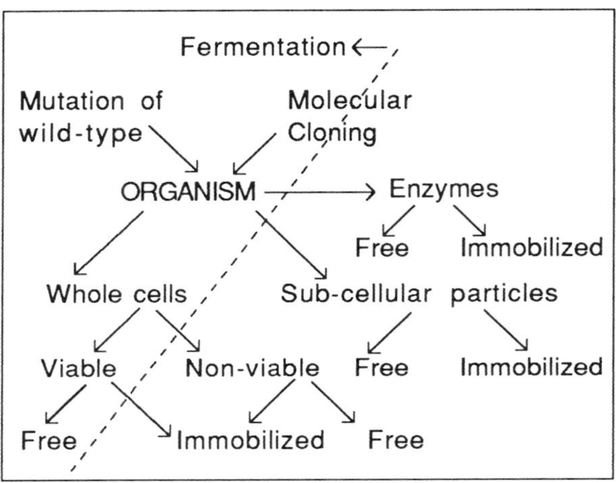

Figure 1.1 Network of bioconversion.

For instance, although only recently appearing on the scene, enzymatic processes *per se* have already found large scale industrial application in the food and pharmaceutical industries (Chibata et al., 1986) and a wide variety of smaller scale applications in medical diagnostics and therapeutics (Pedersen and Horvath, 1981; Bernath et al., 1976). Barker and Petch (1985) elaborate on the enzymatic process for high fructose corn syrup, while Tanaka and Fukui (1985) describe the bioconversion of lipophilic compounds by immobilized biocatalysts in the presence of organic solvents.

Chibata et al. (1986) list some examples of current large scale processes as follows:

- Production of L-amino acids from acetyl-DL-amino acids using immobilized aminoacylase.
- Production of 6-aminopenicillanic acid from penicillin G using immobilized penicillin amidase.
- Production of high fructose syrup using immobilized glucose isomerase.

- Production of L-aspartic acid using immobilized microbial cells.
- Production of L-malic acid using immobilized microbial cells.
- Production of L-alanine using immobilized microbial cells.
- Hydrolysis of lactose in milk using immobilized β-galactosidase.

Recently, Tosa et al. (1988) described an economically attractive slurry bioreactor-crystallizer scheme for the large-scale production of L-malic acid. In the crystallizer, a supersaturated product leaves solution, while remaining unreacted solid substrates dissolve. In this way, a substrate in slurry form is completely converted to a product in slurry form.

In scaled-down operations, Pedersen and Horvath (1981) point out that the most widespread early use of immobilized enzymes in analytical applications (e.g., medical diagnostics) is the employment of enzyme tubes in continuous-flow analyzers of the Technicon type. Highly active enzymic layers are deposited on the inner walls of plastic tubes, which are coiled into modular forms of open tubular heterogeneous enzyme reactors (dubbed "others"). Under the conditions of operation, the reactors are diffusion controlled, displaying "linear chemistry;" i.e., the pseudo first order kinetic conditions assure chemical analytical linearity.

Intermediate between fermentation and enzyme technology lies the immobilized whole cell regime. Examining the character of cells employed in fermentation, it is clear that they possess the desired enzymatic machinery in a highly structured form. The controlled conditions of fermentation permit retention of this structural integrity, but the resulting cellular suspensions are usually at low concentrations. It is possible to concentrate the free enzymes derived from these cells by extraction processes, but without the ancillary structure that stabilizes them in the cell, they are relatively unstable. Some structural reconstitution is possible by immobilization, which leads to higher concentration and better stability, but one is then constrained by economics to consider chiefly single step or two step reactions. With immobilized cells, one has the concentrated form: entire enzymatic pathways remain interconnected and viable, cellular microstructural characteristics are preserved and stabilized and the possibility of improved reactor design is opened up, based upon the characteristics of the carrier.

1.1 ROLE OF KREBS CYCLE IN PRIMARY METABOLITE SYNTHESIS

Salient examples of the above are afforded by careful examination of the tricarboxylic acid (TCA) cycle, when it is functioning in the following ways:

i. microbial respiration;
ii. production of monosodium glutamate;
iii. production of citric acid.

Aerobic cells obtain most of their energy by respiration, the transfer of electrons from organic fuel molecules to molecular oxygen (Fig. 1.2). The final common pathway into which all the fuel molecules - carbohydrates, fatty acids and amino acids - are ultimately degraded in catabolism is the Krebs tricarboxylic acid cycle. A large fraction of the energy released is through this pathway. The overall reaction catalyzed by the cycle may be considered as:

$$CH_3COOH + 2\,H_2O \rightarrow 2\,CO_2 + 8\,H \qquad [1.1]$$

Acetic acid enters by condensation with four-carbon oxaloacetic acid to form citric acid. For each complete turn of the cycle, two molecules of CO_2 are produced with regeneration of the consumed oxaloacetic acid. The oxidations in the cycle are carried out by certain nucleotides (NAD^+, $NADP^+$ and FAD^{++}). The electrons derived from intermediates of the cycle flow down a multimembered chain of electron-carrier enzymes of successively lower energy level until they reduce molecular oxygen, the ultimate electron acceptor in respiration. During this process, much of the free energy of these electrons is conserved in the form of the phosphate-bond energy of ATP, by the process known as oxidative phosphorylation. All the enzymes or enzyme-complexes catalyzing reactions of the tricarboxylic acid cycle, electron transfer and oxidative phosphorylation are located within the mitochondrion, a respiratory organelle in some microbial cells.

It has been known for some time that the process of oxidative phosphorylation can be uncoupled from electron transfer by addition of specific uncoupling agents. Some examples of these compounds are 2,4 dinitrophenol, dicumarol and carbonylcyanide phenylhydrazone. Carbonylcyanide phenylhydrazone is the most powerful compound and completely stops cells' metabolic activities. The mechanism of action of uncoupling agents is not known exactly but it is believed that uncouplers cause structural damage to the inner membrane of the mitochondrion, bringing about alteration of the proton gradient.

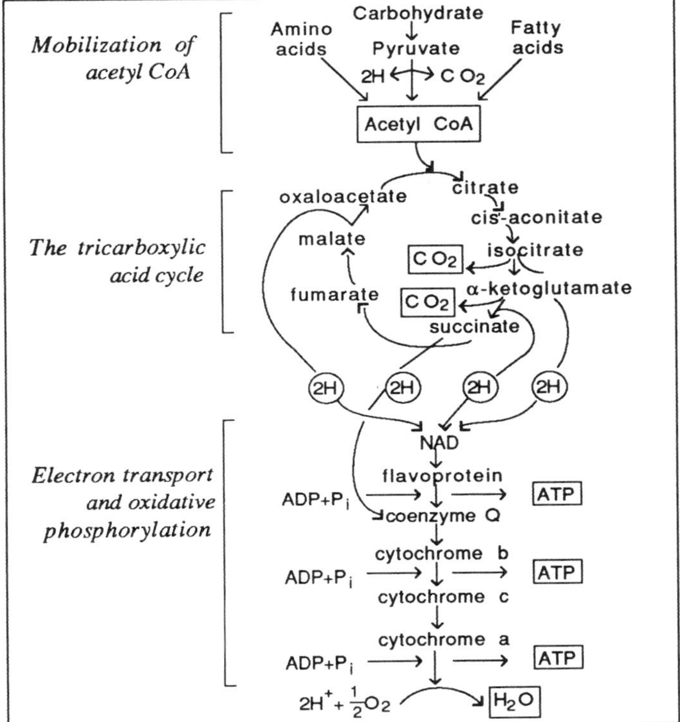

Figure 1.2 Flowsheet of respiration.

Induced microbial respiration refers to the increase in the rate of respiration of cells (i.e., oxygen uptake rate) when oxidative phosphorylation (energy conservation as ATP) is uncoupled from the process of electron transfer. Additionally, reactions involving ATP as energy source - such as synthesis of cellular material - are inhibited to a large extent.

Theoretically, one expects an optimum level of ATP (or uncoupler concentration) to exist just sufficient to carry out the catabolic reactions of the substrate without inhibition. Higher concentrations of ATP would cause feedback inhibition, and lower concentrations would result in exhaustion of the cells by not providing energy to consume the substrate. This phenomenon has been observed by Poe and Estabrook (1968) while measuring enthalpy of oxidations of succinate by the rat liver mitochondrion, and also analyzed in our laboratory (Fig. 1.3) in studies with intact yeast cells (Shah, 1975; Vieth, 1978).

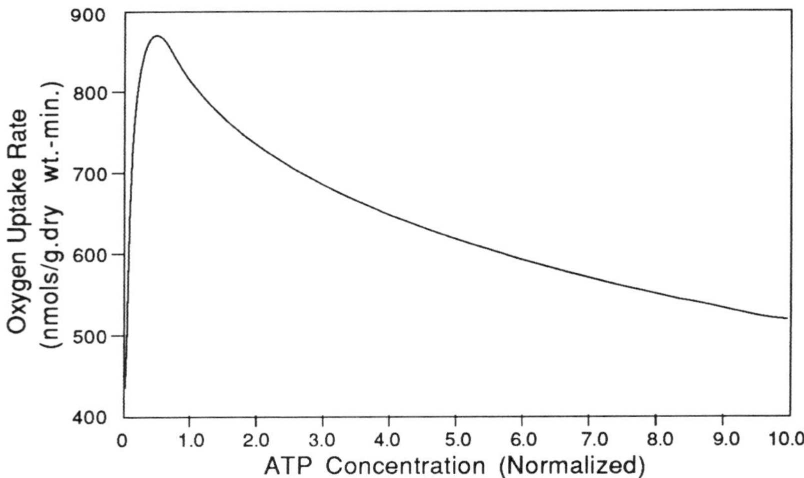

Figure 1.3 Effect of uncoupling on ATP level.

Galazzo, Schlosser and Bailey (1991) examined sensitivity of the flux from glucose to ethanol in nongrowing *Saccharomyces cerevisiae* to changes in the activities of different steps in the pathway based upon a previously presented kinetic model and measurements of intracellular concentrations. Genetic manipulation of the pathway to ethanol has been implemented by transforming *S. cerevisiae* AMW-13C with plasmid pGSF-D1.2 which contains the cloned *PFK1* and *PFK2* genes. These strains show the applicability of the kinetic model over different strains and the high sensitivity in this case of ethanol production to PFK activity.

When a plasmid is added, pathways are added; parallel pathways occur; e.g., ATP-consuming as well as producing steps. Additional message reading steps introduce time lags. In general, as the Sphinx is reported to have said to a delegation of visitors, "Don't expect *too* much!" Overall, perhaps a more effective strategy might be to engineer strains in which competing pathways are eliminated or susceptible to blockage. Recently, the glucose metabolization of an *E. coli* strain bearing mutations abolishing both acetyl phosphotransferase (PTA) and acetate kinase (ACK) activities was studied under aerobic and anaerobic conditions (Diaz-Ricci et al., 1991). These studies were conducted in a complex medium with the mutant carrying no plasmid and the mutant carring a plasmid bearing the "*pet*" operon that encodes *Zymomonas mobilis* pyruvate decarboxylase and alcohol dehydrogenase activities. Expression of the *pet* operon overcame the metabolic stress caused by the plasmid, enhancing growth and glucose uptake rates to the values observed in the plasmid-free mutant. Also, expression of the *pet* operon allowed consumption of pyruvate accumulation during the first hours of fermentation.

Malmberg et al. (1991) performed a detailed kinetic analysis of cephalosporin biosynthesis. Many secondary metabolic pathways are less subject to feedback regulatory events; thus, they are good models for engineering analysis. An example of such a pathway is the biosynthesis of β-lactam antibiotics. Sensitivity analysis revealed that δ-(L-α-aminoadipyl)-L-cysteinyl-D-valine (ACV) synthetase is the rate limiting enzyme. The effects of amplifying ACV synthetase on the specific production rate were analyzed theoretically. It was determined that increasing ACV synthetase enhances the production rate initially until deacetoxycepyhalosporin C (DOAC) hydroxylase becomes rate limiting.

Returning now to the production of primary metabolites, the stoichiometry of glutamic acid synthesis is represented as follows (Constantinides and Shao, 1981):

$$C_6H_{12}O_6 + 1.5\ O_2 + NH_3 \rightarrow C_5H_9O_4N + CO_2 + 3H_2O \qquad [1.2]$$

The overall reaction is depicted in eqn. [1.3]:

$$\begin{aligned} a\ C_6H_{12}O_6 + (b(1-g) + 1.5ag)\ O_2 + (c(1-g) + ag)\ NH_3 \rightarrow \\ (d(1-g))\ C_wH_xO_yN_z + (ag)\ C_5H_9O_4N \\ (e(1-g) + ag)\ CO_2 + (f(1-g) - 3ag)\ H_2O \qquad [1.3] \end{aligned}$$

where a, b, c, d, e, f are stoichiometric coefficients and g is the fraction of substrate used for glutamic acid formation.

Comparison of the predicted and assayed results for glutamic acid synthesis are shown in Fig. 1.4.

Within the fermentation cycle, Constantinides et al. (1981) showed that the specific activity of *Corynebacterium flavum* cells toward glutamic acid synthesis peaked at about 18 hours in fed batch culture (Fig. 1.5). Subsequently, the cells were harvested and immobilized on a collagen membrane carrier which was packed in chipped form into a lab-scale "flashlight" size bioreactor. As shown in Fig. 1.6, the glutamic acid producing ability of the cells was thereby stabilized and expressed for periods of five to nine days. Li et al. (1990) immobilized *Corynebacterium glutamicus* T6-13 on a complex of cellulose acetate and

"Eucheuma gel," obtaining an intragel concentration of 3.0 x 10^9 cells per ml of gel. Glutamic acid concentrations in the range of 5.0 to 6.0 g/L are reported; the bioreactor could be operated for up to 20 cycles of 36 hours each.

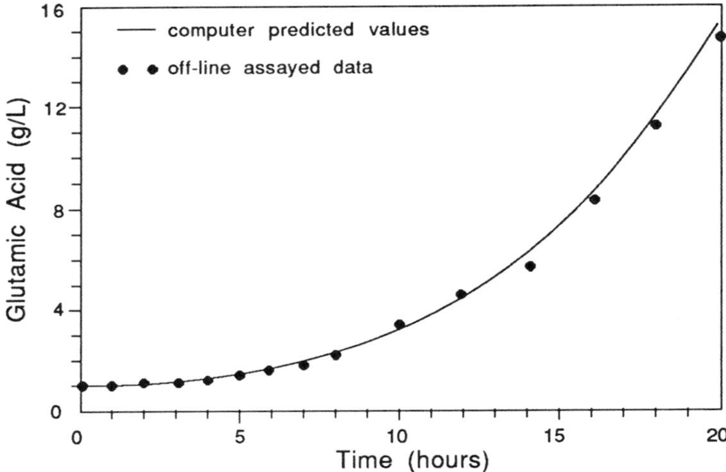

Figure 1.4 Comparison of the predicted and assayed results in glutamic acid.

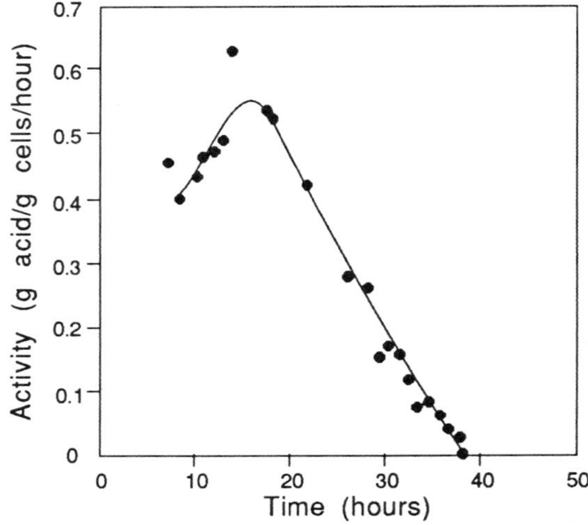

Figure 1.5 Specific activity versus time plot for constant glucose fermentation at 600 RPM.

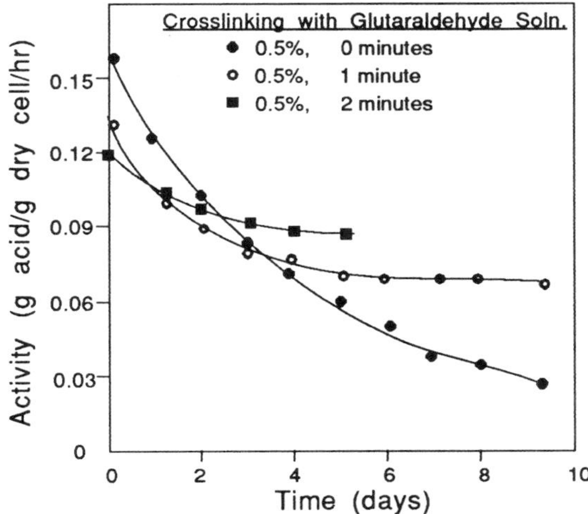

Figure 1.6 Specific activity versus time plots for continuous column reactor system.

In contrast to the case of continuous operation, as for instance the one just described, there are productivity interruptions associated with batch operation. Yet, in the opinion of Heijnen et al. (1992), continuous culture is still currently applied more as a research tool and less as a production process. Fundamental obstacles in continuous culture are discussed to help shed light on this apparent contradiction. Based on a discussion of technical, process-related, and economic/market bottlenecks, the authors conclude that the often mentioned productivity argument in favor of continuous processing is oversimplified. The optimal choice of a process mode is determined by a full understanding of the equipment and production plant factors and of the economic/market factors. Often the resulting choice will be a fed batch and/or the cell retention process mode, which is characterized by low growth rates. Therefore, more research toward product formation at low growth rates (<0.05/h) is needed.

Cells often retain a significant portion of their catalytic activity at the time of nutrient depletion. This catalytically active cell mass is ordinarily discarded upon termination of the fermentation and represents a significant cost in the process. It has been recognized for some time that these difficulties and inefficiencies might be overcome by using immobilized live cell bioreactors. Immobilization decouples the specific growth rate of cells and the dilution rate of the reactor. Thus continuous steady state or dynamic operation of the reactor is possible in a stable fashion. Immobilized live cells frequently offer better utilization of cellular catalytic activity along with reduction of nonproductive growth phases. This is especially advantageous in the synthesis of secondary metabolites, whose production is nongrowth associated. Additionally, they offer the possibility of continuous operation at high dilution rates without the problems of washout. Higher volumetric productivities are possible with increased cell densities and higher yields can be achieved with easier rheological control (Atkinson, 1986; Chibata and Tosa, 1980; Venkatasubramanian, 1979; Vandamme, 1983).

Immobilized cell bioreactors, however, have several operational problems of their own (Emery and Mitchell, 1986). These include the loss of mechanical integrity of the support matrix by dissolution, compaction, and abrasion. The disruption of reactor hydrodynamics due to gas holdup has been reported (Cho et al., 1982). One major problem for the long term operation of immobilized bioreactors is the loss of catalytic activity. The cause of this problem may be enzyme deactivation (Venkatasubramanian and Harrow, 1979) or, in the case of immobilized live cell reactors (Karkare, 1983), the loss of cells by leakage from the support structure or loss of viability. A constant supply of growth promoting nutrients would reduce the specific productivity for nongrowth associated compounds and could eventually also cause cell leakage.

The problem of decaying catalytic activity can be resolved by the use of sequential operation of multiple reactor systems, together with variable flow rates and temperatures (Venkatasubramanian and Harrow, 1979) or by periodic operation of a single reactor (Mehta, 1988).

By way of further illustration, let us consider now the case of a filamentous microorganism for the production of citric acid. To reiterate, cell immobilization is most effective when the cells are harvested at their peak activity in the fermentation. For citric acid production with *A. niger*, 72 to 96 hours are required to arrive at this peak activity (Fig. 1.7). In a batch fermentation process one would have to repeat this pattern each time. A better alternative, it would seem, would be to harvest the cells at their peak activity, then immobilize them so they remain viable for reuse until their stability has decreased to an uneconomical point.

One carrier of choice is the collagen membrane which we have studied extensively since the early 1970s. After loading with cells, the carrier must be crosslinked to make it strong enough to withstand the shear forces in normal contact. It was determined that crosslinking could best be carried out by immersing the collagen cell membrane in a 5% glutaraldehyde solution for one minute. Overall, immobilized whole cells contained within the collagen carrier system represented a significant departure from the collagen-enzyme technology which had already been developed in our laboratory, and, naturally, many questions arose (Vieth, 1979).

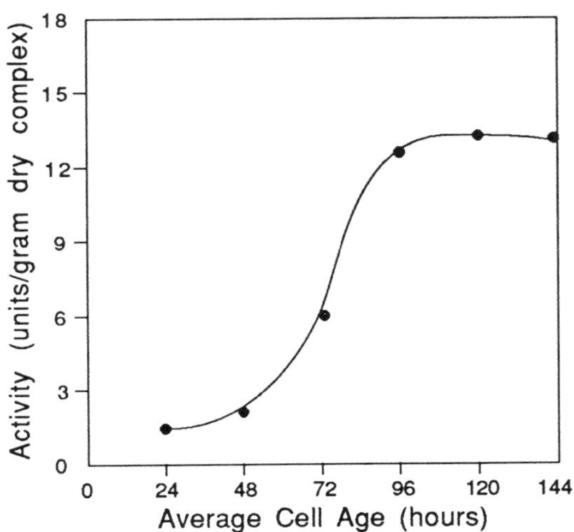

Figure 1.7 Results of batch fermentation.

How can the cells be retained in a viable state without further proliferation, with the attendant degradation of the collagen carrier? One way to accomplish this is to limit one essential nutrient, e.g., the nitrogen concentration (Fig. 1.8). When the concentration of nitrogen is as low as 0.3 grams per liter, it was found that additional cell growth did not occur to any significant extent.

Having a method of preparing controlled catalytic biomass, in what sort of configuration should it be employed? In other words, looking at this question on a scale-up basis, what are some of the important parameters? One approach in our laboratory is to form the membrane into a spirally wound reactor configuration, or "rolled-up fermentor." To do this, the collagen carrier is wound together with a polyolefin Vexar® spacer material. The resulting open multichannel system promotes plug flow contact with very low pressure drop even when operating with particulate substrate matter that would cause plugging problems in the conventional type of fixed bed operation. Fermentation substrates are often quite crude, so this is a plus factor for this type of design. Furthermore, it is possible to design-in high activity per unit volume as a result of the coiling of a large amount of membrane into a confined volume.

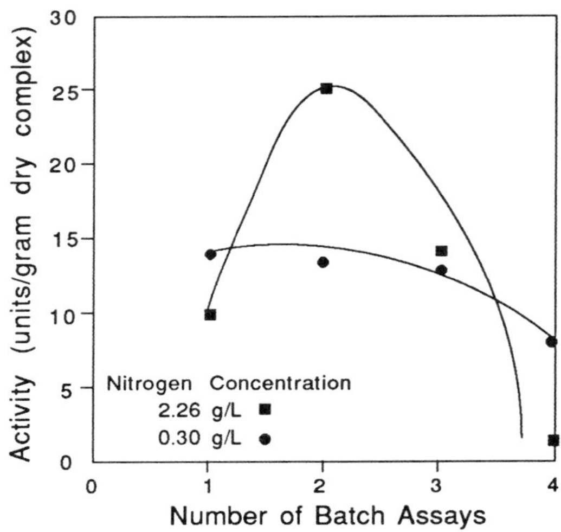

Figure 1.8 Effect of nitrogen limitation.

When examining the kinetic behavior of citric acid in this system, it was found that the activity profile as a function of substrate concentration is essentially linear; this meant that the reactor was operating in a pseudo first order regime.

With regard to other significant factors, oxygen transfer can be singled out as of paramount importance. The spirally wound multipore reactor configuration was used in all reactor studies. Since oxygen transfer is a crucial variable, a special provision was incorporated into the reactor design to allow flow of pure oxygen countercurrent to the flow of substrate; in this sense, the overall system operated as a combined absorber-reactor. Dissolved oxygen concentrations of 80 to 90% of saturation value were maintained throughout the course of the reactor runs.

Presented in Figs. 1.9 and 1.10 are data relating to external and internal mass transfer for the case of citric acid synthesis. The linear effect of fluid velocity on the observed reaction rate (Fig. 1.9) shows, for this case, the presence of a significant boundary layer resistance below a flow rate of 235 ml/min. The existence of nonnegligible pore diffusional resistance is deducible from Fig. 1.10, in which the dependence of observed reaction rate on film thickness is depicted.

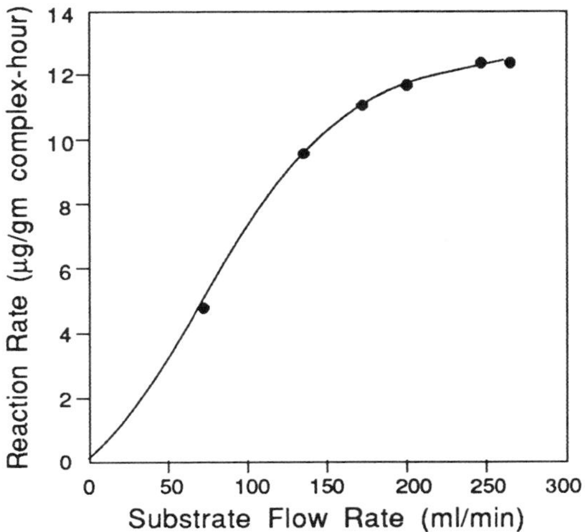

Figure 1.9 Dependence of reaction rate on linear velocity.

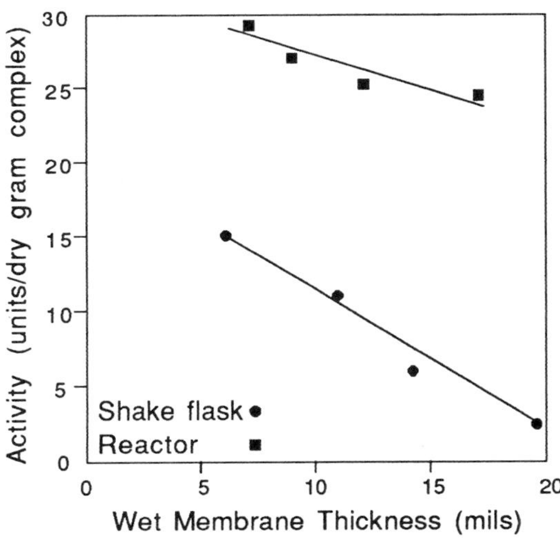

Figure 1.10 Effect of membrane thickness on citric acid production rate.

Overall, the immobilized cells exhibited about 50% of the specific activity of the free cells (in fermentation) toward the production of citric acid. Referring to Fig. 1.10, the substantially increased specific activity of the catalyst observed in the reactor compared to that in a shake flask is attributable, at least in part, to improved oxygen transfer in the reactor, especially when the catalyst elements are made thin. Thus, the simple, effective, flexible use of the membrane form in this type of reactor has demonstrated several positive features.

In any of these immobilized cell processes, the superiority of the heterogeneous form of the biocatalyst rests on three facts (Klein and Wagner, 1987):

i. catalyst retention, i.e., the ease of separation of the catalyst phase and the reaction medium, as well as the uncoupling of the respective characteristic process times (i.e., residence time and cell doubling time);

ii. biomass concentration, i.e., the high cell density in the matrix as the basis for high reactor productivities;

iii. stabilization of the catalytic function, i.e., obtainment of half-life times on the order of weeks and months.

1.2 SYNTHESIS OF SECONDARY METABOLITES

To carry the ideas outlined above a bit further, consider now the synthesis of an antibiotic, candicidin, along a secondary pathway. The employment of resting (i.e., immobilized) cells is potentially still more attractive in this case because of the increased unit value of the product. It was found that candicidin productivity could be increased several-fold by using a nongrowth medium; the catalyst productivity was higher than the free cell productivity as well. An analysis of the carbon conversion profile showed that during cell growth, a major portion of the substrate is routed towards growth, thus reducing the carbon conversion efficiency of the system (Karkare, 1983). The results are presented and analyzed later in this book, and shown in Tables 5.5 and 5.6.

It was also shown that IMC reactors could be operated for several days under nongrowth conditions with consistently high productivity. Furthermore, the catalytic activity of the reactors could be completely rejuvenated by inducing a brief period of cell growth. In the growth mode, the productivity per cell was found to be approximately constant (regardless of growth rate). Thus, the metabolism was shown to switch from growth to production and vice versa by adjusting the phosphate concentration in the medium. This metabolic switch concept appears to be an important operating strategy for IMC systems to increase the carbon conversion efficiency and productivity of secondary metabolite fermentations.

Through the method of protein engineering, a recombinant trypsin was recently designed, the catalytic activity of which can be regulated by varying the concentration of cupric ion in solution (Higaki et al., 1991). A computer program was developed to scan the three dimensional structure of trypsin to find pairs of amino acid residues that, when substituted with a histidine residue (His), could form a stable metal-binding site. Substitution of an arginine residue (Arg96) with a His in rat trypsin (trypsinR96H) places a new imidazole group on the surface of the enzyme near the essential active site, His 57. The unique spatial orientation of these His side chains creates a metal-binding site that chelates divalent first row transition metal ions. Occupancy of this site by a divalent first row transition metal ion prevents the imidazole group of His 57 from participating as a general base in catalysis. As a consequence, the primary effect of the bound metal ion is to inhibit reversibly the esterase and amidase activities of trypsinR96H. The apparent equilibrium constant for this inhibition is in the micromolar range for copper, nickel and zinc, the tightest binding being to cupric ion at 21μM. Trypsin R96H activity can be fully restored by removing the bound cupric ion with EDTA. Multiple cycles of inhibition by cupric ions and reactivation by EDTA demonstrate that reversible regulatory control has been introduced into the enzyme. The authors note that these results describe a novel mode of regulating the activity of a serine protease that may also prove applicable to other proteins. Many such examples of engineered proteins appear throughout this book; e.g., in the area of immediate utility, protein engineering of milk clotting aspartic proteinases is currently underway (Beppu, 1990). Production of active chymosin by recombinant *E. coli* resulted in the accumulation of polypeptides as inclusion bodies. Mucor renin suffered activity losses due to hyperglycosylation in excretion processes. However, these problems were solved and protein engineering is progressing.

ENGINEERING MODELS FOR OVERALL REACTIONS

It is obvious that simple growth kinetics and product formation models are not sufficient to describe the kinds of metabolic shifts that could occur due to nutrient changes and cell age differences. It is necessary to model the biochemistry in greater detail so that one can obtain a better insight into the nutrient changes necessary to bring about the desired metabolic shifts.

One pragmatic approach to the above problem is to develop simplified practical engineering models describing the biochemistry of the metabolic pathways. The idea is to lump chains of reactions into a single reaction concept and focus mainly on certain key intermediates that control the channeling of the substrate into different pathways. Figure 1.11 shows a simplified version of the biochemical pathways involved in the production of polyene macrolides (secondary metabolites) and cell maintenance. We can readily recognize acetyl CoA as one of the key intermediates. The fate of this intermediate can result in a shift in metabolism of the cells. Therefore, the activity of the various enzymes acting on this intermediate is of particular interest. Another possible way of shifting the metabolism is to control the maintenance energy requirement of the cells by attacking the sources of this requirement, such as osmotic pressure, protein turnover rate, pH, etc. (Karkare, 1983).

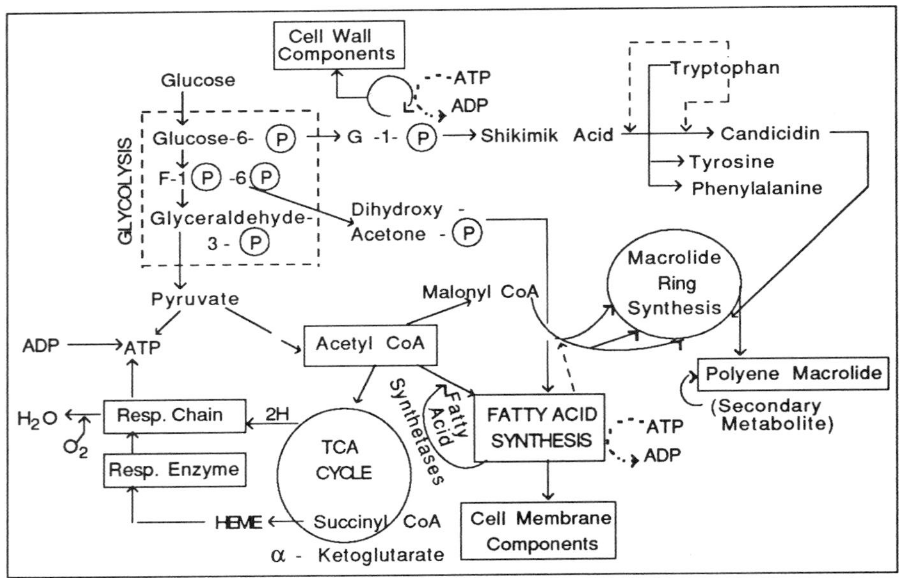

Figure 1.11 Biosynthesis pathways for secondary metabolites (e.g., candicidin and phenylalanine).

1.3 RECOMBINANT CELLS

To conclude this introduction, let us examine briefly the effect of molecular cloning on biocatalyst efficiency/pathway productivity for an inducible enzyme. Figure 1.12 shows steady state bioreactor performance in the production of the enzyme, β-galactosidase, from wild-type *E. coli*, growing on lactose (substrate/inducer) in a chemostat culture.

Note that maximal expressed activities of the product enzyme are a little less than 1,000 (units/per g dry cell). Contrast this with the results for a recombinant strain of *E. coli* MBM7061 (z^-, y^+), harboring the plasmid pMLB1108, which is *lac* i^q z^+, where a maximal activity of 5,000 in comparable units was produced (Vieth, 1988). The principal reason for this result is simply the incorporation of multiple gene copies. (See also discussion in Chapter 7.)

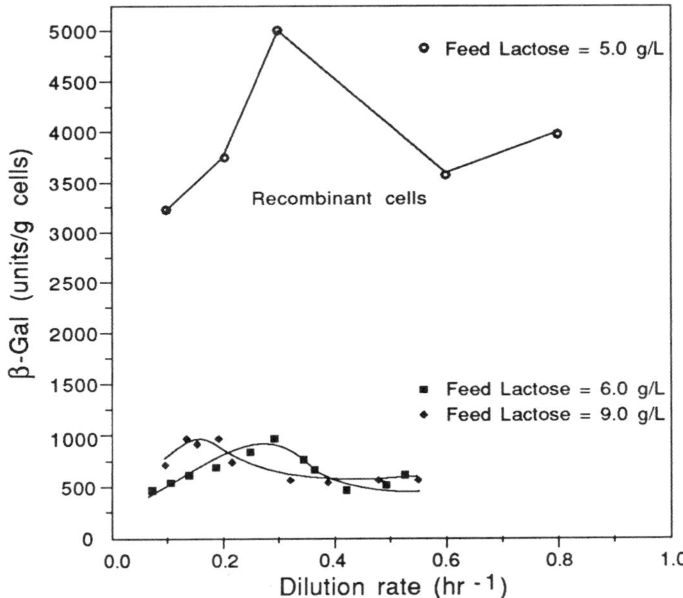

Figure 1.12 Effects of dilution rate and feed lactose concentration on steady state enzyme specific activity in chemostat cultures.

As with the above case, expression of the *araB* gene in *E. coli* is tightly regulated in a rather sophisticated manner: it is subjected to catabolite repression (through cAMP-CAP interaction) and inducer exclusion by glucose, as well as to the specific repression by the *araC* protein. (Process models for the interactions of catabolite repression and inducer exclusion are presented later in this book, in Chapters 7 and 8.)

In another study, both the *araBAD* promoter and the *araC* gene were subcloned from the plasmid pING1; several levels of expression (up to a thousand-fold the repressed level) can be achieved with this system, if advantage is taken of the regulation phenomena by using different types of carbon sources or their combinations (Simoes et al., 1992).

Schofield et al. (1988) note that the use of heat treatment to purify enzymes by selective denaturation and then by the subsequent precipitation of denatured protein is a simple, rapid and well established procedure. Successful applications are limited to those few enzymes that possess a thermostability considerably higher than the majority of cell proteins. The introduction of thermostable enzymes into the protein population of a mesophile by cloning offers a clear opportunity to employ a heat-treatment method of purification to its full advantage. In light of the difficulties involved in purifying bacterial cellulases, the cloning of some of the cellulase and hemicellulase genes of *Caldocellum saccharolyticum* into *Escherichia coli* provided a welcome alternative procedure for obtaining pure enzymes.

It would be inappropriate, however, to leave the impression that creative solutions are always this direct. Wang et al. (1990) point out that, although it is now possible to synthesize a variety of valuable biologically active secondary metabolites from microbial sources using protein-engineered enzymes, e.g., β-lactamase, product screening and recovery pose complex problems. The whole culture broth is an extremely heterogeneous aqueous mixture of biomass, nutrients, residual substrates and other additives. In some cases, the desired product is generally present in very low concentrations (<0.1 wt %) in this heterogeneous colloidal suspension. Also, bioproducts are usually unstable and are prone to physical, chemical or enzymatic degradation that limits the choice for separation and screening methods. Accordingly, the authors have developed the use of membrane-encapsulated affinity and other semispecific adsorbents for directly extracting the bioproduct from the crude solutions, thereby bypassing the solids removal step.

1.4 SCOPE

There is, of course, much that could be added to this book which would be relevant and useful. However, that would once again expand this project to several volumes, which is more than the author could hope to accomplish in the available time interval. The author apologizes in advance for any omissions.

There are current biotechnology meetings and congresses which will, in their proceedings, deal more effectively than the author could here with rapidly developing topics such as: protein post-translational modifications, advances in protein purification methodologies, protein folding and recovery, synthetic methods for production of biological macromolecules, treatment of hazardous wastes, and bioremediation of soils. In particular, the subject of downstream processing deserves a volume of its own. Coming to the fore are chromatographic techniques such as the following: perfusion chromatography, immunoadsorption chromatography, and fiber-based chromatography for protein purification; electrochemically-based techniques such as: affinity capillary electrophoresis, enhanced protein resolution in charged membranes through electrical means, and field flow fractionation in purification processes; and novel solvent-based methods such as: affinity reversed micellar protein purification and methods associated with intracellular and extracellular redox potentials in protein production.

1.5 REFERENCES

Atkinson, B., In "Process Engineering Aspects of Immobilized Cell Systems," C. Webb, G. M. Black and B. Atkinson, Eds., pp. 3-9. The Institution of Chemical Engineers: U. K. (1986).

Barker, S. A. and G. S. Petch, In "Enzymes and Immobilized Cells in Biotechnology," A. I. Laskin, Ed. Benjamin Cummings Publishing Co.: London (1985).

Beppu, T., In "Enzyme Engineering 10," H. Okada, A. Tanaka and H. W. Blanch, Eds. *Ann. N. Y. Acad. Sci.*, **613**, 14-26 (1990).

Bernath, F. R., L. S. Olanoff and W.R. Vieth, In "Biomedical Applications of Immobilized Enzymes and Proteins," p.351. Plenum Press: New York (1976).

Chibata, I. and T. Tosa, *Trends in Biochem. Sci.*, **4**, 88-90 (1980).

Chibata, I., T. Tosa and T. Sato, *J. Mol. Catalysis*, **37**, 1 (1986).

Cho, G. H., C. Y. Choi, Y. D. Choi and M. H. Han, *J. Chem. Technol. & Biotechnol.*, **32**, 959-967 (1982).

Constantinides, A., D. Bhatia and W. R. Vieth, *Biotechnol. Bioeng.*, **23**, 899 (1981).

Constantinides, A. and P. Shao, In "Biochemical Engineering," A. Constantinides, W. R. Vieth and K. Venkatasubramanian, Eds. *Ann. N. Y. Acad. Sci.*, **369**, 167 (1981).

Diaz-Ricci J. C., L. Regan and J. E. Bailey, *Biotechnol. Bioeng.*, **38**, 1318-1324 (1991).

Emery, A. N. and D. A. Mitchell, In "Process Engineering Aspects of Immobilized Cell Systems," C. Webb, G. M. Black and B. Atkinson, Eds., pp. 87-99. The Institution of Chemical Engineers: U. K. (1986).

Gaden, E. L., Jr., *Chem. Ind. (Rev.)*, 154 (1955).

Galazzo, J., P. Schlosser and J. Bailey, Abstract, Biochemical Engineering VII, Engineering Foundation Conf., Santa Barbara, CA, March 3-8 (1991).

Heijnen, J. J., A. H. Terwisscha van Scheltinga and A. J. Straathof, *J. Biotechnol.*, **22**, 3-20 (1992).

Higaki, J. N., B. L. Haymore, S. Chen, R. J. Fletterick and C. S. Craik, Abstract, Biochemical Technology Div., A C S Mtg., New York, Dec. (1991).

Karkare, S. B., Ph. D. Thesis, Dept. of Chemical and Biochemical Engineering, Rutgers University (1983).

Klein, J. and F. Wagner, In "Enzyme Engineering 8," A. I. Laskin, K. Mosbach, D. Thomas and L. B. Wingard, Eds. *Ann. N. Y. Acad. Sci.*, **501**, 306-317 (1987).

Lee, J. M., "Biochemical Engineering." Prentice Hall:Englewood Cliffs, NJ (1992).

Li, Y-F, Y. Huang, L-F. Ye, P. Sui and Q-Q. Wen, In "Enzyme Engineering," H. Okada, A. Tanaka and H. Blanch, Eds. *Ann. N. Y. Acad. Sci.*, **613**, 883 (1990).

Malmberg, L.-H., D. H. Sherman and W.-S. Hu, Abstract, Biochemical Engineering VII, Engineering Foundation Conf., Santa Barbara, CA, March 3-8 (1991).

Mehta, N., Ph. D. Thesis, Dept. of Chemical and Biochemical Engineering, Rutgers University (1988).

Pedersen, H. and C. Horvath, In "Applied Biochemistry and Bioengineering," **3**, p.1. Academic Press: New York (1981).

Poe, M. and R. W. Estabrook, *Arch. Biochem. Biophys.*, **126**, 320 (1968).

Rai, V., Ph. D. Thesis, Dept. of Chemical and Biochemical Engineering, Rutgers University (1973).

Schofield, L. R., T. L. Neal, M. L. Patchett, R. C. Strange, R. M. Daniel and H. W. Morgan, In "Enzyme Engineering," H. W. Blanch and A. M. Klibanov, Eds. *Ann. N. Y. Acad. Sci.*, **542**, 240-243 (1988).

Shah, P., Ph. D. Thesis, Dept. of Chemical and Biochemical Engineering, Rutgers University (1975).

Simoes, D. A, M. D. Jensen, E. Dreveton, M.-O. Loret, S. Blanchin-Roland, J.-L. Uribelarrea and J.-M. Masson, In "Recombinant DNA Technology I," A. Prokop and R. K. Bajpai, Eds. *Ann. N. Y. Acad. Sci.*, **646**, 254-258 (1992).

Tanaka, A. and S. Fukui, In "Enzymes and Immobilized Cells in Biotechnology," A. I. Laskin, Ed. Benjamin Cummings Publishing Co.: London (1985).

Tosa, T., M. Furui, N. Sakata, O. Otsuki and I. Chibata, In "Enzyme Engineering 9," H. W. Blanch and A. M. Klibanov, Eds. *Ann. N. Y. Acad. Sci.*, **542**, 440-444 (1988).

Vandamme, E. J., *Enzyme and Microbial Tech.*, **5**, 403-416 (1983).

Venkatasubramanian, K., "Immobilized Microbial Cells," Academic Press: New York (1979).

Venkatasubramanian, K. and L. S. Harrow, In "Biochemical Engineering," W. R. Vieth, K. Venkatasubramanian and A. Constantinides, Eds. *Ann. N. Y. Acad. Sci.*, **326**, 141-153 (1979).

Vieth, W. R., Abstract, Biochemical Engineering I, Engineering Foundation Conf., Henniker, NH (1978).

Vieth, W. R., In "Biochemical Engineering," W. R. Vieth, K. Venkatasubramanian and A. Constantinides, Eds. *Ann. N. Y. Acad. Sci.*, **326**, 1-7 (1979).

Vieth, W. R., "Membrane Systems: Analysis and Design." Hanser Publishers: Munich; Dist. in U. S. by Oxford University Press: New York (1988).

Wang, H. Y., M. Zhang and T. Imanaka, In "Enzyme Engineering," H. Okada, A. Tanaka and H. W. Blanch, Eds. *Ann. N. Y. Acad. Sci.*, **613**, 376-384 (1990).

2

ENZYME TECHNOLOGY

2.0 GENERAL CONSIDERATIONS FOR ENZYME SYSTEMS

Enzymes are protein biocatalysts which govern the rates of the many biochemical reactions occurring in living things. Unlike ordinary chemical catalysts, enzymes characteristically have the ability to catalyze a reaction under very mild conditions in neutral aqueous solution at normal temperature and pressure, and with very high specificity. However, enzymes cannot be considered ideal catalysts for practical application because they are relatively unstable, and usually cannot be used in organic solvents or at elevated temperature. As one of the methods employed to make enzymes more suitable for the production of useful compounds by biocatalysis, their immobilization has been extensively studied since the late 1960s, and several new techniques have emerged to stimulate the utilization of biocatalysts. One of these techniques amounts to self-immobilization in the form of whole cells colonized within carriers. Therefore, in the discussion that follows, results of some studies with immobilized whole cells are interwoven with those for immobilized enzymes.

The term "immobilized enzyme" refers to a system or preparation in which an enzyme is confined or localized in a relatively defined region of space. The enzyme can actually be attached by one of several mechanisms to a supporting solid surface or simply physically confined within a surrounding solid or liquid barrier. This is a rather simple definition of a catalyst system that can be easily separated from a reaction mixture, and can be used continuously in flow processes or repetitively in batch contacts as long as the enzyme preparation is active. Bu'lock and Kristiansen (1987) and Laskin (1985) provide useful overview treatments.

There are several ways of classifying the various types of immobilized biocatalysts that fall within the above definition. These include (a) the type of interaction responsible for immobilization, (b) the nature of the support, and (c) the nature of the resulting complex (Zaborsky, 1973; 1976a). The author recommends classification based on the nature of the interaction responsible for immobilization; i.e., via either chemical methods which depend on the formation of at least one covalent bond per molecule, or physical methods which utilize noncovalent bond formation, or simple entrapment. Sundaram and Pye (1974) recommend classifying immobilized enzymes as either entrapped or bound. The former group includes matrix-entrapped and microencapsulated enzymes, while the latter includes adsorbed, covalently bound and crosslinked enzymes. From a practical standpoint, the method of classification makes little difference since generally the same subgroups are recognized. Also, many of the more successful techniques have evolved as combinations of the basic methods.

By the early 1980s there existed in the literature over 1,400 reports concerning the immobilization and characterization of more than 200 different enzymes. Well over 100 different supports had been utilized, and possibly an equal number of specific immobilization techniques had been applied. The latter are usually divided under the general methods listed previously. The large number of carriers that have been used are often broadly placed into two groups, i.e., organic or inorganic (Messing, 1975). Organic supports are often classified as vinyl polymers, polysaccharides, proteins or polyamino acids and polyamides (Goldstein and Manecke, 1976). Other useful subclassifications include natural or synthetic, neutral or polyelectrolyte and porous or nonporous (Bernath et al., 1977).

Included among inorganic carriers are glass, alumina, clays, sand, colloidal silica, stainless steel, metal oxides, carbon, ceramics and a host of others. Modified hydrophobic silica gels (e.g., C_{18}-silica gel), which are common packing material for HPLC columns, have been employed as supports for enzyme immobilization (Harada et al. 1990). Goetz et al. (1991) recently developed a versatile magnetic silica support which can be derivatized readily for both adsorption chromatography and enzyme immobilization for use in magnetically stabilized fluid bed applications. The synthesis of branched cyclodextrine was carried out very effectively utilizing pullulanase immobilized on a porous ceramic carrier, prepared by sintering smectite at high temperatures (Yoshida et al. 1990a; b). The stability of immobilized pullulanase was dependent not only on the operating temperature but also the substrate concentration. Highly concentrated substrate was preferable for the pullulanase stability. Kawase et al. (1989) carried out the immobilization of invertase on sepiolite and claim several advantages over immobilization on ceramics of the prior art. The immobilized enzyme was found to be nearly 30-fold more active than invertase immobilized on cordierite.

An additional method of grouping immobilized enzymes is, of course, by the type of reaction the enzyme catalyzes. Here the classification of immobilized enzymes parallels that of their soluble counterparts, i.e., all enzymatic reactions fall into one of six different categories: oxidoreductases, transferases, hydrolases, lyases, isomerases or ligases.

Luck and Bauer (1989) applied immobilized lipases for the interesterification of edible fats. The influence of internal and external mass transfer limitations on the conversion of the reaction was tested in stirred tank reactors and fixed bed reactors and it was found that the reaction rate is influenced by internal diffusion effects. Likewise, the *Candida rugosa* lipase-catalyzed esterification of decanoic acid with glycerol is described for an emulsion system and for a hydrophilic membrane bioreactor (Van der Padt et al., 1990). The activity is a function of the enzyme load. The optimum load in a hydrophilic hollow fiber membrane reactor is one to three times the amount of a monolayer, while in an emulsion system, several times this amount. This could indicate that in the emulsion system the adsorption is in a dynamic state while at the membrane surface the adsorption had reached its equilibrium state.

An augmented reactor, consisting of a stirred vessel, a hydrophilic membrane loop and a hydrophobic membrane loop, is described for the continuous enzymic hydrolysis of soybean oil in an emulsion (Pronk et al., 1991). An important advantage of this system is that it combines the high surface area characteristic of an emulsion with the containment of lipase in a membrane reactor. The composition of the emulsion appears to influence the flux of the membranes; the flux of the hydrophobic membrane increases with an increasing oil fraction of the emulsion, while the flux of the hydrophilic membrane has an optimum for two different oil fractions, 0 and 0.55 (by volume).

Woodley et al. (1991) measured substrate transfer rates from organic to aqueous phases in the presence and absence of biocatalyst in a reaction medium, using modified Lewis cells. These measurements, in combination with intrinsic aqueous phase biocatalytic reaction kinetics, were used to confirm that benzyl acetate hydrolysis by pig liver esterase and toluene oxidation by a strain of *Pseudomonas putida* occur uniformly throughout the bulk of the aqueous phase. Such data may be used to provide a basis for two-liquid-phase biocatalytic reactor design.

As an example of the hydrolase reaction, continuous resolution of acetyl-DL-methionine and acetyl-DL-phenylalanine was carried out rather recently in an ultrafiltration membrane reactor containing kidney or mold aminoacylases (Leuchtenberger et al., 1984). Also, wheat germ acid phosphatase was immobilized in κ-carrageenan gel, followed by crosslinking with glutaraldehyde, and used for the preparative resolution of O-phospho-DL-threonine (Scollar et al., 1985). Whole cells of *Pleurotus ostreatus* immobilized by entrapment into chitosan have been used for the production of 6-amino penicillanic acid from penicillin V (Kluge et al., 1982). Barenschee et al. (1989) describe an integrated process for the production and biotransformation of penicillin G in a liquid membrane system containing penicillin acylase in the aqueous phase. Factors affecting permeation rates and membrane stability were analyzed. In a related contact mode, the production of aspartame precursor in a semicontinuous pulsed extraction column bioreactor retaining free enzyme in an aqueous phase is reported by Hirata et al. (1991). The reactor made it possible to use free enzyme continuously and to carry out stable production and extraction by pulsation.

In the early phases, the bulk of the work conducted in the field had primarily involved hydrolases and simple isomerases. It is clear that when the full potential of immobilized enzymes is realized, all six categories will be amply represented. Other useful classifications under this heading include single or multienzyme systems, and cofactor or noncofactor requiring enzymes. Yet, there exist relatively few comparative studies aimed at making a critical evaluation of the many different techniques available for preparing immobilized enzymes.

It is the purpose of this chapter to review the current status of enzyme technology and to enunciate and discuss its principles as they have evolved over the last several decades.

2.1 HISTORICAL PERSPECTIVE

In order to comprehend the current status of immobilized enzyme technology one needs to have an appreciation of its roots. Goldstein and Katchalski-Katzir (1976) provided an excellent detailed survey of the field from its rather diffuse and unspectacular beginnings to its status as a sophisticated technology with the potential for dramatically affecting future life styles. The first immobilized enzyme was probably prepared by Nelson and Griffin (1916) over seventy five years ago. The first group of scientists to make use of this phenomenon was the immunologists who initially adsorbed and later covalently bound antigens to solid supports for the isolation of specific antibodies (Isliker, 1957). This period, which spanned the late 1930s to the early 1950s, saw the development of some important early techniques including the chemical modification of available carriers, such as polystyrene and cellulose, for the purpose of improving immunoadsorbent preparations. These same modified carriers were applied to enzyme immobilization in the mid 1950s with little success, probably due to their hydrophobic nature. Shortly thereafter, however, in the early 1960s, a number of carriers with varying degrees of hydrophilicity were developed which produced conjugates having higher levels of bound protein and much improved activity and stability characteristics. This important early work has been summarized in reviews by Manecke (1962), Katchalski (1962) and Silman and Katchalski (1966).

These early workers, for the most part biochemists, biophysicists and other basic scientists, began to appreciate that immobilized enzymes could serve not only as highly specific water-insoluble reagents, but also as models for bound enzymes in living systems, and perhaps even as a new class of highly specific and efficient industrial catalysts. From the mid 1960s, research in the field proceeded along at least three parallel paths. It was clear by this time that the chemical and mechanical nature of the support material was extremely important in determining the characteristics of the immobilized enzyme, and some workers began attempts to predesign carriers in order to achieve optimal binding and enzyme stability. Simultaneously, others sought to develop milder, more general techniques as an alternative to covalent methods (Tosa et al., 1966; Berfield and Wan, 1963; Chang, 1964). As experimental data were collected for these many different carrier systems, it became apparent that the behavior could vary considerably for the same enzyme on different carriers. It was discovered that this variable behavior could be predicted, even if no intrinsic changes occurred in the enzyme molecule during immobilization, by so-called "microenvironmental effects." Hence began intensive experimental and theoretical analyses of partitioning effects (electrostatic and physicochemical) and mass transfer effects (Nernst layer and pore diffusion), coupled with enzymatic reaction (Goldman et al., 1971; Katchalski et al., 1971). The third area of development was the study of immobilized enzymes in continuous flow packed bed and continuous stirred tank reactors (Lilly et al., 1966; Lilly and Sharp, 1968). This marked the initial contributions of chemical engineers who could apply their knowledge and experience in the design and analysis of continuous heterogeneous catalytic reactors.

By the end of the 1960s, several of the fundamental principles of immobilized enzyme technology had been developed and emphasis began to shift towards utilization in practical applications. The beginning of the new decade saw a tremendous increase in research activity in the field as reflected by the large number of publications that resulted (O'Driscoll and Mercer, 1976). By 1970, there had accumulated perhaps 300 papers in the literature concerning enzyme immobilization. By 1975, the number had increased to approximately 1,000. Unfortunately, this period was characterized to a degree by a lack of any specific direction. Criteria for evaluation and comparison of various systems and methods were conspicuously absent, and critical analysis was often overlooked. Nevertheless, a considerable amount of experience with immobilized enzyme systems was accumulated during this time as researchers continued to explore new carriers and methods, new enzymes, combinations of enzymes in multienzyme systems, cofactor regeneration, reactor design and analysis and other important aspects.

Shortly after this flurry of activity began in the early 1970s, many began to recognize the need for more organization and specific direction in the field. In August 1971, the first Enzyme Engineering Conference was held in Henniker, New Hampshire, for the purpose of bringing together researchers who shared the interests and goals of developing immobilized enzyme tehnology, to discuss the status of the field and present guidelines for its future. The meeting, as a reflection of the field itself, brought together chemical engineers, organic and physical chemists, biochemists, biologists, microbiologists and food scientists, all contributing their own special expertise to the common goal of utilizing enzyme technology and making its potential a reality. This combination of disciplines with a common interest defined a new field, "enzyme engineering," of which immobilized enzymes was a major but not the sole component. Identification of new sources for enzymes, isolation and purification of enzymes, stabilization of enzymes and other aspects important in the overall develpment of practical applications became an integral part of the field.

Initially, the most active researchers and major proponents of this new technology were from the academic community, and much of the work was supported by the National Science Foundation under its RANN (Research Applied to National Needs) program. From 1971 to 1975, the NSF granted 93 awards totaling approximately $8.6 million (Zaborsky, 1976b), which contributed greatly to advancement of the field. As work progressed, however, and as more conferences and symposia were held (Enzyme Engineering II, 1973, and III, 1975; Gordon Conferences in 1974 and 1976; and many others) at which researchers could touch base, evaluate new data and re-evaluate former guidelines, the field gradually gained organization, direction and credibility. Industrial participation both in research and in the public gatherings increased continuously; criteria for evaluation of systems were developed; economic evaluations were conducted; pilot plant studies were initiated; and, in general, the field became focused on the actual attainment or specific practical applications for immobilized enzymes.

By the late 1970s, attention was shifting to bioprocess applications involving controlled catalytic biomass in predesigned bioreactor configurations. The Biochemical Engineering Conferences were begun in 1978, in parallel and alternating with the Enzyme Engineering Conferences, to address these issues. Consequently, literature concentrations reflect this shift, with a heavier emphasis on the enzyme technology of the 1970s giving way a bit to the biochemical engineering literature of the 1980s and 1990s.

2.2 COENZYMES AND MULTIENZYMES

IMMOBILIZED COENZYMES

Some of the more dramatic applications of enzyme technology occur when the more complicated cofactor-requiring enzymes are utilized. An important group of cofactors are the stoichiometric coenzymes such as NADH(H) and NADP(H), ATP, ADP, AMP and coenzyme A, which actually act as cosubstrates in that their activity must be regenerated after involvement in the enzyme reaction. The use of these coenzymes in practical applications has been limited by at least three problems. First of all, the coenzyme must retain its diffusibility since it acts as a group carrier. Secondly, the coenzyme must be maintained within the system in order to prevent the continuous requirement of an expensive compound. Finally, in order to continuously reuse a given amount of coenzyme, it must be regenerated within the reactor system.

Miyawaki et al. (1990) describe an affinity chromatographic reactor which utilizes the dynamic affinity between immobilized enzymes and the free coenzyme, leading to dynamic immobilization of the latter. Beijer et al. (1990) have developed an Amber molecular mechanics model to investigate the interactions of NAD^+ with horse liver alcohol dehydrogenase (HLADH). The authors found that the reactivity of the coenzyme derivatives can be directly related to their analog geometries in the ternary complex. Larsson et al. (1990) concluded that the inactivation of HLADH in microemulsions corresponds to a progressive loss of functional sites, whereas the properties of the remaining functional sites are unchanged. Yomo et al. (1990) prepared redox enzyme system mimics based on conjugates of 5-ethylphenazine-, polyethylene glycol and NAD, which show the rate acceleration effect of the linking of two reactants by PEG. Kulbe et al. (1987) prepared PEG-coupled N(6)-succinyl-NAD(H) and demonstrated the feasibility of its functional efficacy with glucose dehydrogenase and mannitol dehydrogenase in a membrane reactor.

Brevibacterium ammoniagenes cells having high polyphosphate NAD kinase activity were immobilized into a polyacrylamide gel. By using a column packed with the immobilized cells, highly pure NADP was continuously produced from NAD and metaphosphate, the phosphoryl donor, in high yield. In addition, *Saccharomyces cerevisiae* cells for high ATP-regenerating activity and *B. ammoniagenes* cells for high NAD kinase activity were co-immobilized by entrapment in polyacrylamide gel (Murata et al., 1981) or by microcapsulation with cellulose acetate butylate (Ado et al., 1980), and the co-immobilized cells were also used for production of NADP.

IMMOBILIZED MULTIENZYME SYSTEMS

The immobilization on the same carrier matrix of two or more enzymes that catalyze sequential sets of reactions has generated considerable interest both from the standpoint of representing actual *in vivo* multienzyme systems and also for the development of practical processes. Although the basic carriers and immobilization techniques are no different than for single enzymes, there are a number of important considerations that are unique to multienzyme systems. For example, an extremely important parameter is the molar ratio of one enzyme to another on the carrier matrix. Another important consideration is the determination of the relationship between the molar ratio of enzymes, the mass transfer resistances in the reactor and conversion to the final product (Fernandes et al., 1975). Multienzyme systems have been reviewed by Mosbach and Mattiasson (1976) and Vieth et al. (1976). Experimental data for some representative multienzyme systems that have potential for practical application have been provided for the glucose oxidase-catalase system (Bouin et al., 1976), for a β-amylase-pullulanase system (conversion of starch to maltose) (Martensson, 1974; Nakajima et al., 1990) and for a β-galactosidase-glucose oxidase system (Bjorck and Rosen, 1976). Srere et al. (1973) and Mattiasson and Mosbach (1971) have described the preparation of immobilized 3-step enzyme systems.

Bülow and Mosbach (1987) used gene fusion techniques in the preparation of bifunctional enzyme complexes which catalyze the first two steps in lactose metabolism. Their strategy included construction of a protein with a thiol-containing tail, which is allowed to covalently attach to a chromatographic support for ease of purification.

A direct approach to obviating the costs of multienzyme reconstitution consists in the employment of immobilized whole cells. Whole cells of *C. simplex* have been entrapped into collagen membranes for the production of prednisolone from hydrocortisone (Constantinides, 1980). Similarly, the two-step bioconversion of cortexolone to prednisolone was successfully achieved by the combined use of *Curvularia lunata* mycelia and immobilized *Arthrobacter simplex* cells (Omata et al., 1980; Mazumder et al., 1985). Immobilized living mycelia of *C. lunata* having high 11β-hydroxylation activity were prepared by spores entrapped in photo-crosslinkable resin. Transformations of hydrocortisone (Koshcheenko et al., 1981) and of progesterone (Sonomoto et al., 1982) were carried out by *Arthrobacter globiformis* cells together with spores of *Aspergillus ochraceus*, immobilized in polyacrylamide.

Several steroid transformation reactions have been performed in organic solvents using immobilized microbial cells. *Arthrobacter simplex* cells immobilized with urethane polymer and *Nocardia rhodocrous* cells immobilized with photo-crosslinkable resin were used in non-aqueous solvent for conversion of hydrocortisone and testosterone (Yokozeki et al., 1982). As another example of the oxidoreductase reaction, the pharmaceutical intermediate dihydroxyacetone was obtained via the oxidation of glycerol using *Gluconobacter oxydans* cells immobilized with calcium alginate (Adlercreutz et al., 1985).

In general, Chibata (1986) has supplied an excellent compendium of immobilized whole cell (or enzyme) biocatalytic processes. Elements of it are shown in Table 2.1

Table 2.1 Application of Immobilized Biocatalysts for Production of Useful Compounds

Enzymes or microbial cells	Matrices for immobilization	Products	References
Oxidoreductase Reactions:			
Arthrobacter globiformis (D^1-dehydrogenase)	polyacrylamide	prednisolone	Koshcheenko et al., 1981
Arthrobacter simplex (D^1-dehydrogenase)	photo-crosslinkable resin	prednisolone	Mazumder et al., 1985
Aspergillus niger (glucose oxidase)	calcium alginate	gluconic acid	Linko, 1981
Aspergillus ochraceus (11α-hydroxylase)	polyacrylamide	11α-hydroxy-progesterone	Koshcheenko et al., 1981
Brevibacterium fuscum (reductase)	carrageenan	12-ketochenodeoxycholic acid	Sawanda et al., 1981
Candida tropicalis (oxidase)	carrageenan	α, ω-dodecanedioic acid	Yi & Rehm, 1982
		α ω-tridecanedioic acid	Yi & Rehm, 1982

Table 2.1 (Cont.)

Chlorella vulgaris and Anacystis nidulans (L-amino acid oxidase)	agarose	α-keto acid	Wiskström et al., 1982
Cornybacterium sp (9α-hydroxylase)	photo-crosslinkable resin	9α-hydroxy-4-androstene-3, 17-dione	Sonomoto et al., 1983
Curvularia lunata (11β-hydroxylase)	photo-crosslinkable resin	hydrocortisone	Mazumder et al., 1985
Flavobacterium sp (halohydrin epoxidase)	polyacrylamide	propylene oxide	Niedleman & Geigert, 1983
Gluconobacter oxydans (glycerol oxidase)	calcium alginate	dihydroxyacetone	Adlercreutz et al., 1985
Nocardia rhodochrous (Δ^1-dehydrogenase)	photo-crosslinkable resin	4-androstene-3, 17-dione	Yokozeki et al., 1982
Rhizopus stolonifer 11α-hydroxylase)	photo-crosslinkable resin	11α-hydroxy-progesterone	Sonomoto et al., 1982
Transferase Reactions:			
Arthrobacter oxydans	polyvinylalcohol	FAD	Yamada et al., 1980
Enterobacter aerogenes (transglucosylase)	photo-crosslinkable resin	adenine arabinoside	Yokoseki et al., 1982
Hydrolase Reactions:			
acid phosphatase	carrageenan	L-threonine	Scollar et al., 1985
aminoacylase	ultrafiltration membrane	L-phenylalanine	Leuchtenberger et al., 1984
		L-methionine	Leuchtenberger et al., 1984
Bacillus sp (Hydantoinase)	polyacrylamide	D-α-phenylglycine	Yamada et al., 1980
		D-α-p-hydroxy-phenylglycine	Yamada et al., 1980
Nocardia sp (nitrilase)	polyacrylamide	acrylamide	Watanabe et al., 1983
Pleurotus ostreatus (penicillin amidase)	chitosan	6-APA	Kluge et al., 1982
Lyase Reactions:			
aspartase	Duolite A-7	L-aspartic acid	Kimura et al., 1981
aspartase & L-aspartate 4-decarboxylase	ultrafiltration membrane	L-alanine	Jandel et al., 1982; Wandrey et al., 1982
Brevibacterium flavum (fumarase)	carrageenan	L-malic acid	Takata et al., 1980
Escherichia coli (aspartase)	polyurethane	L-aspartic acid	Fusee et al., 1981
Escherichia coli (aspartase) & Pseudomonas dacunhae (L-aspartate 4-decarboxylase)	carrageenan	L-alanine	Tosa et al., 1984
Escherichia coli (tryptophan synthase)	chitosan	L-tryptophan	Verlop & Klein, 1981
Pseudomonas dacunhae (L-aspartate 4-decarboxylase)	carrageenan	L-alanine	Yamamoto et al., 1980
Thermusrubens nov. sp. (fumarase)	Duolite A-7	L-malic acid	Ada et al., 1982

2.3 KINETIC BEHAVIOR OF IMMOBILIZED ENZYME SYSTEMS

INTRODUCTION

As mentioned earlier, the expressed activity and kinetic parameters of an immobilized enzyme are highly dependent on such factors as the physical dimensions and chemical nature of the carrier, the mixing or flow characteristics in the reaction system and the concentrations of substrates, products and effectors in the surrounding solution or macroenvironment. As such, these properties have often been labeled as "apparent" parameters since their values depend so strongly on reaction conditions. Unfortunately, "apparent" rate parameters are of little value unless there is some knowledge and understanding of how they are affected by important environmental variables. A considerable effort has been directed toward the understanding and modeling of the heterogeneous nature of immobilized enzyme catalysts in order to improve the analysis and use of rate data from these systems.

Due to the heterogeneous nature of bound enzymes, the catalytic reaction itself is only one in a sequence of important steps that result in the overall transformation of substrate to product. First, the substrate, as well as ions, inhibitors and cofactors, must diffuse through the stagnant fluid layer surrounding the solid carrier (external diffusion). Once at the surface, affinity or repulsive effects of a chemical or electrostatic nature may cause a partitioning of these species. This step is followed by reaction with surface bound enzyme or simultaneous diffusion (internal) and reaction within the carrier matrix. Finally, the products must diffuse through the carrier and stagnant fluid layer to the bulk solution surrounding the catalyst. With these steps in mind, a number of factors have been identified as affecting the observed or apparent kinetics of bound enzymes; i.e., factors which may cause the bound enzyme to behave much differently than its soluble counterpart. These factors can usually be grouped under two general headings which are: (1) effects of immobilization on the enzyme molecule itself, and (2) effects on the microenvironment of the bound enzyme. Under the former heading we can consider such factors as conformational changes, matrix interactions and steric effects which may affect not only the rate of reaction but also the enzyme's catalytic mechanism, specificity, and stability. The second group of factors, often called "microenvironmental effects," includes electrostatic and chemical partitioning effects and external and internal diffusional resistances which result in differences between the macro- and microenvironmental concentrations of substrates, products or other effectors.

By far the greatest attention has been given to the interaction of enzymic reaction and diffusional resistances. It has been shown that mass transfer limitations can cause alterations in the apparent temperature dependence of the bound enzyme (Buchholz and Ruth, 1976); in the apparent variation of reaction rate with substrate, product, buffer and inhibitor concentrations (Engasser and Horvath, 1976); and even in the apparent operational stability of the enzyme (Bailey and Ollis, 1986). All of these effects are reversed at high substrate concentrations. It is obvious that in interpreting kinetic data for design purposes, it is extremely important to uncouple mass transfer effects from reaction effects, since these two phenomena are affected differently by environmental conditions. In order to facilitate the theoretical treatment of mass transfer effects, Engasser and Horvath (1976) suggest that the terms intrinsic, inherent and effective rates be used in place of the traditional "apparent" activity. The intrinsic kinetic parameters of the bound enzyme refer to its behavior at the substrate, product and effector concentrations of the macroenvironment. The intrinsic parameters of the bound enzyme can differ from those of the soluble enzyme due to conformational changes, matrix interactions and steric effects. The inherent rate refers to the rate that would be observed in the absence of diffusional resistances, and the effective rate is equivalent to the observed or apparent activity. These definitions should facilitate both the theoretical and experimental analysis of immobilized enzyme kinetics by distinguishing between the different factors that affect the kinetics of bound enzymes. These authors also suggest that the term "apparent K_m" be abandoned for all systems that do not follow Michaelis-Menten kinetics, specifically for systems that are limited by external and internal diffusion. In a similar light, Buchholz (1982) has more recently summarized diffusion and kinetics with immobilized enzymes. Moser (1989) has recently synthesized the several points of view into a description of biokinetics via the macro- versus micro-approach.

2.4 KINETIC MODELS FOR IMMOBILIZED ENZYMES

The simplest model of conversion of substrate (S) to product (P) catalyzed by *unsupported* enzyme (E) is:

$$E + S \underset{k_{-1}}{\overset{k_1}{\Leftrightarrow}} ES \overset{k_2}{\Rightarrow} P + E$$

where k_1, k_{-1} and k_2 are kinetic constants. If one assumes that a steady state exists in which the concentration of intermediate (ES) does not vary with time, the Michaelis-Menten relationship can be developed:

$$V = \frac{k_2 E_0 S}{K_m + S} \qquad [2.1]$$

where V = velocity of the enzyme reaction; K_m = Michaelis-Menten constant = $(k_{-1} + k_2)/k_1$; E_0 = total enzyme concentration; and S = substrate concentration. If the concentration of substrate is large relative to K_m, then $V = k_2E_0 = V_m$ and the rate of reaction is at its maximum.

When an enzyme is attached to a solid support, the kinetic pattern of reaction changes considerably, leading to changes in the values of the kinetic parameters K_m and V_m. The kinetics of such reactions are obscured by several factors and the observed kinetic parameters are only "apparent." As mentioned before, these factors are: (a) change in enzyme conformation, (b) steric effects, (c) microenvironmental effects and (d) bulk and internal diffusional effects (A portion of the material from here to the end of this chapter is excerpted with permission from W. R. Vieth and K. Venkatasubramanian, *Chem. Tech, Part I*, **Nov.**, 1973. Copyright 1973 American Chemical Society).

Partitioning effects have frequently been documented in the literature, often revealing themselves through altered pH-activity optima with charged supports and through altered apparent Michaelis constants for charged or hydrophobic carriers. Positively charged carriers repel hydrogen ions, resulting in a microenvironmental pH that is higher than the pH of the bulk solution and a shift in pH-activity optima to more acidic values. Polyanionic carriers have the opposite effect. By the same mechanism, charged carriers can cause increases in apparent K_m values for substrates carrying like charges, and decreases in apparent K_m's for oppositely charged substrates. Displacements of pH-activity optima of from 1-2.5 pH units and alterations in K_m's of more than an order of magnitude have been attributed to electrostatic effects (Goldstein, 1976). These effects are neutralized at high levels of ionic strength. Experimental evidence for hydrophobic partitioning of substrate and the resulting effect on the apparent K_m has been provided by Johansson and Mosbach (1974) for the reaction of n-butanol by liver alcohol dehydrogenase bound to an acrylamide/methacrylate copolymer.

A close correlation between conformational changes, as measured by fluorescent properties, and enzyme activity was obtained in one instance (Gabel and Hofsten, 1971). Sun and Faust (1990) have recently reported the on-line determination of enzymatic catalytic activity and activation energy by a precise photometric method. Designed for use on an industrial scale, it can be adapted for research and to optimize the parameters of catalytic reactions carried out in the laboratory. Combes et al. (1988) proposed the modification of the biocatalyst environment with additives as a viable approach to the stabilization of enzymes. Polyols and simple salts of the lyotropic series have been shown to have a stabilizing influence on hydrolytic enzymes.

Clark et al. (1988) report on the conformational analyses of enzymes and substrates in processing environments where the enzyme may be in an immobilized form in reverse micelles, or in solvent systems that do not solubilize proteins. In each of two model cases, kinetic data analyzed in conjunction with structural information provided by electron paramagnetic resonance (EPR) spectroscopy have shed considerable light on the factors that govern enzyme function.

Kirchner, Scollar and Klibanov (1985), Reslow, Adlercreutz and Mattiasson (1988) and Khmelnitsky et al. (1987) show that enzymes can be used in reaction systems with very low water contents, which favors mass transfer of hydrophobic substrates to the biocatalyst phase and thermodynamic equilibrium in reversed hydrolytic reactions. This was demonstrated in work on the transesterification of triglycerides (Aldercreutz et al., 1988). Paul et al. (1988) rather recently demonstrated the direct enzymatic synthesis of aspartame using a new peptidase activity from *M. caseolyticus.* The use of low water-content media containing concentrated nonhydroxylated polyethers such as triglyme results in an important shift in the reaction equilibrium toward synthesis: a twenty-fold increase in the equilibrium constant K_{syn} is obtained by using a mixture containing 60% triglyme.

In this connection, Ishikawa et al. (1990) report on the kinetic properties of enzymes in reversed micelles. They concluded that micelle size is not a main controlling factor of reaction rates for solubilized enzymes, but that the rates depended, rather, on the water and surfactant concentrations. Maximal reaction rates were found to exceed those in bulk aqueous solution. Miller et al. (1990) studied the enzymatic interesterification of triglycerides in supercritical carbon dioxide. They found that the interesterification reaction is not influenced by external mass transfer in the range of flow rates studied. The rate of interesterification increases with rising pressure as a result of enhanced substrate solubilities at higher pressures. Interesterification is favored over hydrolysis to a greater degree in dry CO_2 than in water-saturated CO_2, leading to a higher production rate of interesterified triglycerides.

In the case of enzymes covalently attached to solid supports, the covalent binding reaction may modify the innate reactivity of the enzyme. However, such effects occur concomitantly with the electrostatic effects induced by the carrier and therefore it is difficult to study these effects separately (Sri Ram et al., 1954; Axen et al., 1970; Goldstein et al., 1970).

Steric limitations imposed on the enzyme by immobilization sometimes impede the expression of enzymatic activity, particularly in the case of high molecular weight substrates. Data on steric hindrances are scarce, and it is difficult to separate steric effects from diffusional resistance experienced by the substrate, both in the bulk phase as well as in the carrier phase. Immobilized proteolytic enzymes exhibit decreased proteolytic activity toward high molecular weight protein substrates, as compared to their activity toward low molecular weight model substrates such as benzoyl-L-arginine ethyl ester (Silman et al., 1966; Lilly et al., 1966; Mitz and Sumaria, 1961; Wharton et al., 1968a). Steric restrictions have been implicated as a factor in impeding the lysis of the cell walls of the bacterium *Micrococcus lysodeikticus* by lysozyme immobilized on collagen (Venkatasubramanian et al., 1972). Choi et al. (1991) carried out partial characterization and localization of cell wall fraction-bound enzyme (CWE) in *A. oryzae* IAM 2640. Multiple enzymes such as acid proteinase (APase), acid carboxypeptidase (ACPase) and glutaminase needed for protein digestion were bound on the cell wall fraction. CWEs were found to be tightly and hydrophobically associated with the cell wall through the membranes.

Lowering the initial rate of hydrolysis of proteins by bound proteases is often accompanied by a decrease in the total number of peptide bonds susceptible to hydrolysis (Goldstein, 1969). In addition, the peptide maps of the digests of pepsinogen obtained by its hydrolysis by native and immobilized trypsins were different. In a similar manner, the protein fragments obtained on limited hydrolysis of heavy meromyosin and myosin with soluble and bound trypsin were completely different (Lowey et al., 1967). These results allude to steric interferences in modified kinetic behavior.

Work in our laboratory on lysozyme immobilized in micelles of a nonionic surfactant revealed a sterically favorable orientation of the enzyme to the approach of its substrate, thereby enhancing activity (Bernath and Vieth, 1972). F. Davis et al. (1979) successfully launched a high technology enterprise (ENZON, Inc., No. Plainflield, NJ) on the basis of the protective qualities of poly(ethylene glycols) toward polypeptides, against immune response. Andersson and Hahn-Hägerdal (1987) investigated penicillin acylase activity and stability in PEG solutions. PEG 600, which is the strongest water activity-decreasing polymer, does not increase the stability of PA, but PEG 8,000, which decreases the water activity very little, increases the stability by a factor of three. In a recent paper, Asenjo and Andrews (1989) describe the characteristics, production and use of enzyme systems with lytic activity against yeast and bacterial cells. Factors affecting enzymatic lysis, including the sequential protease and glucanase degradation of the two-layered yeast wall and the obtainment of sufficiently high lytic activities, are analyzed. The advantages of using enzyme systems, as well as their design and actual processes for the controlled lysis and selective product release (concept of a "Biochemical Cell Refinery"), are discussed.

MICROENVIRONMENTAL EFFECTS

The environment in the vicinity of an immobilized enzyme can be quite different from that in the bulk phase. The modified microenvironment has been shown to be responsible for the changes in pH-activity profiles observed in the case of several enzymes bound to polyelectrolytic carriers (Fig. 2.1). pH displacements of 1-2.5 pH units

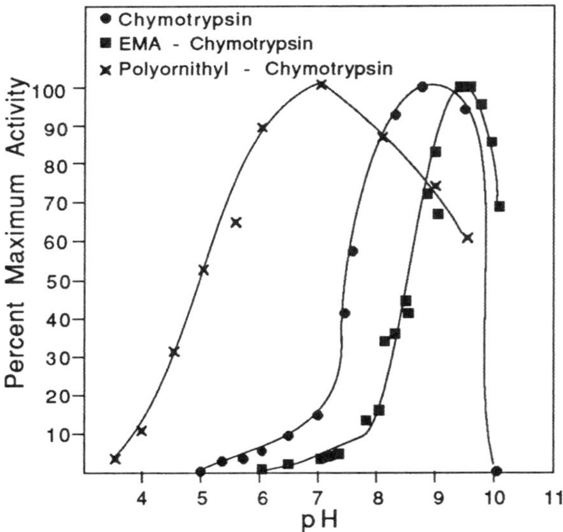

Figure 2.1 pH-Activity profiles at low ionic strength (1 = 0.008) for supported and unsupported chymotrypsin, using acetyl-L-tyrosine ethyl ester as substrate (Goldstein and Katchalski, 1968).

have been observed (Goldstein and Katchalski, 1968) at low ionic strengths. However, at high ionic strengths no shift was found. Correspondingly, the apparent Michaelis constant K'_m decreases at low ionic strengths (Table 2.2).

Table 2.2 Apparent Michaelis Constants for Soluble and Immobilized Trypsins: Benzoyl-L-Arginine Substrate (Goldstein, 1970)

	Ionic strength	103 K'm (moles/liter)[a]
Unsupported native trypsin	0.04	6.9
Ethylenemaleic acid-trypsin	0.04	0.2
EMA-trypsin	0.5	5.2

[a]Michaelis-Menten constants

Unequal distribution of hydrogen ions, hydroxyl ions, and charged substrates between the carrier phase and bulk phase leads to an alteration in the microenvironment, so that hydrogen ion concentration in the locale of the enzyme bound to a negatively charged (polyanionic) carrier is higher than in the external solution. Similarly, it is lower in a positively charged (polycationic) carrier. This accounts for the shift of the pH-activity profile of a bound enzyme. These effects have been correlated with a Boltzmann-type distribution (Goldstein et al., 1965).

In the case of a charged substrate acted upon by an enzyme attached to a polyelectrolytic carrier, a similar unequal distribution of charges occurs between the matrix phase and the external solution phase. Substrates carrying the same charge as the matrix are repelled from the carrier surface, resulting in a lower concentration in the vicinity of the enzyme and vice versa, so that the substrate concentration at which maximal reaction velocity (V_m) is achieved is lower for an oppositely charged substrate-carrier-bound enzyme system. Conversely, a lower rate is obtained for an immobilized enzyme-substrate system where both carrier and substrate carry the same charge (Goldman et al., 1971). Goldstein and Katchalski (1968) have derived a quantitative expression for the electrostatic partitioning of the substrate between the bulk and the carrier phases, and this analysis was extended by Wharton et al. (1968b). The validity of this prediction is shown in Fig. 2.2 for the hydrolysis of benzoyl-L-arginine ethyl ester by carboxymethyl cellulose-bromelain.

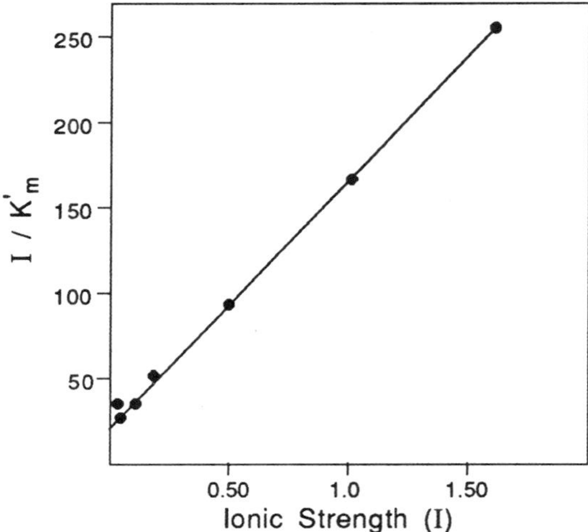

Figure 2.2 Dependence of the values of the apparent Michaelis-Menten constant, K'_m, on ionic strength, I, for the hydrolysis of benzoyl-L-arginine ethyl ester by CMCellulose-bromelain (Wharton et al., 1968b).

In some cases, displacements in pH-activity profiles have been observed that are not explicable from the foregoing considerations. For example, Bernfeld et al. (1969) report a pH shift toward the acid side for phosphoglycerate mutase entrapped on an electrically ***neutral*** carrier, polyacrylamide. Accumulation of one of the reaction products in the microenvironment contributes to altered pH-activity profiles in some instances.

DIFFUSIONAL LIMITATIONS

Unlike soluble enzyme, matrix-supported enzyme has to exercise its catalytic action in a heterogeneous environment. As in conventional heterogeneous catalysis, at least four distinct steps in the overall enzymatic process can be identified: (1) diffusion of the substrate from the bulk phase to the carrier surface, (2) transport of the substrate from the carrier surface to the domain of the enzyme, (3) enzyme-catalyzed conversion of substrate, and (4) reversal of steps (2) and (1) for product. Steps 1 and 3 are external or bulk diffusional effects, while steps 2 and 4 include internal or pore diffusional effects.

External diffusional resistance results from the presence of a near-stagnant layer of fluid around the immobilized enzyme particle surface. Transport of substrate through this region - known as the Nernst diffusion layer - occurs mainly by molecular diffusion, and if this rate is slow, reaction can be slow. External film diffusional resistances have been experienced in a number of immobilized enzyme systems.

The influence of combined bulk diffusional and electrostatic effects was extended to a nonporous immobilized enzyme system by Shuler et al. (1972). An effectiveness factor η_{DE} may be defined as the ratio of the actual reaction rate to that which would be obtained in the absence of diffusive and electrostatic effects. The effectiveness factor is a function of three parameters: (a) the electrostatic parameter λ, (b) the ratio of the maximum reaction rate to the maximum mass transfer rate μ, and (c) the Michaelis constant. When μ is small, the effectiveness factor can be approximated as:

$$\eta_{DE} = \frac{1+\gamma}{1+\gamma e^{\lambda}\left[1+\dfrac{\mu}{1+\gamma e^{\lambda}}\right]} \qquad [2.2]$$

where γ = dimensionless Michaelis constant = K_m/S_B; $\lambda = ZF\psi(o)/RT$; $\mu = V_m\delta/MD_sS_B$; M = electrostatic potential distribution defined by:

$$M = \frac{\delta}{\int_0^{\delta} \exp\left[\lambda \dfrac{\psi(x)}{\psi(o)}\right] dx}$$

The modified Michaelis constant K'_m is given by:

$$K'_m = K_m e^{\lambda}\left[1+\frac{V_m\delta}{MD_s(S_B+K_m e^{\lambda})}\right] \qquad [2.3]$$

By assuming the Gouy diffuse double-layer potential distribution, calculations were presented that show modification of K'_m by the diffusional and electrostatic effects (Shuler et al., 1972).

Often the immobilized enzyme is investigated in small packed columns. In most cases, a substantial bulk diffusional limitation has been observed (Allison et al., 1972; O'Neill et al., 1971). Kinetic parameters vary with substrate flow rate, and for a column operating at steady state an integrated form of the Michaelis-Menten kinetic expression may be used to estimate the kinetic constants (Lilly and Sharp, 1968).

$$XS_o - K'_m \ln(1 - X) = \frac{k_2 E_t \beta}{Q} \qquad [2.4]$$

where X = fractional conversion = (S_o - S_e/S_o); S_o = inlet substrate concentration; S_e = outlet substrate concentration; E_t = total amount of enzyme in the column; β = porosity of the column; and Q = flow rate through the column.

Eqn. [2.4] may be rewritten as:

$$XS_o = K'_m \ln(1 - X) + C/Q \qquad [2.5]$$

where $C = k_2E_t\beta$ = reaction capacity of the column.

At constant flow rate a plot of XS_o against ln (1 - X) should yield a straight line whose slope is K'_m and whose intercept is C/Q. Tests of eqn. [2.5] and the variation of the apparent Michaelis constant (K'_m) with flow rate are shown in Figs. 2.3 and 2.4. K'_m decreases asymptotically with increasing flow rate, reaching a minimum at high flow. The observed dependence of K'_m on linear velocity is the result of bulk diffusion limitation of transport of the substrate.

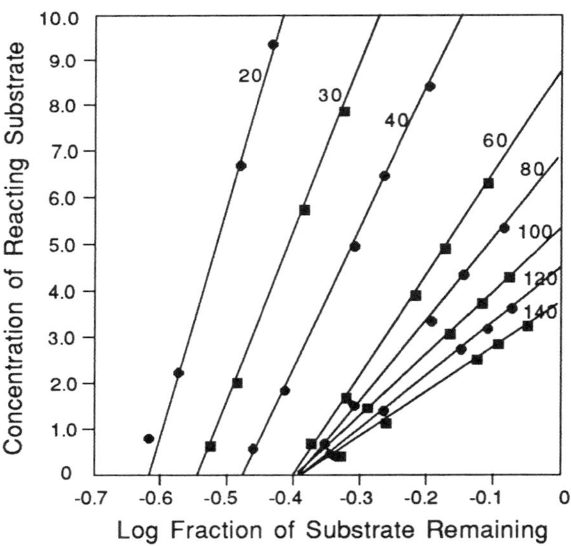

Figure 2.3 The hydrolysis of benzoyl-L-arginine ethyl ester by a column of CMCellulose-ficin. The numbers on the curves indicate the flow rate (ml/h) (Hornby et al., 1966).

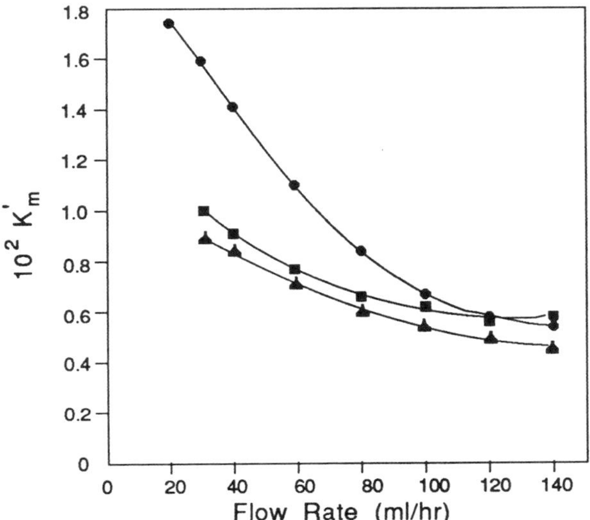

Figure 2.4 Dependence of K'_m on flow rate for columns containing different preparations of CMCellulose-ficin acting on benzoyl-L-arginine ethyl ester (Hornby et al., 1966).

Intraparticle or internal diffusional impedances also influence the kinetics of immobilized enzyme reactions. The enzyme is usually distributed within a porous carrier matrix, and substrate has to diffuse through the carrier to reach the reaction site. In many cases, internal diffusional resistance substantially reduces the expressed catalytic activity of the enzyme. Enolase covalently attached to a porous carrier exhibits an enzymatic activity only a fourth that of its "potential activity" because only that portion of the aldolase located at or near the surface of the carrier particle displays activity (Bernfeld et al., 1968).

If one considers an enzyme membrane of nearly infinite expanse and of thickness 2L, the steady state mass balance on substrate diffusing and reacting in the membrane is:

$$D_e \frac{d^2S}{dx^2} - \frac{V_m S}{K_m + S} = 0 \qquad [2.6]$$

which can be solved for a few limiting cases. When $S >> K_m$, then reaction becomes first order in S,

$$V = kS \qquad [2.7]$$

where k = first order reaction rate constant.

Analytical solution of eqn. [2.6] for first order kinetics with appropriate boundary conditions leads to eqn. [2.8], which describes the local substrate concentration as:

$$S = S_o \frac{\cosh \phi\left(1 - \frac{X}{L}\right)}{\cosh \phi} \qquad [2.8]$$

where ϕ = Thiele modulus = $L[k/D_e]^{0.5}$.

An effectiveness factor, η, defined as the ratio of actual reaction rate in the matrix to the (maximum) rate obtainable without diffusional influences, is given by:

$$\eta = \frac{\tanh \phi}{\phi} \qquad [2.9]$$

The effectiveness factor η increases with decreasing values of the modulus ϕ (i.e., the diffusion rate is considered rapid with respect to the reaction rate), as shown in Fig. 2.5. An effectiveness factor of unity signifies no diffusional limitation. The effectiveness factor is an important design parameter because it measures the extent of internal diffusional resistance.

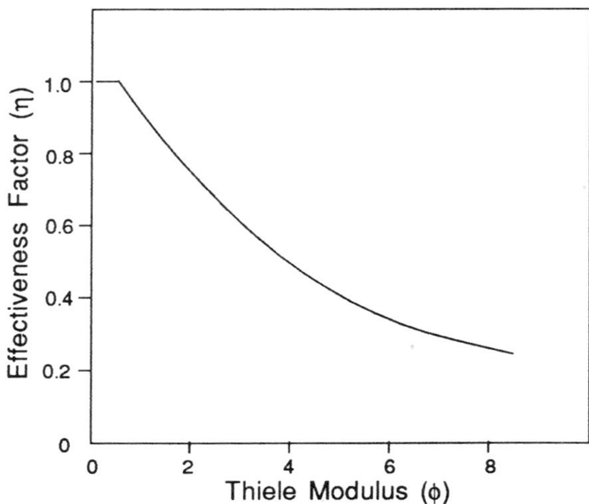

Figure 2.5 Variation of effectiveness factor with Thiele modulus.

The variation of enzymatic activity as a function of substrate concentration for different moduli values is shown in Fig. 2.6 (Selegney et al., 1971a). When $\phi = 0$ (i.e., $D_s = \infty$), there is no diffusional limitation and the rate-substrate concentration profile is of the Michaelis-Menten type. As ϕ increases, the curves become more and more sigmoidal. These results can be interpreted in terms of biological regulatory mechanisms (Thomas et al., 1972).

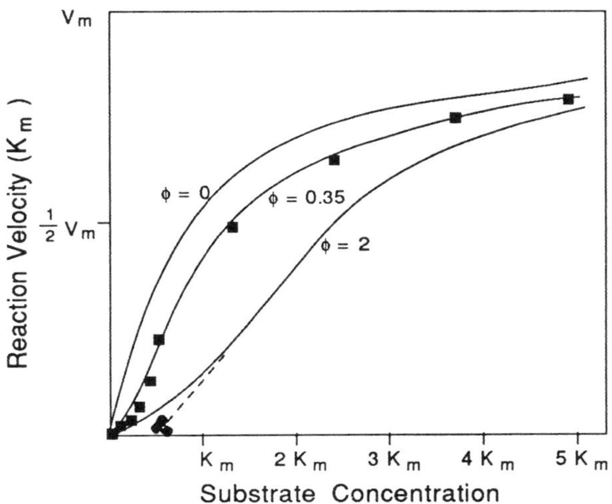

Figure 2.6 Variation of enzymatic activity as a function of substrate concentration for different values of Thiele modulus, ϕ. Experimental points are shown for a glucose oxidase-albumin membrane of $\phi = 0.35$ (Selegny et al., 1971a).

An approximate analytical solution of eqn. [2.6] has been obtained for unsymmetrical systems such as a membrane placed between two different substrate solutions (Selegny et al., 1971a). Effectiveness factors for the general Michaelis-Menten kinetics have been developed by Moo-Young and Kobayashi (1972) in terms of a general modulus m defined as:

$$m = \frac{LV(S_i)}{\sqrt{2}}\left[\int_o^{S_i} D_e V(S)dS\right]^{0.5} \qquad [2.10]$$

where S_i = substrate concentration at the surface; S = substrate concentration anywhere in the membrane; and V(S) = reaction rate as a function of S.

Effectiveness factors for complex kinetic cases involving competitive inhibition have also been developed (Selegny et al., 1971a). Unsteady state mass transfer kinetic models for enzyme-membrane systems have been formulated (Selegny et al., 1971b).

Both external and internal diffusional effects may often be present simultaneously. This situation can be quantitatively treated by incorporating a partition coefficient, P, defined as the ratio of the substrate concentration at the carrier surface (S_i) to that at the bulk (S_B) to account for the bulk diffusion in the above analysis (Kobayashi and Laidler, 1973). Under these conditions, an apparent Michaelis constant can be formulated (Bunting and Laidler, 1972) as:

$$K'_m = \frac{K_m}{P\eta} \qquad [2.11]$$

Experimental results on the kinetic behavior of β-galactosidase entrapped in a polyacrylamide gel corroborate this analysis (Bunting and Laidler, 1972). Czermak et al. (1990) describe a bioreactor for lactose hydrolysis which utilizes a steam-sterilizable dialysis membrane in which the enzyme solution is separated from the milk- or whey-lactose solution, but free diffusion of lactose and the hydrolysis products through the membrane is allowed. Optimization of the process is described.

Analysis of the influence of diffusion on apparent thermal stability of immobilized enzyme has shown (Ollis, 1972) that an apparently more stable enzyme results when the overall reaction rate is diffusion controlled. However, this analysis makes the assumption that V_m and K_m values are essentially the same for soluble and bound enzymes. These ideas are elaborated in Bailey and Ollis (1986).

Diffusional effects in a microencapsulated enzyme system or in hollow fiber-trapped enzymes have an additional feature. Mass transfer resistance to diffusion of the substrate within the enzyme solution may be superimposed on external- and membrane-phase resistances. Kinetic analyses have been reported (Rony, 1971; Mogensen and Vieth, 1973).

Internal diffusional restrictions sometimes lead to a modified pH activity profile for the attached enzyme compared to the soluble form. This is because the reaction products such as acids do not readily diffuse out from the microenvironment. For example, the hydrolysis of esters such as benzoyl arginine ethyl ester by bound proteolytic enzymes leads to the liberation of hydrogen ions. Therefore, the local pH may be much lower than bulk pH value, resulting in a shift of the pH activity curve. Such distortions in pH activity profiles have been observed with collodion-papain (Goldman et al., 1968), collagen-glucose oxidase membranes (Constantinides et al., 1973), and agarose-protocatechuate dioxygenase (Zaborsky and Ogletree, 1972).

Particulate immobilized derivatives of papain exhibit a pH dependence of activity that is very similar to that of the native enzyme (Fritz et al., 1968). Similarly, when collodion-papain membranes were ground to a fine powder and used, the anomaly in the pH dependence of activity is eliminated, implying that diffusional barriers are indeed responsible for the altered pH activity curve of the intact membranes. Further, when the reaction is done in the presence of a strong buffer, no anomalous pH dependence of activity is observed, since the buffers used diffuse into the membrane and increase its inner pH by neutralizing the acid generated by the enzymatic action (Goldman et al., 1968).

The same collodion-papain membranes when acting on an amide substrate (benzoyl-L-arginine amide) produce carboxylate and ammonium ions in the pH range 4.5-8.0. At higher pH the ammonium ions liberate protons and the pH of the membrane is lowered. This leads to a corresponding shift in the pH activity profiles toward more alkaline pH (Goldman et al., 1968).

In some cases, the accumulation of ionic products within a carrier-enzyme conjugate of the same charge causes it to swell, thereby increasing local diffusional rates (Fritz et al., 1968).

TEMPERATURE EFFECTS

Kinetic constants of reaction for both soluble and bound enzyme catalysis have a temperature dependence of the Arrhenius form. However, the apparent activation energy of an immobilized enzyme differs from that of the soluble enzyme. Enzymatic reactions by matrix-supported glucose oxidase (Weetall and Hersh, 1970), enolase (Bernfeld et al., 1969), and lysozyme (Venkatasubramanian et al., 1972) exhibit higher activation energies than those in solution. Bound aminoacylase (Tosa et al., 1967) and chymotrypsin (Sharp et al., 1969) exhibit lower activation energies, while no change occurs in the case of immobilized glycerate mutase (Bernfeld and Bieber, 1969) and asparaginase (Nikolaev, 1962). A possible explanation for these results has been advanced by Melrose (1971) which invokes: (a) the introduction of strain in the enzyme, substrate or the carrier, (b) favorable electron flow between the reactants and the polymer, and (c) favorable changes of polarity or ionic concentration in the vicinity of the active site.

In connection with the above, Kragl et al. (1991) describe an enzyme membrane reactor made of a hydrophobic polymer, especially a polyalkene which is used to prolong the half-life of aldolase. In particular, the reactor surfaces must be free of ether or thioether linkages. Decay curves for enzyme activity in reactors prepared from a number of materials are presented. Immobilization of *Trichoderma viride* cellulase on CNBr-activated agarose beads, and its application to bioreactor operation are reported by Sasaki et al. (1988). Compared with native cellulase, the immobilized cellulase was improved in stability at pH 1.5-2.0. Lin (1991) determined optimal feed temperature for a nonisothermal immobilized enzymic reaction with enzyme deactivation in a packed bed reactor. This was carried out by maximizing the average substrate conversion over a given reaction period. Simulation showed the optimal feed temperature to be strongly dependent on the flow dispersion, the reaction activation energy, the corresponding enzyme inactivation energy, and the enthalpy of reaction. In a plug flow reactor the enzyme reaction generally exhibited a lower optimal feed temperature and higher substrate conversion than in a continuously stirred tank reactor.

2.5 REFERENCES

Ada, T., T. Kawamoto, I. Masunaga, T. Takayama, S. Takasawa and K. Kimura, *Enzyme Eng.*, **6**, 303 (1982).

Ado, Y., K. Kimura and H. Samejima, *Enzyme Eng.*, **5**, 295 (1980).

Aldercreutz, P., K. Larsson and B. Mattiasson, In "Enzyme Engineering 9," H. W. Blanch and A. M. Klibanov, Eds. *Ann. N. Y. Acad. Sci.*, **542**, 270-274 (1988).

Allison, J. P., L. Davidson, A. Gutierrez-Hartman and G. B. Kitto, *Biochem. Biophys. Res. Commun.*, **47**, 66 (1972).

Andersson, E. and B. Hahn-Hägerdal, In "Enzyme Engineering 8," A. I. Laskin, K. Mosbach, D. Thomas and L. B. Wingard, Eds. *Ann. N. Y. Acad. Sci.*, **501**, 85-87 (1987).

Asenjo, J. A. and B. A. Andrews, In "Bioproducts and Bioprocesses," A. Fiechter, H. Okada and R. D. Tanner Eds. Springer-Verlag: Berlin, Heidelberg (1989).

Axen, R., P. A. Myrin and J. C. Janson, *Biopolymers*, **9**, 401 (1970).

Bailey, J. E. and D. F. Ollis, "Biochemical Engineering Fundamentals," McGraw-Hill Book Co.: New York (1986).

Barenschee, T., A. Hasler, T. Scheper and K. Schuegerl, *DECHEMA Biotechnol. Conf.*, **3** (Pt. B, Biochem. Eng., Environ. Biotechnol., Recovery Bio-Prod., Saf. Biotechnol.), 1055-1058 (1989).

Beijer, N. A., H. M. Buck, L. A. Sluyterman and E. M. Meijer, In "Enzyme Engineering 10," H. Okada, A. Tanaka and H. W. Blanch, Eds. *Ann. N. Y. Acad. Sci.*, **613**, 494-501 (1990).

Berfield, P. and J. Wan, *Science*, **142**, 678 (1963).

Bernath, F. R. and W. R. Vieth, *Biotechnol. Bioeng.*,**14**, 737 (1972).

Bernath, F. R., K. Venkatasubramanian and W. R. Vieth, In "Annual Reports of Fermentation Processes," D. Perlman, Ed., 235-266. Academic Press: New York (1977).

Bernfeld, P., R. E. Bieber and P. C. MacDonnell, *Arch. Biochem. Biophys.*, **127**, 779 (1968).

Bernfeld, P., R. E. Bieber and D. M. Watson, *Biochim. Biophys. Acta*, **191**, 570 (1969).

Bernfeld, P. and R. E. Bieber, *Arch. Biochem. Biophys.*, **131**, 587 (1969).

Bjorck, L. and C. G. Rosen, *Biotechnol. Bioeng.*,**18**, 1033 (1976).

Bouin, J. C., M. T. Atallah and H. O. Hultin, *Biotechnol. Bioeng.*, **18**, 179 (1976).

Buchholz, K. and W. Ruth, *Biotechnol. Bioeng.*, **18**, 95 (1976).

Buchholz, K., *Adv. Biochem. Eng.*, **24**, 39 (1982).

Bu'lock, J. and B. Kristiansen, "Basic Biotechnology." Academic Press: London (1987).

Bülow, L. and K. Mosbach, In "Enzyme Engineering 8," A. I. Laskin, K. Mosbach, D. Thomas and L. B. Wingard, Eds. *Ann. N. Y. Acad. Sci.*, **501**, 44-50 (1987).

Bunting, P. and K. J. Laidler, *Biochemistry*, **11**, 4477 (1972).

Chang, T. M. S., *Science*, **146**, 524 (1964).

Chibata, I., T. Tosa and T. Sato, *J. Mol. Catalysis*, **37**, 1 (1986).

Choi, M. R., N. Sato, T. Yamagishi and F. Yamauchi, *J. Ferment. Bioeng.*, **72**, 214-216 (1991).

Clark, D. S., P. S. Skerker, T. W. Randolph, H. W. Blanch and J. M. Prausnitz, In "Enzyme Engineering," H. W. Blanch and A. M. Klibanov, Eds. *Ann. N. Y. Acad. Sci.*, **542**, 16-29 (1988).

Combes, D., W. N. Ye, A. Zwick and P. Monsan, In "Enzyme Engineering," H. W. Blanch and A. M. Klibanov, Eds. *Ann. N. Y. Acad. Sci.*, **542**, 7-10 (1988).

Constantinides, A., W. R. Vieth and P. M. Fernandes, *Mol. Cell. Biochem.*, **1**, 127 (1973).

Constantinides, A, *Biotechnol. Bioeng.*, **22**, 119 (1980).

Czermak, P., D. Bahr and W. Bauer, *Chem.-Ing.-Tech.*, **62**, 678-679 (1990).

Davis, F., T. Van Es and N. Palzuk, US Patent 4179337, Non-Immunogenic Polypeptides (1979).

Engasser, J. M. and C. Horvath, In "Applied Biochemistry and Bioengineering," L. B. Wingard, E. Katchalski-Katzir and L. Goldstein, Eds., Vol. 1, p. 128. Academic Press: New York (1976).

Fernandes, P. M., A. Constantinides, W. R. Vieth and K. Venkatasubramanian, *Chemtech*, **5**, 438 (1975).

Fritz, H., K. Hochstrasser, E. Werle, E. Brey and B. M. Gebhardt, *Z. Anal. Chem.*, **243**, 352 (1968).

Fusee, M. C., W. E. Swann and G. J. Calton, *Appl. Environ. Microbiol.*, **42**, 672 (1981).

Gabel, D. and B. V. Hofsten, *Eur. J. Biochem.*, **15**, 410 (1971).
Goetz, V., M. Remaud and D. J. Graves, *Biotechnol. Bioeng.*, **37**, 614-626 (1991).
Goldman, R., O. Kadem, I. H. Silman, S. R. Caplan and E. Katchalski, *Biochemistry*, **7**, 486 (1968).
Goldman, R., L. Goldstein and E. Katchalski, In "Biochemical Aspects of Reactions on Solid Supports," G. R. Stark, Ed., p. 1. Academic Press: New York (1971).
Goldstein, L., Y. Levin and E. Katchalski, *Biochemistry*, **3**, 1913 (1965).
Goldstein, L. and E. Katchalski, *Z. Anal. Chem.*, **243**, 375 (1968).
Goldstein, L.,In "Fermentation Advances," D. Perlman, Ed., p. 391. Adademic Press: New York (1969).
Goldstein, L., *Methods Enzymol.*, **19**, 935 (1970).
Goldstein, L., M. Pecht., S. Blumberg, D. Atlas and Y. Levin, *Biochemistry*, **19**, 2322 (1970).
Goldstein, L., In "Methods in Enzymology," K. Mosbach, Ed., Vol. 44, p. 397. Academic Press: New York (1976).
Goldstein, L. and G. Manecke, In "Applied Biochemistry and Bioengineering," L. B. Wingard, E. Katchalski-Katzir and L. Goldstein, Eds., Vol. 1, p. 23. Academic Press: New York (1976).
Goldstein, L. and E. Katchalski-Katzir, In "Applied Biochemistry and Bioengineering," L. B. Wingard, E. Katchalski-Katzir and L. Goldstein, Eds., Vol. 1, p. 1. Academic Press: New York (1976).
Harada, F., T. Takahashi and K. Kitagawa, Jpn. Patent JP 02046282 A2, Silica Gel as Support for Enzyme Immobilization in Bioreactor (1990).
Hirata, A., M. Hirata, H. Furuzawa and N. Honda, *Kagaku Kogaku Ronbunshu*, **17**, 586-588 (1991).
Hornby, W. E., M. D. Lilly and E. M. Crook, *Biochem. J.*, **98**, 420 (1966).
Ishikawa, H., K. Noda and T. Oka, In "Enzyme Engineering," H. Okada, A. Tanaka and H. W. Blanch, Eds. *Ann. N. Y. Acad. Sci.*, **613**, 529-534 (1990).
Jandel, A.-S., H. Husted and C. Wandrey, *Eur. J. Appl. Microbiol. Biotechnol.*, **15**, 59 (1982).
Johansson, A. C. and K. Mosbach, *Biochim. Biophys. Acta*, **370**, 339 (1974).
Katchalski, E.,In "Polyamino Acids, Polypeptides, Proteins," M. A. Stahman, Ed., p. 283. Univ. of Wisconsin Press: Madison (1962).
Katchalski, E., I. Silman and R. Goldman, *Adv. Enzymol. Relat. Areas Mol. Biol.*, **34**, 445 (1971).
Kawase, M., Y. Yoshida and H. Yonekawa, Jpn. Patent JP01296975 A2, Immobilization of Enzymes on Sepiolite for Bioreactors (1989).
Khmelnitsky, Y. L., I. N. Zharinova, I. V. Berezin, A. V. Levashov and K. Martinek, In "Enzyme Engineering 8," A. I. Laskin, K. Mosbach, D. Thomas and L. B. Wingard, Eds. *Ann. N. Y. Acad. Sci.*, **501**, 161-165 (1987).
Kimura, K., K. Takayama, T. Ado and I. Masunaga, Jap. Patent, 81-75097 (1981).
Kirchner, G., M. Scollar and A. Klibanov, *J. Am. Chem. Soc.*, **107**, 7072-7076 (1985).
Kluge, M., J. Klein and F. Wagner, *Biotechnol. Lett.*, **4**, 293 (1982).
Kobayashi, T. and K. J. Laidler, *Biochim. Biophys. Acta*, **302**, 1 (1973).

Koshcheenko, K. A., G. V. Sukhodolskaya, V. S. Tyurin and G. K. Skryabin, *Eur. J. Appl. Microbiol. Biotechnol.*, **12**, 161 (1981).

Kragl, U., J. Peters, C. Wandrey and M. R. Kula, German Patent, DE 3937892 A1, Materials for Enzyme Membrane Reactors for Prolonging Enzyme Half Lives (1991).

Kulbe, K. D., U. Schwab and M. Howaldt, In "Enzyme Engineering 8," A. I. Laskin, K. Mosbach, D. Thomas and L. B. Wingard, Eds. *Ann. N. Y. Acad. Sci.*, **501**, 216-224 (1987).

Larsson, K. M., P. Adlercreutz and B. Mattiasson, In "Enzyme Engineering 10," H. Okada, A. Tanaka and H. W. Blanch, Eds. *Ann. N. Y. Acad. Sci.*, **613**, 791-796 (1990).

Laskin, A. I., Ed., "Enzymes and Immobilized Cells in Biotechnology." Benjamin Cummings: London (1985).

Leuchtenberger, P., M. Karrenbauer and U. Plöcker, *Enzyme Eng.*, **7**, 78 (1984).

Lilly, M. D., W. E. Hornby and E. M. Crook, *Biochem. J.*, **100**, 718 (1966).

Lilly, M. D. and A. K. Sharp, *Chem. Eng.* (London), CE 12 (1968).

Lin, S. H., *J. Chem. Technol. Biotechnol.*, **50**, 17-26 (1991).

Linko, P., In "Advances in Biotechnology," M. Moo-Young, Ed., Vol. 1, p.711 (1981).

Lowey, S., L. Goldstein, C. Cohen and S. M. Luck, *J. Mol. Biol.*, **23**, 807 (1967).

Luck, T. and W. Bauer, *GBF Monogr. Ser.*, Vol. Date 1988, **11** 131-142 (1989).

Manecke, G., *Pure Appl. Chem.*, **4**, 507 (1962).

Martensson, K., *Biotechnol. Bioeng.*,**16**, 567, 579 (1974).

Mattiasson, B. and K. Mosbach, *Biochim. Biophys. Acta*, **235**, 253 (1971).

Mazumder, T. K., K. Sonomoto, A. Tanaka and S. Fukui, *Appl. Microbiol. Biotechnol.*, **21**, 154 (1985).

Melrose, G. J. H., *Rev. Pure Appl. Chem.*, **21**, 83 (1971).

Messing, R. A., In "Immobilized Enzymes for Industrial Reactors." Academic Press: New York (1975).

Miller, D. A., H. W. Blanch and J. M. Prausnitz, In "Enzyme Engineering," H. Okada, A. Tanaka and H. W. Blanch, Eds. *Ann. N. Y. Acad. Sci.*, **613**, 534-537 (1990).

Mitz, M. A. and L. J. Sumaria, *Nature*, **189**, 576 (1961).

Miyawaki, O., T. Yano and K. Nakamura, In "Enzyme Engineering 10," H. Okada, A. Tanaka and H. W. Blanch, Eds. *Ann. N. Y. Acad. Sci.*, **613**, 816-820 (1990).

Mogensen, A. O. and W. R.Vieth, *Biotechnol. Bioeng.*, **15**, 467-481 (1973).

Moo-Young, M. and T. Kobayashi, *Can. J. Chem. Eng.*, **50**, 162 (1972).

Mosbach, K. and B. Mattiasson, In "Methods in Enzymology," K. Mosbach, Ed. Vol. 44, p. 453. Academic Press: New York (1976).

Moser, A., *Experientia.*, **45**, 1035-1041 (1989).

Murata, K., K. Tani, J. Kato and I. Chibata, *Enzyme Microb. Technol.*, **3**, 233 (1981).

Nakajima, M., K. Iwasaki, H. Nabetani and A. Watanabe, *Agric. Biol. Chem.*, **54**, 2793-2799 (1990).

Nelson, J. M. and E. G. Griffin, *J. Amer. Chem. Soc.*, **38**, 1109 (1916).

Niedleman, S. L. and J. Geigert, *Trends Biotechnol.*, **1**, 21 (1983).

Nikolaev, A. Y., *Biokhimiya* (Eng. Trans.), **27**, 713 (1962).
O'Driscoll, K. F. and D. G. Mercer, In "Encyclopedia of Polymer Science and Technology," N. M. Bikales, Ed. Wiley (Interscience): New York (1976).
Ollis, D., *Biotechnol. Bioeng.*, **14**, 871 (1972).
Omata, T., A. Tanaka and S. Fukui, *J. Ferment. Technol.*, **58**, 339 (1980).
O'Neill, S. P., M. D. Lilly and P. N. Rowe, *Chem. Eng. Sci.*, **26**, 173 (1971).
Paul, F., D. Auriol and P. Monsan, In "Enzyme Engineering 9," H. W. Blanch and A. M. Klibanov, Eds. *Ann. N. Y. Acad. Sci.*, **542**, 351-356 (1988).
Pronk, W., M. Van der Burgt, G. Boswinkel and K. Van 't Riet, *J. Am. Oil Chem. Soc.*, **68**, 852-856 (1991).
Reslow, M., P. Adlercreutz and B. Mattiasson, In "Enzyme Engineering," H. W. Blanch and A. M. Klibanov, Eds. *Ann. N. Y. Acad. Sci.*, **542**, 250-254 (1988).
Rony, P. R., *Biotechnol. Bioeng.*, **13**, 431 (1971).
Sasaki, T., Y. Kashiwagi and Y. Magae, *Baiomasu Henkan Keikaku Kenkyu Hokoku*, (11), 41-49 (1988).
Scollar, M. P., G. Sigal and A. M. Kibanov, *Biotechnol. Bioeng.*, **27**, 247 (1985).
Selegny, E., G. Broun and D. Thomas, *Physiol. Veg.*, **9**, 25 (1971).
Selegny, E., J. P. Kernevez, G. Broun and D. Thomas, *Physiol. Veg.*, **9**, 51 (1971).
Sharp, A. K., G. Kay and M. D. Lilly, *Biotechnol. Bioeng.*, **11**, 373 (1969).
Shuler, M. L., R. Aris and H. M. Tsuchiya, *J. Theor. Biol.*, **35**, 67 (1972).
Silman, I. H. and E. Katchalski, *Annu. Rev. Biochem.*, **35**, 873 (1966).
Silman, I. H., M. Albu-Weissenberg and E. Katchalski, *Biopolymers*, **4**, 441 (1966).
Sonomoto, K., K. Nomura, A. Tanaka and K. Fukui, *Eur. J. Appl. Microbiol. Biotechnol.*, **16**, 57 (1982).
Sonomoto, K., N. Usui, A. Tanaka and S. Fukui, *Eur. J. Appl. Microbiol. Biotechnol..*, **17**, 203 (1983).
Srere, P. A., B. Mattiasson and K. Mosbach, *Proc. Natl. Acad. Sci. U.S.A.*, **70**, 2534 (1973).
Sri Ram, J., L. Terminiello, M. Bier and F. F. Nord, *Arch. Biochem. Biophys.*, **52**, 464 (1954).
Sun, Z. and U. Faust, *DECHEMA Biotechnol Conf.* **4** (Pt. B. Lect. DECHEMA Annu. Meet. Biotechnol. 8th) 959-962 (1990).
Sundaram, P. V. and E. K. Pye, In "Enzyme Engineering," E. K. Pye and L. B. Wingard, Eds., Vol. 2, p. 449. Plenum Press: New York (1974).
Takata, I., K. Yamamoto, T. Tosa and I. Chibata, *Enzyme Microb. Technol.*, **2**, 30 (1980).
Thomas, D., M. C. Tran, G. Gellf, D. Domurado, B. Paillot, R. Jacobsen and G. Broun, *Biotechnol. Bioeng. Symp. No. 3*, **299** (1972).
Tosa, T., T. Mori, N. Fuse and I. Chibata, *Enzymologia*, **31**, 214 (1966).
Tosa, T., T. Mori, N. Fuse and I. Chibata, *Enzymologia*, **32**, 153 (1967).
Tosa, T., S. Takamatsu, M. Furui and I. Chibata, *Enzyme Eng.*, **7**, 450 (1984).
Van der Padt, A., M. J. Edema, J. J. Sewalt and K. Van't Riet, *J. Am Oil Chem. Soc.*, **67**, 347-352 (1990).
Venkatasubramanian, K., W. R. Vieth and S. S. Wang, *J. Ferm. Technol.*, **50**, 600-614 (1972).
Verlop, K.-D. and J. Klein, *Biotechnol. Lett.*, **3**, 9 (1981).

Vieth, W. R. and K. Venkatasubramanian, *Chem. Tech, Part I*, **Nov.**, 677 (1973).
Vieth, W. R., K. Venkatasubramanian, A. Constantinides and B. Davidson, In "Applied Biochemistry and Bioengineering," L. B. Wingard, E. Katchalski-Katzir and L. A. Goldstein, Eds., Vol. 1, p. 222. Academic Press: New York (1976).
Wandrey, C., R. Wichmann and A.-S. Jandel, *Enzyme Eng.*, **6**, 61 (1982).
Watanabe, E., K. Sakashita and Y. Ogawa, Jap. Pat. 83-35 078 (1983).
Weetall, H. H. and L. S. Hersh, *Biochim. Biophys. Acta*, **206**, 54 (1970).
Wharton, C. W., E. M. Crook and K. Brocklehurst, *Eur. J. Biochem.*, **6**, 565 (1968).
Wharton, C. W., E. M. Crook and K. Brocklehurst, *Eur. J. Biochem.*, **6**, 572 (1968).
Woodley, J. M., A. J. Brazier and M. D. Lilly, *Biotechnol. Bioeng.*, **37**, 133-140 (1991).
Yamada, H., H. Shimizu, Z. Tani and T. Hino, *Enzyme Engineering*, **5**, 405 (1980).
Yamada, H., S. Shimizu, H. Shimada, Y. Tani, S. Takahashi and T. Ohashi, *Biochimie*, **62**, 395 (1980).
Yamamoto, K., T. Tosa and I. Chibata, *Biotechnol. Bioeng.*, **22**, 2045 (1980).
Yi, S.-H. and H. J. Rehm, *Eur. J. Appl. Microbiol. Biotechnol.*, **16**, 1 (1982).
Yokozeki, K., S. Yamanaka, T. Utagawa, K. Takinami, Y. Hirose, A. Tanaka, K. Sonomoto and S. Fukui, *Eur. J. Appl. Microbiol. Biotechnol.*, **14**, 225 (1982).
Yomo, T., I. Urabe and H. Okada, In "Enzyme Engineering 10," H. Okada, A. Tanaka and H. W. Blanch, Eds. *Ann. N. Y. Acad. Sci.*, **613**, 313-319 (1990).
Yoshida, Y., M. Kawase and T. Shiraishi, *Hakko Kogaku Kaishi*, **68**, 197-203 (1990).
Yoshida, Y., M. Kawase, T. Majima and T. Shiraishi, *Hakko Kogaku Kaishi*, **68**, 267-273 (1990).
Zaborsky, O. R. and J. Ogletree, *Biochim. Biophys. Acta*, **289**, 68 (1972).
Zaborsky, O. R., "Immobilized Enzymes," CRC Press: Cleveland (1973).
Zaborsky, O. R., In "Methods in Enzymology," K. Mosbach, Ed., Vol. 44, p. 317. Academic Press: New York (1976).
Zaborsky, O. R., In "Enzyme Technology and Renewable Resources," Proc. of Grantees-Users Conf., sponsored by NSF, RANN at Univ. of Virginia, May 19-21 (1976).

3

ENZYME AND CELL-BASED REACTORS

3.0 INTRODUCTION

Some perspective on the possible impact of developments in bioreactor technology can be gained from a recent report in Genetic Engineering News. Salient features are summarized in Table 3.1. Levels of business activity as well as rates of growth are projected to be quite substantial.

Table 3.1 Bioreactor Market Projections for the U.S. ($ Millions) (after Wrotnowski, 1990)

	1989	1990	1992	1995	2000	%Avg. Ann. Growth Rate
Membrane Bioreactors	4	5	8	15	47	25.0
Specialty Bioreactors	6	7	11	20	56	22.5
Related Devices*	27	29	33	41	57	7.0
Conventional Bioreactors	39	50	67	116	177	14.7
Total	76	91	119	192	337	14.5

*These include shaker flasks, roller bottles and other cell culture bottles.
Source: Business Communications Co. Estimates.

3.1 CARRIER MATRIX AND IMMOBILIZATION METHOD

In much of the work carried out in our laboratory, we have employed reconstituted bovine hide collagen as the carrier matrix of choice. The biomaterial, collagen, offers a number of rather unique advantages as a carrier material for enzyme and whole cell immobilization, as indicated in Table 3.2.

Table 3.2 Collagen as a Carrier for Immobilized Cells and Organelles

Simple immobilization procedure; mild conditions.
High density of reactive groups.
Hydrophilic proteinaceous nature.
High swelling levels in aqueous solutions.
Ease of substrate penetration.
Fibrous nature.
Variety of reactor configurations.
Number of sources.
Inexpensive.
High mechanical strength - improved by varying degrees of crosslinking.
Chemical modifications possible for modifying properties.
Possibility of modifying hydrophile/hydrophobe ratio so as to enhance reactions in non-aqueous media.
Possibility of duplicating *in vivo* function.

3.2 COLLAGEN TECHNOLOGY: FIRST GENERATION

Collagen is a material which has demonstrated versatility as a support matrix for enzymes and whole cells for use in food, pharmaceutical, environmental and medical applications (Vieth and Venkatasubramanian, 1976). Several general methods have been developed, including impregnation, electrocodeposition and complexation (see Fig. 3.1). In all three methods the enzyme interacts with the protein carrier through multiple, cooperative, physicochemical interactions prior to crosslinking with glutaraldehyde. Suzuki et al. (1976) have also prepared collagen-enzyme complexes by electrocodeposition and have studied their use in enzyme fuel cells. Coulet and Gautheron (1976) describe a covalent binding technique ("azide method") which has been successfully applied in the preparation of several different collagen-enzyme complexes utilizing hydrolases, transferases and oxidoreductases. Collagen appears to offer several advantages as an enzyme support material such as a hydrophilic, proteinaceous nature, a fibrous structure, high swellability in aqueous media and a high concentration of binding sites. Furthermore, it is readily available and can be cast into a number of different physical forms including tubes, membranes and particles. In addition, it appears to be generally applicable to a large number of enzymes. Due to its biological derivation, it appears to be particularly useful in applications with animal cells, as will be discussed later, in Chapter 9.

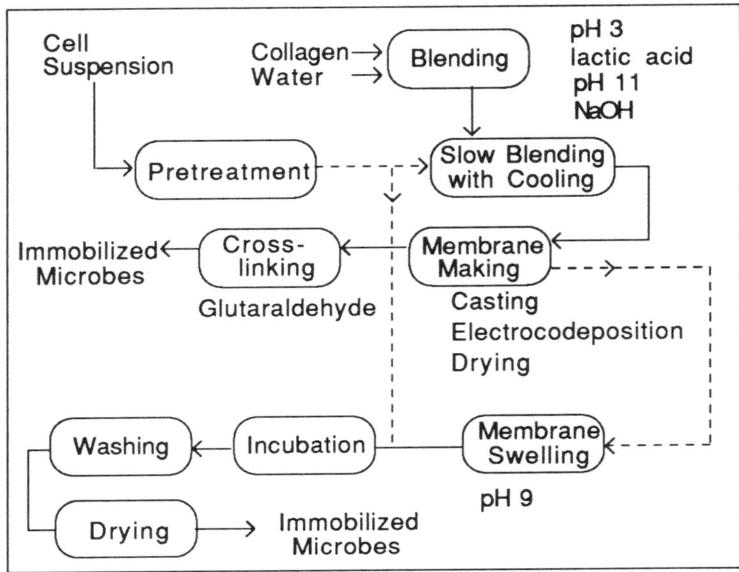

Figure 3.1 Methods of immobilization with collagen.

In mammalian systems, the major role of collagen is in connecting and supporting cell masses, forming tissues. A reconstituted collagen-whole cell structure may be similar in some respects. Thus, the cells are considered bound to collagen by a multiplicity of bonds between chain receptors and whole cell surface moieties. Formation of stable macromolecular complexes by the same type of interactions between enzyme and collagen has been reported for a number of cases (Vieth et al., 1972).

The epsilon-amino group of lysine, which is also implicated in the phenomenon of ageing, has been identified as a principal enzyme receptor site for β-galactosidase in collagen (Luo et al., 1979). At its colloidal degree of subdivision, a cell has a very large surface area. There would thus be a propensity for a large number density of bonds to be formed between the cell and collagen. The cooperative action of this multiplicity of interactions results in a stable cell-collagen complex structure. The structure can then be crosslinked to the desired level of mechanical strength - by brief exposure to a dialdehyde solution such as glutaraldehyde - for prolonged high temperature use (up to 70°C if desired) in a batch or continuous reactor.

Thus the introduction of physicochemical crosslinks in the membrane by cells forming filler-type bridges between the chains complements the action of the crosslinking agent, and the resulting supranetwork structure is stabilized by collagen-collagen, collagen-cell, intercell and intracell linkages acting in concert. The internal architecture of collagen, its hydrophilicity and the high level of surface charge distribution on the tropocollagen molecules in a dispersed phase allow immobilization of a large weight fraction of cells (up to 50% or more on a dry basis) per unit weight of the complex.

Collagen is of interest in and of itself as a major protein constituent of the connective tissue in the vertebrate and invertebrate animal kingdoms. In the body, collagen imparts strength to blood vessels, holds bone crystals together, enables muscles to pull on bones and holds individual cells together in tissues. In the latter case, collagen receptors bind to cells, holding the tissue together, as already mentioned.

Collagen from a large number of animal sources has been extensively studied, and it has been observed that its amino acid composition is distinctive in its very high content of glycyl residues (33% of the total number) and of amino acid prolyl and hydroxyprolyl residues (about 25% of the total number) (see Table 3.3). It is also known that the glycyl residues are evenly spaced in the protein and act as hinges for rotation of the chain, giving rise to a multiplicity of configurations (Howell, 1981).

The fundamental structural element for collagen is a right-handed triple helix, stabilized by interchain hydrogen bonds. Every third residue is a glycine and there is also a substantial representation by proline and hydroxyproline. Ribosomal synthesis of collagen in the endoplasmic reticulum of a cell produces three identical procollagen chains which at their carboxy terminal end form disulfide linkages, enabling correct registry of each of the chains within the triple helix. The enzyme prolyl hydroxylase adds hydroxyl groups to some proline residues after the chains are synthesized and before they twist together. These hydroxyls probably help stabilize the triple helix through hydrogen bonding. The twisted, but still intact, polypeptide is then transported through the cell membrane where enzymes substantially remove the extra portions at each end of the protein (King, 1989).

The structural complexity attainable by a polymer is dependent on the diversity of bonds that hold the constituent atoms together. This property has been christened heterodesmisity. Thus, proteins have a more complex structure (because of the diversity of amino acids available) than any simple synthetic polymer.

Table 3.3 The Amino Acid Composition of Collagen (Human Tendon)

Trivial Name	R in repeating sequence -NH CH R CO-; or amino acid formula	Number of amino acid residues per 1000 total residues
Glycine	-H	324.0
Proline	H_2C—CH_2; H_2C, CH - COOH; ring closed by NH (pyrrolidine ring)	126.4
Alanine	$-CH_3$	110.7
Hydroxyproline	HO - HC—CH_2; H_2C, CH - COOH; ring closed by NH	92.1
Glutamic Acid	$- CH_2 - CH_2 - COOH$	72.3
Arginine	$- CH_2 - CH_2 - CH_2 - NH - C(NH_2)=NH$	49.0
Aspartic Acid	$- CH_2 - COOH$	48.4
Serine	$- CH_2 - OH$	36.9
Leucine	$- CH_2 - CH(CH_3)_2$	26.0
Valine	$- CH(CH_3) - CH_3$	25.4
Lysine	$- CH_2 - CH_2 - CH_2 - CH_2 - NH_2$	21.6
Threonine	$- CH(CH_3) - OH$	18.5
Phenylalanine	$- CH_2 - C_6H_5$	14.2
Isoleucine	$- CH(CH_3) - CH_2 - CH_3$	11.1

Table 3.3 (Cont.)

Trivial Name	R in repeating sequence -NH CH R CO-; or amino acid formula	Number of amino acid residues per 1000 total residues
Hydroxylysine	$-CH_2-CH_2-CH(OH)-CH_2-NH_2$	8.9
Methionine	$-CH_2-CH_2-S-CH_3$	5.7
Histidine	$-H_2C-C=CH-NH-CH=N-$ (imidazole ring: C=N, HC, CH, NH)	5.4
Tyrosine	$-CH_2-C_6H_4-OH$	3.6

The first level of structural hierarchy pertains to the basic amino acid sequence in a protein (Fig. 3.2A). Collagen is made up of 18-19 amino acid residues in relative amounts, which are well documented for several species, but in sequences which are incompletely known for any one species.

Thus, a level of structural complexity arises, due to the arrangement of the amino acid sequence in three-dimensional space. In polypeptides, the option of rotating around skeletal bonds is usually decided in favor of formation of a large number of N-H...O hydrogen bonds, either intramolecularly (α-helix) or intermolecularly (β-structure).

Referring to Table 3.3, it can be seen that proline and hydroxyproline form 25% of the total amount of amino acid residues, i.e., one out of every four residues is a rigid link. [As shown in Fig. 3.2E, at the sequence Gly-Pro-Z, rotation occurs freely at the glycyl residues (arrows in the figure) unless prevented by chain-chain interaction. By contrast, rotation is hindered at the site of the prolyl residue (bonds 1, 2, 3 in Fig. 3.2E), which act as rigid links.] Instead, collagen forms a left-handed version of poly-proline II with a pitch of 9Å (Fig. 3.2B).

The presence of glycyl residues (1 out of 3 amino acid residues) ensures the capacity of extensive sterically permissible hydrogen bonding. Individually considered, therefore, the chains of collagen are helical macromolecules with a strong propensity for interchain rather than intrachain hydrogen bonding interactions.

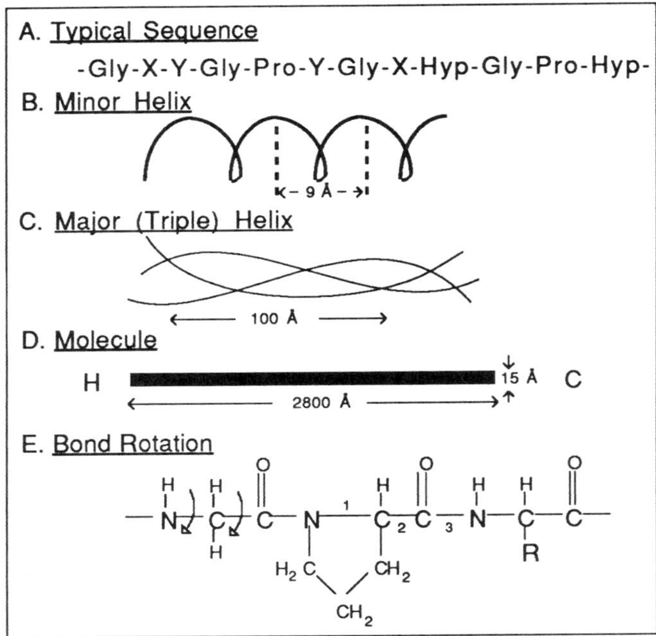

Figure 3.2 Collagen structural hierarchy.

The next level of structural hierarchy refers to the large scale folding, helicity and other convoluted forms assumed by the totality of the polymeric chains which constitute the fundamental molecular unit. In collagen, the fundamental unit is the tropocollagen (TC) molecule, a right-handed super helix or coiled element with a repeat distance of 100Å (Fig. 3.2C), made up of three left-handed strands (Fig. 3.3) supported by interchain N-H...O hydrogen bonds.

In dilute solutions, the tropocollagen molecule behaves like a rod 2800Å in length and 15Å in diameter (Fig. 3.2D), with a molecular weight of about 300,000. Each such α-helix extends the entire length of the tropocollagen molecule and has a molecular weight of ca. 100,000; i.e., about 1,000 amino acid residues.

All three constituent helices are not identical. Of the three helices, two have almost identical composition, *vis-a´-vis* amino acid residues, and are called the α1-helices. The third, or so-called α2-helix, appears to be a bit different. One end of the tropocollagen molecule, involving 10-20 amino acid residues at the N-terminal ends of each of the α-chains, is probably non-helical and contains the sites (probably including lysyl residues) where covalent crosslinking can occur.

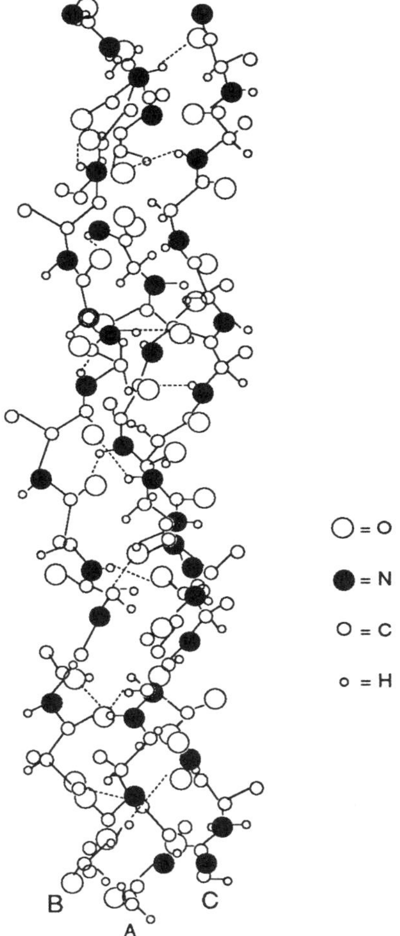

Figure 3.3 Tropocollagen (after Ramachandran and Ramakrishnan, 1976).

This level of complexity refers to the three-dimensional arrangements of tropocollagen molecules. In most synthetic polymers, structural order at this level is usually limited to folded-chain lamella. In collagen, because of a high degree of heterodesmisity, the tropocollagen molecules aggregate to form complex microfibrils. These are several hundred angstroms in diameter and many microns in length.

Relatively less is known about the existence of the next higher order of structuring in collagen. Its existence is necessitated because a macroscopic collagen tendon is not a random conglomerate of quarternary structure crystallites, but a conglomerate of highly oriented entities of crystallites along the fiber axis. This has been proven by x-ray scans of collagen tendon.

Grant and co-workers (1967) regard the tropocollagen molecule as composed of five so-called bonding zones (a zones), separated by four so-called non-bonding zones (b zones). The model does not require that each zone, even of a given type, have components of the same chemical structure, but it does postulate that aggregation is determined by the tendency for the a bonding zones of any one molecule to become associated with like bonding zones on adjacent molecules. Individual chains may twist and rotate and even cross over one another in order to maximize the bonding. The model also provides for considerably more free volume, resulting primarily from the smaller areal density of macromolecules in the aggregated b, or non-bonding zones. Thus, surface area of fibrils should be considerable and diffusion of even large molecules, like enzymes, to appropriate binding sites should be facilitated. The expressed potency of such immobilized biocatalysts in several reactor configurations is taken up directly, in the next section.

3.3 ENZYME REACTORS

A source of contact inefficiency in fixed bed reactors is the presence of a distribution of nonuniform flow channels arising from packing imperfections. These channels allow bypassing of uncontacted fluid elements and may support large radial velocity gradients. One way to circumvent this problem is to deliberately design a system with uniform fine-capillaric channels, which meter out fluid elements with small radial velocity gradients and uniform local mean flow velocities, thereby minimizing bypassing effects. Such predesigned systems (e.g., the catalytic muffler) are occasionally referred to as "monoliths."

For a fixed bed reactor containing an enzyme-membrane, Vieth et al. (1976) developed expressions to describe different reaction kinetic schemes. The physical system modeled is the spiral-wound multichannel bioreactor shown in Fig. 3.4. Considering the membrane to be a semi-infinite plate of thickness 2ℓ, the reactor can be construed to be made up of a set of membranes. Thus, it can be approximated by a parallel membrane model as shown in Fig. 3.5. At steady state, the mass flux through the boundary layer is equal to that into the membrane, i.e.:

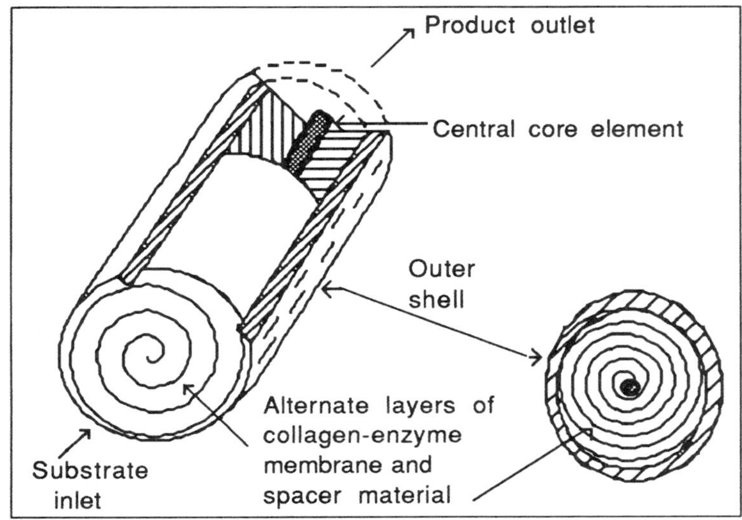

Figure 3.4 Spiral-wound multichannel bioreactor.

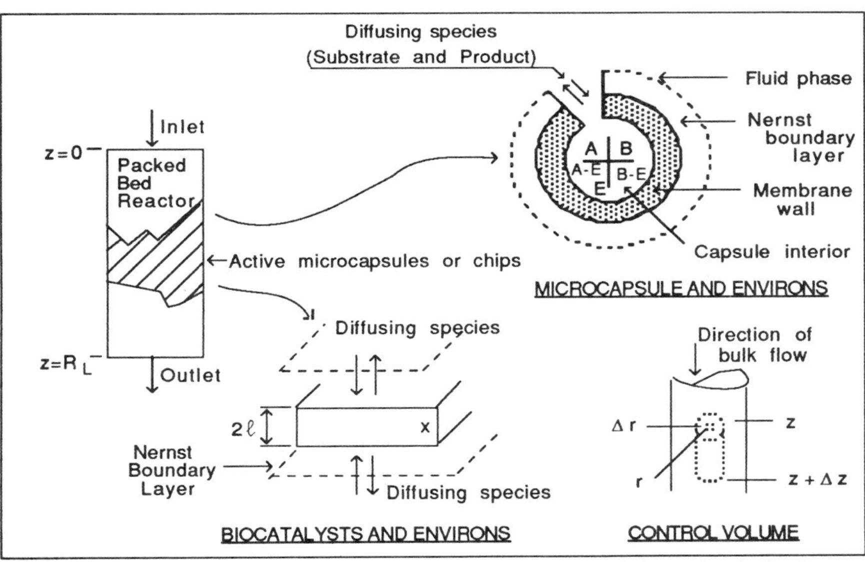

Figure 3.5 Schematic diagram of a packed bed reactor containing spherical microcapsules or collagen-enzyme membrane in the form of chips. z is the direction of fluid flow along the length of the reactor. Membrane chips are considered to be semi-infinite slabs of thickness 2ℓ. Substrate diffusion into the catalyst chips occurs in the x direction. The control volume shell is also shown.

$$J = k_L [S_F - S_s] = \eta \ell k_f S_s \qquad [3.1]$$

where J is the steady state flux, S_f is the bulk substrate concentration, S_s is the substrate concentration at the membrane surface, and k_f is the pseudo first order constant.

Rewriting eqn. [3.1],

$$[S_F - S_s] / J = 1 / k_L \quad \text{and} \quad S_s / J = 1 / \eta k_f \ell \qquad [3.2]$$

which leads to:

$$S_F / J = [1 / k_L] + [S_s / J] = [1 / k_L] + [1 / \eta k_f \ell] \qquad [3.3]$$

Defining a combined mass transfer-kinetic coefficient, K' ($\equiv S_F / J$), eqn. [3.3] can be written as:

$$1 / K' = [1 / k_L] + [1 / \eta k_f \ell] \qquad [3.4]$$

Equation [3.4] represents a series of resistances, uncoupling the effects of diffusion and reaction. This is similar to the analysis of Aris (1957) for heterogeneous catalytic systems in which the reciprocal of an overall effectiveness factor is expressed as a series of resistances. In the case of spherical beads, eqn. [3.4] becomes:

$$1 / K' = [1 / k_L] + [3 / \eta k_f R] \qquad [3.5]$$

where R is the radius of the particle. For microcapsules having an aqueous phase inside a semipermeable membrane wall, the wall permeability must also be accounted for. The wall resistance ($1/k_w$) is included as shown below (Mogensen and Vieth, 1973):

$$1 / K' = [1 / k_L] + [1 / k_w] + [3 / \eta k_f R] \qquad [3.6]$$

The results obtained above - even though based on an idealized solid phase geometry - can be readily extended to immobilized enzyme reactors of practical importance, such as the spiral wound, multichannel biocatalytic modules already mentioned. Integrating the steady state material balance on the substrate passing through a differential reactor element, we obtain:

$$\ln [1 - \chi] = - K'a\tau' \qquad [3.7]$$

where a is the catalyst surface per unit of reactor fluid volume.

It is worth noting that K' can be evaluated experimentally by knowing χ and τ' for steady state reactor operation. The mass transfer coefficient k_L is dependent on fluid velocity. Therefore, the combined coefficient K' would also be a function of fluid velocity. Experimental correlations of (K'a) with flow rate (or linear velocity) can be readily developed. An example of such a correlation is shown in Fig. 3.6 for a collagen-lactase reactor system (Eskamani, 1972). The overall resistance (1/K'a) was found to vary linearly with the reciprocal of the flow rate. From eqn. [3.4] it is seen that the effectiveness factor η can be evaluated from the intercept of Fig. 3.6. For instance, the effectiveness factor for the collagen-β-galactosidase system was thus evaluated to be 0.47. Correlations of this type are useful for scale-up purposes for this particular system. They can be used to design an immobilized enzyme reactor for other combinations of throughput and conversion.

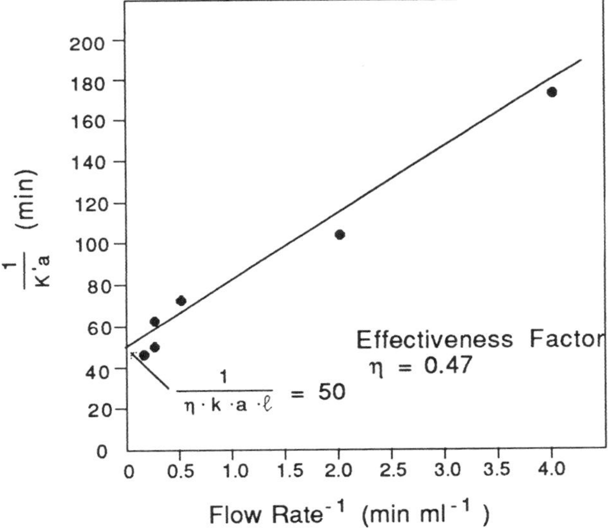

Figure 3.6 Correlation of combined mass transfer-kinetic coefficient with flow rate, based on equation [3.4]. Experimental system studied was collagen-lactase multichannel reactor. This system obeys pseudo first order kinetics.

The reactor performance equation (eqn. [3.7]) can be written as:

$$\ln [1 - \chi] = - [k_2E_0\tau' / K_m'] \eta\bar{P} [k_L / (k_L + [k_2E_0 / K_m'] \eta \ell)] \quad [3.8]$$

$$\ln [1 - \chi] = - [k_2E_0\tau' / K_m'] \eta\bar{P} [k_L / (k_L + [k_2E_0 / K_m'] \eta [R / 3))] \quad [3.9]$$

where $\bar{P} = \ell a$ for flat sheets and (R/3)a for beads or microcapsules; it is the carrier packing factor expressed in cc of carrier per cc of fluid. Equations [3.8] and [3.9] refer to flat sheets and beads or microcapsules, respectively. The term $k_2 E_0$ is the catalytic potency of the reactor; for purposes of simplicity, let us examine the case where the dimensionless parameter $k_2E_0\tau/K'_m$ is assigned the value of unity for the free enzyme in a CSTR/UF system (see column C in Table 3.4).

The quantity $k_2 E_0 \bar{P}$ can be expressed as the product of the enzyme loading factor per cc of carrier, and the reactor loading factor ($\bar{P}$) expressed in cc of carrier per cc of fluid. The values of the latter factor are shown in column D in Table 3.4. Values of the effectiveness factor η which defines the microdiffusional efficiency of the system are tabulated in column F.

In eqns. [3.8] and [3.9], the fourth term on the right-hand side represents the bulk phase transfer or "macrodiffusional" efficiency (ν) of the substrate. Column G (Table 3.4) shows this effect. Now, eqns. [3.8] and [3.9] can be rearranged in terms of the different efficiency factors as:

$$\ln (1 / [1 - \chi]) = [(1.0) \bar{P}\eta\nu] \quad [3.10]$$

where (1.0) = unit value of $[k_2 E_0 \tau' / K'_m]$.

For the combined CSTR/ultrafiltration (UF) membrane reactors employing the free enzyme, a back-correction has been used; i.e., at the same level of conversion (χ = 0.9), the CSTR will require approximately two times the space-time value (τ') required for the packed bed. Therefore, relative to the latter, the CSTR has a space-time factor of 0.5, as shown in column H.

Yet, relatively high values for all the above efficiency factors can be negated simply by a poor integral mean value of the operational stability of the enzyme. This occurs when E_0 decays over more extended operations (e.g., 1000 reactor space-times). Examining columns I and J (Table 3.4), it can be noted that in the case of a free enzyme CSTR/UF process, the serious drawback of the system arises from its poor

Table 3.4 Estimate of Maximum Relative Expressed Potency of Several Single-Enzyme Reactor Configurations[a]

A Enzyme form	B Enzyme configuration	C $\frac{k_2 E_0 \tau'}{K'_m}$	D Carrier packing factor	E Max. relative efficiency, or C x D	F Effectiveness factor	G Macrodiffusional efficiency v	H Contact efficiency: residence time factor relative to plug flow 90% conversion	I Enzyme stability factor	J Maximum Expressed potency relative to Case i. E x F x G x H x I
i. Free Enzyme	CSTR/UF[b]	1.0	1.0	1.0	1.0	1.0	0.5	0.01 (denaturation)	≡1.0
ii. Micro-capsules	Column	1.0	0.5	0.5	0.5	1.0	1.0	0.05 (capsule breakage)	2.5
iii. Ion Exchange Beads	Column	0.02 - 0.5	0.5	0.25	0.5	1.0	1.0	0.5 (slow leaching)	12.5
iv. Collagen Membrane	Micro-Channel Module	0.1 - 1.0	0.5	0.5	0.5	1.0	1.0	0.5 (slow denaturation)	25.0

[a] Basis: rate expression considered pseudo first-order with respect to substrate.
[b] Continuous-flow stirred-tank reactor/ultrafiltration system.

operational stability. Enzyme stability could conceivably be increased by attaching the enzyme to soluble high molecular weight supports. On the subject of stability it must also be pointed out that the operational life and characteristics of the ultrafiltration membrane itself need to be scrutinized carefully. Since the enzyme could be adsorbed on the ultrafiltration membrane, the filtration rate and the rejection efficiency could be significantly altered.

Among the immobilized enzyme reactor systems listed in Table 3.4, microcapsule reactors seem to suffer from poor stability (column I), mainly owing to breakage of capsules. More rigidity of microcapsules could possibly reduce this problem, but may increase the transport impedances. In a more recent paper (Arbeloa et al., 1986), the feasibility of containing microencapsulated urease within a fluidized bed reactor, eliminating problems of membrane rupture, was demonstrated. Hydrolysis of urea under conditions simulating that of an artificial kidney device was measured as a function of reactor residence time, microcapsule diameter, volume of microcapsules, urea concentration in the feed, and enzyme activity. Empirical correlations were developed based on dimensional analysis, which may be used to predict urea conversion within the range of experimental operating conditions. Results under transient state conditions better represent the operation of the reactor in treatment of uremic patients.

Returning to the consideration of Table 3.4, between packed columns and collagen multichannel modules, the main difference arises from the superior enzyme loading capacity of the latter. Thus, on the basis of the data appearing in column J (Table 3.4), it is clear that immobilized enzyme reactor systems can be technically advantageous as compared to free or microencapsulated systems, for process scale conversion of substrates. Within the former, the biocatalytic modules employing collagen-enzyme complex membranes exhibited definite advantages of overall reactor performance, within the limited scope of the comparative study.

Marconi et al. (1978) have described a lactose hydrolysis reactor comprised of cellulose fibers, in which lactase is entrapped, wound around a perforated tube. A schematic diagram of this reactor is presented in Fig. 3.7. This radial flow reactor has been used to produce low lactose milk at an operating temperature of 7°C (Pastore et al., 1976; Marconi and Bartoli, 1981). The same system has been used to achieve 85-90% hydrolysis of whole whey in a semi-batch process at operating temperature between 5°C and 25°C (Marconi et al., 1978).

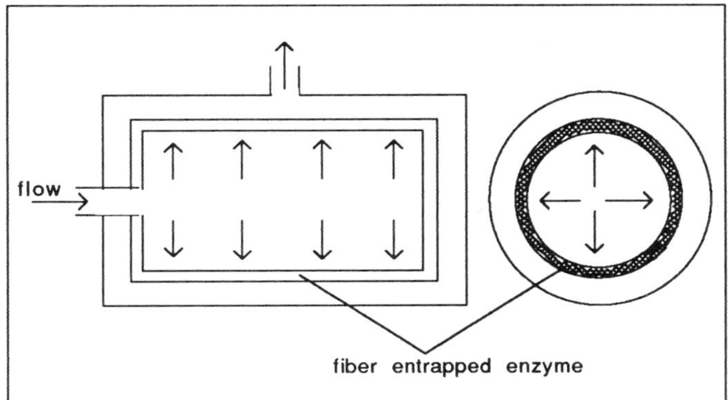

Figure 3.7 Radial fiber immobilized enzyme reactor.

Ichijo et al. (1990) have developed a new fibrous support module made of braided poly(vinyl alcohol) superfine fibers. Continuous sucrose hydrolysis and alcohol fermentation were carried out by invertase and bakers' yeast immobilized on the braid. Immobilized invertase showed high stability, and no pressure increase was observed after repeated usage. Yeast immobilized on the braid grew rapidly and fermented honey to ethanol. Likewise, Inushima (1991) spun fibers of cellulose triacetate entrapping *Saccharomyces cerevisiae.* Cultivation of the immobilized yeast successfully produced alcohol.

The composition and kinetic properties of *S. cerevisiae* were found to be substantially different when the cells were immobilized on gelatin, a hydrolyzed form of collagen. Batch fermentation experiments conducted in a gradientless reaction system allowed comparison of immobilized cell and suspended cell performance. In complete nutrient medium, the specific rate of ethanol production by the immobilized cells was 40-50% greater than for the suspended yeast. The immobilized cells consumed glucose twice as fast as the suspended cells, but their specific growth rate was reduced by 45%. Yields of biomass from the immobilized cell population were lower, at one-third the value for the suspended cells.

Cellular composition was also affected by immobilization. Measurements of intracellular polysaccharide levels showed that the immobilized yeast stored larger quantities of reserve carbohydrates and contained more structural polysaccharide than suspended cells. Flow cytometry was used to obtain DNA, RNA and protein frequency functions for suspended and immobilized cell populations. These data showed that the immobilized cells have higher chromosome concentration than cells in suspension. The level of stable, double-stranded RNA in immobilized cells was only one-quarter that measured for suspended cells. The observed changes in immobilized cell metabolism and composition may have arisen from disturbance to the yeast cell cycle by cell attachment, causing alterations in the normal patterns of yeast bud development, DNA replication, and synthesis of cell wall components (Doran and Bailey, 1986; 1987). Experiments with biotin starvation show that the trend toward reduced stable RNA content in immobilized cells is independent of the response that causes accumulation of DNA.

Gist-Brocades has put into practice an industrial process employing a lower cost collagen variation (i.e., gelatin-immobilized whole cells with glucose isomerase activity, in bead form). Use of collagen enzyme systems for hydrolysis of vegetable protein (Constantinides and Adu-Amankwa, 1980) and for enhanced starch conversion efficiency (Ram and Venkatasubramanian, 1982) are other examples appearing in the literature.

3.4 LACTOSE PROCESSING APPLICATION

Goldberg (1984) earlier had examined hydrolysis of lactose in a spiral wound microporous, PVC-based membrane reactor system. Goldberg and Chen (1985) used a spiral flow reactor, which contains an enzyme immobilized on a porous 2-surface (ribs or net-like sheet) medium coiled in a jelly roll-like configuration. If the feed is free of particulate material (e.g., if it is permeate from an ultrafiltration unit), the feed can flow both in the direction of the spiral (axial flow) and radially through the medium (see Fig. 3.8). If the feed contains particulate matter (e.g., if it is whole whey), the particulates will block the pores and prevent radial flow; in this case, only axial flow will occur. Lactose conversions greater than 90% are easily obtained with this reactor configuration for permeate, whole whey and skim milk.

Peterson (1987) completed a thorough investigation of the system described above. Details of the fine structure of the process flow are indicated in Fig. 3.9. Peterson showed on the basis of dimensionless group analysis (Damköhler number $<<1$) that the reaction is kinetically limited; i.e., that mass transfer rates are indeed rapid in the fine capillaries. Employing a kinetic model involving separate competitive inhibition by α-galactose and β-galactose which was shown to be valid over the range of substrate and product concentrations, Peterson was able to fit the data successfully.

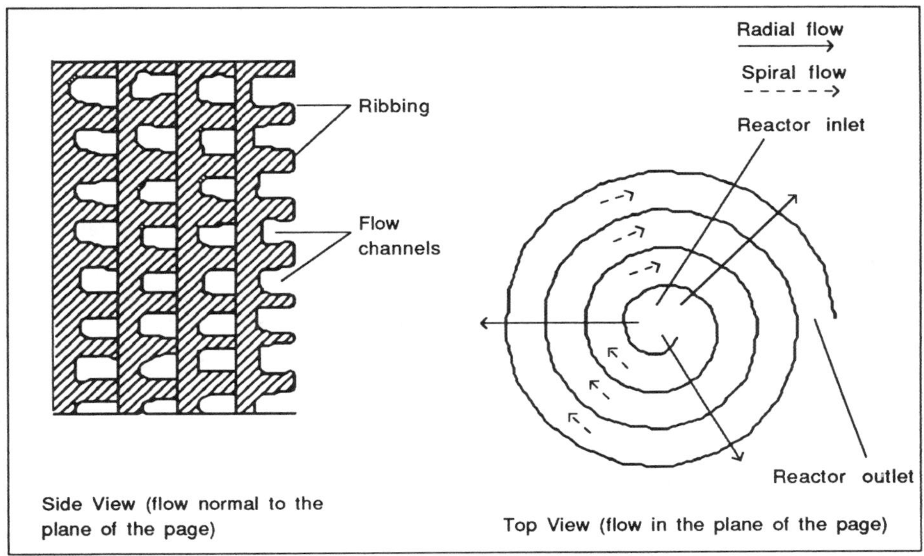

Figure 3.8 Spiral flow reactor.

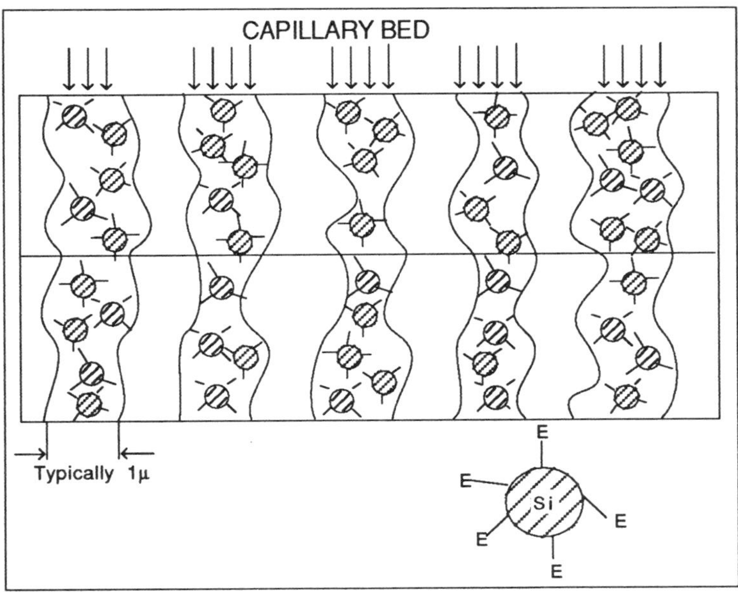

Figure 3.9 Elements of process flow.

Relevant associated kinetic studies include those of Hill and coworkers (Scott et al., 1985; Bakken et al., 1991; 1992) on the determination of the steady state behavior of immobilized β-galactosidase, utilizing an integral reactor scheme constructed of a ribbed membrane made from PVC and silica. Yang and Okos (1989) contributed a new graphical method for determining parameters in Michaelis-Menten type kinetics for enzymatic lactose hydrolysis.

The dispersion coefficient may be related to changes in variances (σ^2) of a pulse entering and exiting a reactor and the mean residence time (Levenspiel, 1973). Peterson was able to confirm the validity of the plug flow postulate, i.e., from the information presented in Fig. 3.10, it is clear that axial dispersion is minimal. On the basis of an axial dispersion coefficient estimated from the correlation in Fig. 3.11, he determined that there was only a 1.5% difference between the conversions predicted by the axial dispersion model and the plug flow model. The recent book by Van't Riet and Tramper (1991) discusses the subjects of mixing and residence time distribution specifically for bioreactors, in some detail.

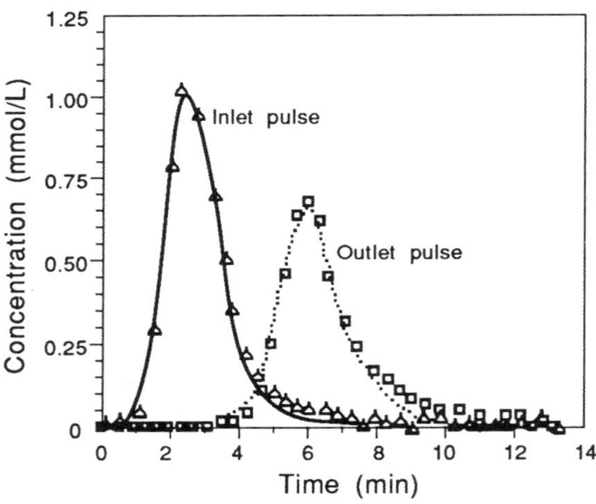

Figure 3.10 Pulse profiles.

Rai (1984) has described an adaptation of the module approach for bioseparations via "cartridge chromatography." The spirally wound cartridge contains a plurality of flow compartments, providing large surface area for efficient mass interchange for protein molecules with an ion exchange matrix in membrane form. Blood proteins and enzymes have been efficiently recovered with this device. Millipore's MemSep cartridges for convective chromatography are another case in point. Ganetsos et al. (1990) report on preparative-scale chromatographic systems as combined biochemical reactor-separators, with application to the biosynthesis of dextran.

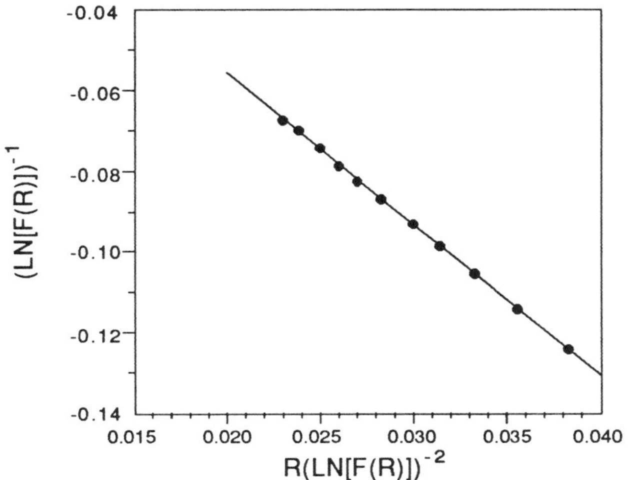

Figure 3.11 Dispersion correlation.

Amerace Esna has announced its modular bioreactor and separation support media. In addition to its basic flow-through modular designs based upon a microporous PVC-silica sheet, the reactor incorporates a spiral design which can be operated in a flow-over mode for processing difficult feedstocks.

3.5 BIOSENSORS

An important special type of bioreactor is the biosensor, which, in a convenient mode, monitors the presence of biologically interesting species by monitoring their rates of appearance at a catalytically active surface. In general, to investigate a penetrant diffusion rate under the influence of an activity gradient, the time lag method may be used to advantage. The method was first proposed by Daynes (1920), and refined by Barrer (1939). It allows a monitoring of diffusion from "time zero" until the diffusing species attains a constant permeation rate.

Under most measurement conditions, the subject membrane (or diffusion zone) is initially free of penetrant. Penetrant is then introduced and equilibrated at the upstream position and the accumulated amount of transported species (kept very small in concentration by using a sweep fluid or a large collection volume) is measured on the downstream side. In this manner, transient diffusion as well as steady state diffusion can be studied.

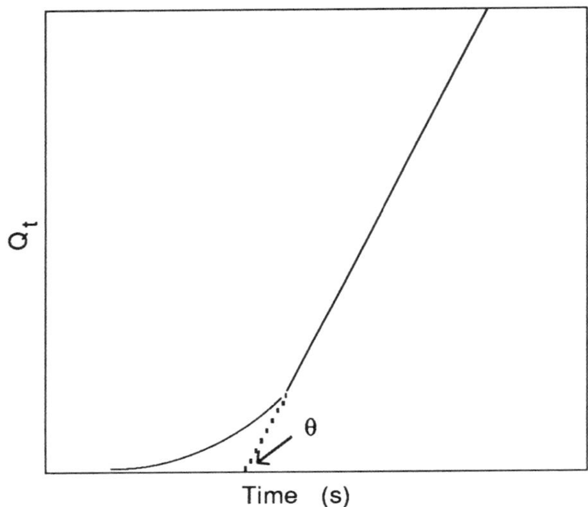

Figure 3.12 Time lag method.

Figure 3.12 is a typical example of a time lag experiment where the early time segment is featured to illustrate the transient; i.e., accumulated diffusive flux is plotted as a function of time. The initial curve shows the changing diffusion rate in the unsteady state and the beginning of the trend toward linearization which becomes much more apparent at longer times (not shown). Linearity demonstrates constant flux conditions when steady state is reached. Extrapolation from the steady state portion of such a figure will result in an intersection of the time axis; this time is referred to as the time lag and is denoted here by θ (Crank and Park, 1968; Crank, 1975).

For diffusion in a plane sheet, the accumulated amount of diffusing substance Q_t which has passed through it in time t is given by:

$$Q_t = \int_0^t J \,|_{x=\ell} \cdot dt = \int_0^t - D\,[\partial C / \partial x] \,|_{x=\ell} \cdot dt \qquad [3.11]$$

where ℓ is the sheet thickness, D is the diffusion coefficient, C is the concentration, and J is local diffusive flux as the penetrant emerges from unit area of the downstream face, $x = \ell$, of the membrane. A simple relationship for the time lag and diffusion coefficient can be obtained:

$$\theta = \ell^2 / 6D \qquad [3.12]$$

where θ is the time lag. According to Crank (1975), within the accuracy of plotting, the steady state is reached when $t = 0.45\ (\ell^2/D)$. Siegel and Coughlin (1972) estimated the relationship between time lag and the time required for a Fickian diffusion process to reach steady state. They concluded that the proper experimental data which should be analyzed for steady state diffusion are those obtained after experimental times larger than 3θ. These results imply that the time for a specific diffusion process to reach steady state bears a fixed relation to the measured time lag.

For a more complicated sorption and diffusion mechanism than simple Fick's law diffusion of a single Henrian species, the time lag function may not be so readily obtained, nor will it always be possible to calculate precisely the time to reach steady state. However, an experimental time lag can always be obtained from its definition; that is, to extrapolate the accumulated amount (Q_t) from the steady state region until it crosses the time axis. While the numerical ratio between time lag and the time required to reach steady state varies from case to case, for each specific system there should exist a certain fixed ratio for purposes of approximation.

3.6 THE PENETRANT TIME LAG IN A MEMBRANE BIOSENSOR

Ludolph et al. (1979) derived an expression for the time lag in a system influenced by the effects of a linear irreversible reaction. The transient equation becomes:

$$\frac{\partial C}{\partial t} = D\frac{\partial^2 C}{\partial x^2} - kC \qquad [3.13]$$

with $C(x,0) = 0$, $C(0,t) = C_0$ and $C(\ell, t) = 0$ as the boundary conditions.

By manipulating eqn. [3.13] and its steady state solution, $C_s(x)$, it is possible to obtain an expression for the time lag, θ, without solving the full transient problem. The expression is:

$$\theta = \frac{\int_0^{\ell} C_s\,dx + k\int_0^{\ell} u\,[x]\,dx - \frac{1}{C_0}\int_0^{\ell} C_s^2\,dx}{J_s\,[\ell]} \qquad [3.14]$$

where u(x) satisfies the boundary value problem:

$$D\,\frac{d^2u}{dx^2} - ku = C_s \qquad [3.15]$$

$$u\,[0] = u\,[\ell\,] = 0 \qquad [3.16]$$

Upon integration, a rather detailed expression results which may be summarized as:

$$\frac{\theta D}{\ell^2} = f\,(\phi) \qquad [3.17]$$

where $\phi = R\ell$ and $R = (k/D)^{1/2}$.

Leypoldt and Gough (1980) presented a solution for the concentration field by using finite Fourier transforms, leading more directly to the useful form:

$$\frac{\theta D}{\ell^2} = \frac{1}{2}\,[\coth\phi\,/\,\phi - 1\,/\,\phi^2] \qquad [3.18]$$

where: $$\phi^2 = \frac{k\,\ell^2}{D} = R^2\,\ell^2$$

Substantial simplification revealed that eqn. [3.17] reduces to this form, reminiscent of the familiar effectiveness factor-Thiele modulus correlation which describes the effect of diffusion on catalytic reaction in a microporous catalyst particle, just described in the previous section.

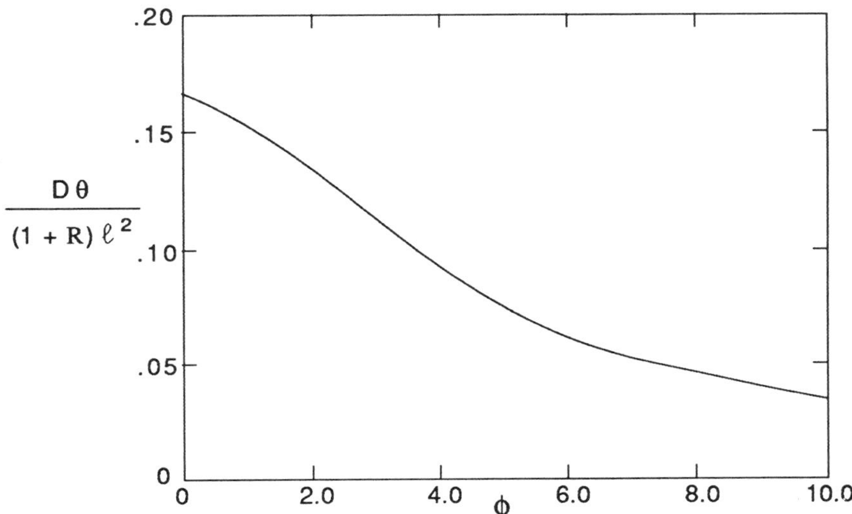

Figure 3.13 The non-dimensional time lag as a function of ϕ.

The main conclusion is that the effect of an irreversible first order reaction is to *decrease the time lag* (Fig. 3.13). This is in contrast to the effect of equilibrium reversible penetrant immobilization, which is to increase θ. Further results are obtainable in a paper by Siegel (1986). A simple formula for diffusional time lags is derived for systems governed by linear equations with source terms. This formula is used to calculate the time lag for systems in which diffusion, dissolution and reversible or irreversible reactions occur. It is also shown that time lags can be calculated for nondiffusional systems. The application of time lags to pharmacokinetics is discussed in particular.

3.7 WHOLE CELL IMMOBILIZATION

Events of the last two decades have borne witness to a surging interest in the field of enzyme and whole cell immobilization for carrying out a variety of bioconversion processes. In addition to the studies introduced above, Radovich (1985) reports on mass transfer limitations in immobilized cells, while Enfors and Mattiasson (1983) describe oxygenation in processes involving immobilized cells. Adams et al. (1988) discuss a range of ceramic biosupports, which are becoming ever more popular. For instance, Sueki et al. (1991) report on the production of a marketable grade of vinegar, employing *Acetobacter* cells immobilized on a new Aphrocell ceramic carrier which has a shape designed for effective liquid/gas exchange. Nillson (1987) covers entrapment of cultured cells in agarose beads. Several comprehensive compendia have rather recently appeared, such as "Enzymes and Immobilized Cells in Biotechnology" (Laskin, 1985) and "Process Engineering Aspects of Immobilized Cell Systems" (Webb et al., 1986). The latter contain detailed treatments of the enzymatic process for high fructose corn syrup as well as an assessment of the industrial potential for application of immobilized cells.

A variety of immobilization techniques ranging from gel entrapment and physical adsorption to covalent attachment are available (e.g., see Fig. 3.14). Depending on the method and the application, the immobilized cells may be growing, live but resting, or in the dead (intact or autolyzed) state. Thus, one encounters systems of varying degrees of complexity - starting from simple single enzyme systems to more complex systems of cells growing in the immobilized state - while performing the necessary bioconversions. Hence, the analysis of an immobilized whole cell reactor would depend on the type of immobilization and the specific application.

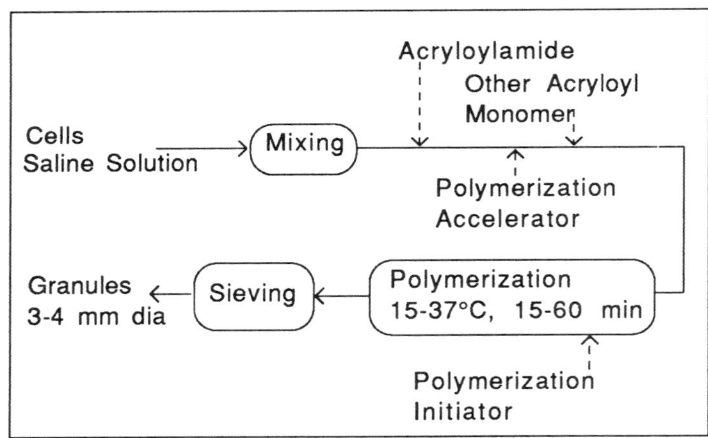

Figure 3.14 Immobilization with polyacrylamide.

Much of the earlier literature on the design and performance analysis of immobilized cell reactors focuses on single enzyme kinetics. Detailed chemical engineering analyses for such systems are now well documented for a variety of reactor configurations. However, the concept of using live microorganisms in the immobilized state has emerged only within about the last ten years. It has been convincingly demonstrated that gel entrapped microorganisms can replicate within the gel matrix (e. g., Larretta Garde et al., 1981; Jirku et al., 1981).

In this section some basic considerations for different types of immobilized cell systems are reviewed. We continue then, by introducing reactor kinetic equations for the case of whole cell-immobilized single enzyme systems and discuss their applications to reactor design. Later, we present the analysis of immobilized cell systems in the context of primary and secondary metabolite synthesis (Chapters 4 and 5). The approach suggests a direction for further analysis which would include the effects of external and internal mass transfer resistances, built upon first generation models introduced near the end of this chapter. Total integration of the concepts then takes place in succeeding chapters (e.g., Chapters 7 and 8) relative to specific case studies.

Technology has been developed to a point where it is capable of accomplishing the design and operation of quite sophisticated enzyme reactors, but the costs are still too high in a number of cases. These costs arise from such aspects as purification of enzymes, replenishment of cofactors and loss of enzyme activity during immobilization and operation. One approach to improve the economics of enzyme processes is to continue to work on purification methods, on cofactor immobilization and regeneration, and on ways of artificially improving storage and operational stabilities of enzymes. An alternate approach is immobilization of whole microbial cells containing the desired enzymic activities. The latter possibility has given rise to the development of immmobilized microbial cell systems or "controlled catalytic biomass."

The first generation of catalytic processes mediated by immobilized microbial cells has already been translated into commercial realities. However, many other processes have been practiced historically which embody the basic principle of enzymic catalysis by microorganisms bound to surfaces. Waste treatment in trickling filters and activated sludge, ethanol oxidation to produce vinegar, microbial activity in soil and marine environments, and leaching of mineral ores are but a few examples of such processes (Abbott, 1977). It is well known that several classes of microorganisms also adhere to solid surfaces and form microbial films (Atkinson, 1974). Tissue culture growth of animal cells is also routinely carried out on solid matrices as a surface monolayer. The term "immobilized microbial cells" as used herein is defined as a system or preparation in which microbial cells are attached to - and confined in - a relatively constricted region of space and employed specifically to effect heterogeneous catalysis in a reactor system. The catalytically active surface should be capable of both easy separation from a reaction mixture and ready use in either continuous flow processes or in repeated batch contacts.

Our purpose here is not only to report further on the recent developments mentioned at the outset of this chapter, but also to draw upon our own work in this area to examine recent developments critically. In this manner, we can attempt to achieve a degree of internal self-consistency possibly unattainable otherwise, relative to: the catalytic role of such systems *vis-a´-vis* immobilized (isolated) enzymes and classical submerged fermentations; biochemical, physiological and process engineering considerations; and delineation of potential technological advances.

The Rutgers Biochemical Engineering Laboratories have devoted a rather concerted effort to the study of immobilized cell systems in the past two decades. Work began by initiating studies on immobilized whole cells with simple systems such as glucose isomerization, which involve a single enzymatic reaction. The excellent results obtained in this case, and our strong belief that several fundamental elements of immobilized cell systems could be understood only by examining more complex reaction sequences characterized by entire metabolic pathways, have steered our efforts to the study of such systems.

Table 3.5 summarizes several immobilized cell processes which have been investigated. It is worth noting that the types of cells bound to collagen range from bacterial cells all the way up to mammalian cells. The complexity of these systems in terms of the nature of catalytic reaction can be categorized as follows:

1. Reactions involving single enzymes.
2. Reactions with multi-enzyme systems, but no specific cofactor requirement.
3. Cofactor-requiring systems (single or multiple enzymes).
4. Processes which involve a complete metabolic pathway (primary or secondary metabolite).
5. Immobilization of cell organelles.

The difficulty of the immobilization task compounds as one proceeds from category 1 to category 5.

It is obvious that the systems detailed in Table 3.5 represent a wide range of bound cell-mediated catalysis and provide a firm basis to discuss the objectives outlined in the previous paragraph. We have followed the sequence of Table 3.5 closely in developing these themes.

Immobilized cell systems are, of course, also subjected to a number of disadvantages. These are considered in some detail in a later section.

Table 3.5 Collagen-Immobilized Whole Cell Systems

Microbe/Organelle	Substrate	Product	Comments
Streptomyces venezuelae Bacillus coagulans	Glucose	Fructose	Glucose isomerization; single enzyme process
Saccharomyces cerevisiae	Sucrose	Invert sugar	Single enzyme
Escherichia coli	Fumaric acid	L-Aspartic acid	Single enzyme
Nocardia erythropolis Mycobacterium rhodochorus	Cholesterol	Δ^4-Cholestenone	Non-aqueous medium; single enzyme
Corynebacterium simplex	Hydrocortisone	Prednisolone	Steroid modification; single enzyme w. cofactor reqiorement; non-aqueous medium
Serratia marcesçens	Glucose	2-Keto gluconic acid	Multi-enzyme
Acetobacter sp.	Ethanol	Acetic acid	Multi-enzyme; cofactor
Corynebacterium lilium	Glucose	Glutamic acid	Pathway (primary metabolite)
Aspergillus niger	Sucrose	Citric acid	Primary metabolite
Chloroplast	Water	Oxygen	Immobi.organelle; first step in biophotolysis of water
Anacystis nidulans	Water	Oxygen	Immobilized algal cells
Anacystis nidulans	Nitrate	Ammonia	Biological nitrogen fixation
Streptomyces griseus	Glucose	Candicidin	Antibiotic synthesis; secondary metabolite
Pseudomonas aeruginosa	- - -	- - -	Concent.of plutonium from faste waters (bioadsorp.)
Klebsiella pneumoniae	Nitrogen	Ammonia	Microbial fixation of atmospheric nitrogen
Mammalian erythrocyte	- - -	- - -	Model studies of *in vivo* enzyme action

3.8 ACTIVITY AND STABILITY OF IMMOBILIZED CELLS

From a practical standpoint, the two most important characteristics of an immobilized cell catalyst are its activity and its operational stability. The latter parameter is usually expressed as catalyst half-life. The amount of activity would be a function of cell-to-carrier loading ratio. Membranes containing up to 50% cells (by dry weight) are found to possess an excellent combination of mechanical and catalytic properties. The apparent initial specific activities of several bound cell membranes catalyzing simple conversions (mono-enzyme reactions) are listed in Table 3.6. In all these cases, 98 to 99% of theoretical maximum conversion of the substrate was achieved. The high half-life values obtained lend credence to the speculation that the enzyme in question is more stable when left in the cell. This is particularly exemplified by the glucose isomerase system. The bound cells yielded a half-life of as high as 1800 hours. Translation to the commercial scale may reduce this value. Kalevi (1989) recently reported on manufacture of crosslinked water-insoluble glucose isomerase with dialdehydes and amines. The half-life of this crosslinked enzyme was about three-fold greater than that of DEAE cellulose-immobilized enzyme when used in a reactor. The transport rate for the crosslinked crystals was five- to ten-fold greater than that for columns containing another commercially available immobilized enzyme preparation.

Table 3.6 Activity and Operational Half-Life of Immobilized Cell Systems Catalyzing Single Reactions

Reaction	Organism	Apparent specific activity (I.U./g catalyst)	Operational half-life (h)	Reference
Glucose isomerization	*S. venezuelae*	28	500	Saini & Vieth,1975
Glucose isomerization	*B. coagulans*	70	1800	Vieth & Venkatasubramanian,1976
Fumarate to aspartate	*E. coli*	400	2000	Venkatasubramanian et al., 1977
Sucrose inversion	*S. cerevisiae*	73	-	Goldstein et al.,1977
Penicillin modification	*E. coli*	26	350[a]	Venkatasubramanian et al., 1977

[a]Extrapolated from 60 h of actual reactor operation.

Dehydrogenation of the steroid hydrocortisone to prednisolone - pharmacologically a more active steroid - involves a single enzyme. However, this reaction is quite complex on two counts: insolubility of the substrate in water and requirement of menadione as a cofactor to effect the overall dehydrogenation reaction. The reaction was therefore carried out in a non-aqueous medium with ethanol (at 15% level) as the solvent. Similarly, in the case of cholesterol oxidation by fixed cells of *Mycobacterium rhodochorus*, carbon tetrachloride or tetradecane was used as the solvent. Maximum achievable levels of conversion were 80 and 98% for hydrocortisone dehydrogenation and cholesterol oxidation, respectively. Specific reaction rates and half-life values for these systems are reported in Table 3.7. While these results are not spectacular, they appear at least to demonstrate the feasibility of the immobilized cell approach for reactions conducted in non-aqueous media. Furthermore, the conversion levels and half-life values obtained in these experiments are better than those in a number of fermentation processes. It is well known that many microbial steroid transformation processes are often hampered by substrate and/or product inhibitions. However, the results seem to indicate that the collagen-immobilized cell process does not suffer from such inhibitions (Montana, 1977).

Table 3.7 Steroid Modifications by Immobilized Cells

Reaction	Organism	Substrate Concent. (g/liter)	Max. Conver. (%)	Specific reaction rate[a]	Half-life (h)
Hydrocortisone dehydrogenation	*C. simplex*	0.67	80	0.17	130
Cholesterol oxidation	*M. rhodochorus*	20	82	800	95

[a]Milligram substrate converted per gram catalyst per h.

Ikemi et al. (1990a; b) describe sorbitol production in a charged membrane bioreactor with coenzyme regeneration. A reactor equipped with a charged membrane as the coenzyme separator module was constructed and used in the continuous production of sorbitol. NADPH-dependent aldose reductase was used for the production of sorbitol from glucose. The oxidized enzyme was enzymically regenerated by conjugation with glucose dehydrogenase, together with the coproduction of gluconic acid from glucose. With a substrate conversion of 85%, 100 g/L sorbitol was produced and equimolar gluconic acid was coproduced for more than 800 hours, indicating that the reaction was efficiently coupled to the enzymic regeneration.

Chithra and Baradarajan (1990) carried out a mathematical analysis of the performance of a packed bed co-immobilized bioreactor. Each step in the consecutive reaction is assumed to follow Michaelis-Menten type kinetics. The model includes all the possible limiting steps controlling the rate of reaction (as well as the additional effect of axial dispersion in the bulk liquid).

3.9 IMMOBILIZATION OF ISOLATED ORGANELLES

In the cell, organelles represent highly ordered compartments where specific reaction sequences are carried out. The spatial and conformational arrangements of such multienzyme complexes have been recognized as crucially important factors in mediating these reactions very efficiently. Respiration and phosphorylation in the mitochondria are examples of such reaction schemes. Unlike the intact cells, isolated organelles are not protected by a thick cell wall and are therefore very fragile. Organelle immobilization thus presents additional problems. Special techniques must be used to preserve not only the activities of the individual enzymes but also their structural and conformational integrity. As an example of organelle immobilization, fixation of chloroplasts is discussed here briefly.

Splitting of water biologically to hydrogen and oxygen in the presence of sunlight is referred to as biophotolysis of water. Conceptually it is an appealing means for capturing solar energy to produce a valuable fuel, namely, hydrogen. The reaction is highly complex, involving two photosynthetic electron transport systems (PETS), as well as electron acceptors such as ferrodoxin or methyl viologen. A second enzyme, hydrogenase, is required to complete the reaction scheme to produce hydrogen. Chloroplasts isolated from spinach leaves contain the PETS system but do not have the hydrogenase, which must be supplied from a different source. Certain blue-green algae also have the PETS system in their cells.

Activities of free and immobilized chloroplasts and algal cells (*A. nidulans*) obtained in preliminary experiments are shown in Table 3.8. Immobilized chloroplasts retain about half of the activity of free chloroplasts, while the activity retention is about 27% in the case of bound algae cells. The normal photosynthetic rate in an actively growing plant is about 450 μmol O_2/mg chlorophyll-h. The activities observed in this early study represented an energy recovery efficiency of 1.5 to 3.0%

Table 3.8 Typical Activities of Collagen-Chloroplast and Collagen-*A. nidulans* Membranes

Sample	Chlorophyll Content	Specific activity (mmol O_2/mg chlorophyll-h)
Stripped chloroplast preparation	2.5 mg chlorophyll/ml	35.4
Collagen-chloroplast membrane	11.8 mg chlorophyll/g	18.5
Algal cells	11.9 mg chlorophyll/g	29.1
Collagen-algal cells membrane	3.6 mg chlorophyll/g	7.7

Assay conditions: 25°C; light intensity 40,000 Lux; ca. 4-mil-thick membranes; 70% of dry weight of the membrane was collagen; assay medium consisted of pH 7.4, 0.34 M sucrose or sorbitol, 0.01 M sodium chloride, 0.6 mM phenylene diamine and 1.5 to 3.0 mM potassium ferricyanide. All assays were short-term (3 to 8 min.).

when compared with a growing plant. Intermittent batch assays conducted at 25°C using potassium ferricyanide as electron acceptor indicate that the fixed chloroplast preparation is active after 15 days of storage. These subjects are taken up in much greater detail in Chapter 10, and a kinetic model for photosynthetic electron transport is presented there.

3.10 REACTORS WITH WHOLE CELLS

In approaching the analysis of immobilized whole cell reactors, appropriate consideration must be given to the two broad classifications of immobilized cell processes. Non-viable cells which express the activity of intracellular enzyme can be treated essentially as if they were carriers of a single bound enzyme system. The analysis of such a system necessitates integration of enzyme kinetics with appropriate catalyst environmental quantitative indices, (e.g., effectiveness factor) which account for external and internal mass transfer resistances. In contrast, for viable cells which may be resting or growing in the immobilized state, proper allowances must also be made for cell growth kinetics and cell maintenance requirements, as are discussed in the next chapter.

3.11 SINGLE ENZYME TYPE IMC REACTORS

The reactor spacetime τ is defined as:

$$\tau = V_R / Q \qquad [3.19]$$

where V_R is the volume of the reactor and Q is the flow rate through the reactor.

Fractional conversion χ is expressed by:

$$\chi = [S_0 - S] / S_0 \qquad [3.20]$$

where S_0 and S are the inlet and outlet substrate concentrations respectively; reactor (volumetric) productivity Pr is defined as:

$$Pr = \chi S_0 / \tau \qquad [3.21]$$

The enzymatic rate of reaction is commonly expressed by the Briggs-Haldane monoenzyme, monosubstrate, stationary state model:

$$r = -\,[dS / dt] = \frac{k_2 ES}{K_m + S} \qquad [3.22]$$

where r is the reaction rate, S is the substrate concentration, K_m is the Briggs-Haldane (popularly known as the Michaelis-Menten) constant, and $k_2 E$ is the maximum reaction rate (V_m) for that system. Occasionally K_m and V_m are replaced by apparent constants K'_m and V'_m to account for external influences on intrinsic kinetics.

Enzyme inactivation kinetics can be adequately described as a pseudo first order process.

$$-\,[dE / dt] = k_d E \qquad [3.23]$$

where E is the effective enzyme concentration in the reactor at time t and k_d is the first order decay constant.

In analyzing the industrial process for production of L-aspartic acid using immobilized *E. coli* cells in a packed bed reactor, Seko et al. (1990) report that the reactor could best be modeled as a multistage continuous stirred tank reactor. A simulation model enabled prediction of temperature and concentration profiles along the bed height, leading to the establishment of an optimal policy for temperature control to maximize the operational life of the reactor. The policy was implemented through the use of a heat exchanger tube bundle of four tubes incorporated into the reactor shell.

3.12 EFFECT OF MASS TRANSFER ON THE PERFORMANCE OF IMMOBILIZED CELL REACTORS

External (film) diffusion, diffusive and electrostatic effects, internal (pore) diffusion, and combinations of these effects constitute the array of possible influences which may be encountered.

EXTERNAL (FILM) DIFFUSION

Mass transfer of the substrate from the fluid to the catalyst surface can be represented by:

$$r_m = k_L\, a_m\, [S_F - S_S] \qquad [3.24]$$

where k_L is the mass transfer coefficient, a_m is the surface area for mass transfer, and S_F and S_S are the substrate concentrations in the bulk and at the surface, respectively. Correlations may be consulted to estimate k_L for different particle geometries and operating conditions (Vieth et al., 1976).

First order kinetics constitute a reasonable approximation to enzyme reaction behavior for many engineering calculations. For packed bed geometries, an equation of the type:

$$\tau' = k_f\,(-\ln\,[1 - \chi]\,) \qquad [3.25]$$

can be employed where $k_f = V'_m/K'_m$, the pseudo first order rate constant, and τ' is the reactor spacetime based on reactor fluid volume:

$$\tau' = [V_R\varepsilon]\,/\,Q \qquad [3.26]$$

The constant k_f is a limiting form of k'_f, the modified pseudo first order constant incorporating both kinetic and diffusional resistances.

$$k'_f = \frac{k_f\, k_L\, a_m}{k_f + k_L} \qquad [3.27]$$

DIFFUSIVE AND ELECTROSTATIC EFFECTS

Boundary layer diffusional resistance can appear together with substrate partitioning by electrostatic forces. Surface immobilized whole cells might be expected to display the net negative or positive charge borne on the cell walls. Perhaps it will prove possible to carry over the analysis of Hamilton et al. (1973) who examined a wide range of surface potentials using the Gouy-Chapman potential distribution.

INTERNAL (PORE) DIFFUSION

The classical approach to this problem is through the use of an effectiveness factor, η, which compactly expresses the ratio of the observed reaction rate to that which would apply if the enzyme particle were gradientless in substrate concentration. The actual situation for a single catalyst element can be described by a second order differential equation:

$$D_e \nabla^2 S - r = 0 \qquad [3.28]$$

The solution of this equation can be put in the form of a relation for the effectiveness factor in terms of the Thiele modulus ϕ:

$$\phi = \ell \, [k_{true} \, S_2^{m-1} / D_e]^{0.5} \qquad [3.29]$$

where z is the distance from the center of the catalyst particle, ℓ is the characteristic particle dimension, D_e is the effective diffusivity, m is the reaction order and k_{true} is the undisguised or true kinetic constant.

The effectiveness factor for spherical particles in a packed bed is then:

$$\eta = [1 / \phi] \, [1 / \tanh 3 \phi - 1 / 3\phi] \qquad [3.30]$$

while that for a packed bed of membrane-like chips is:

$$\eta = [\tanh \phi / \phi] \qquad [3.31]$$

Once the effectiveness factor is calculated, the actual rate expression is multiplied by this factor and used in the reactor performance equation. More intricate examples, including hollow fiber ones, are described in detail in another publication (Venkatasubramanian, 1979).

3.13 FLUIDIZED BEDS

The foregoing discussion pertains chiefly to fixed bed reactors. Until recently, such reactors have been used almost exclusively in processing with immobilized enzymes, even with fairly rapidly deactivating catalysts. However, in an important departure, a fluidized bed of alginate-immobilized yeast cells for the production of ethanol from glucose has been successfully operated on the pilot scale (4,000 liters) by Kyowa Hakko Kogyo Co. (Nagashima et al., 1984). In related work, ethanol fermentation was demonstrated in a multistage, fluidized bed bioreactor with glucose as the main substrate, using yeast cells immobilized in calcium alginate particles (Tzeng et al., 1991). The bioreactor alleviated problems associated with CO_2 evolution and provided good mixing of liquid and solid phases; no gas clogging and channeling were observed. Enhanced volumetric productivity and conversion were obtained, compared to a single-stage bioreactor. The volumetric productivity ranged from 14.90 to 17.41 g/L-h with 87% to 97% conversion and 66.8 to 93.3 g/L outlet ethanol concentrations. Ethanol conversion in the multistage bioreactor was adequately simulated by a model which considers a complete-mixing flow pattern in each stage, free-cell kinetics, liquid-solid mass transfer and intraparticle diffusion.

Phase holdup and axial dispersion measurements were made in a model fluidized bed reactor (FBR) using small low-density beads as the solid phase. Previous results from a one inch internal diameter columnar fluidized bed reactor were extended to a larger three inch column. Davison (1990; Davison and Scott, 1988) found that larger diameter systems for high productivity bioconversions demonstrate improved operability over the smaller diameter fermentation FBRs that had been successfully tested earlier. Slugging behavior is not a problem in three to six inch diameter columns because the bubbles remain small, relative to the column diameter. Likewise, wall effects should be negligible, with the result that the hydrodynamics should scale to even larger systems. In addition, if the limit on liquid fraction holds for a broad range of gas and liquid velocities, the gas hold-up (less than 3%) can be virtually neglected in a larger fermenting FBR. This will simplify the modeling and scale-up of the fluidized bed fermentor system. Nonetheless, the gas flow will have an effect on the overall system behavior due, in part, to bed expansion, lowering the solids fraction (Bly and Worden, 1990; Bajpai et al., 1990).

Recently, Petersen and Davison (1991) developed a model of a three-phase, tapered fluidized bed bioreactor which includes a variable dispersion coefficient, the concentration profile inside the biocatalyst bead and the effects of tapering. Analyzing reaction rates within the bed, the model was found to predict experimentally obtained concentration profiles quite accurately; it also demonstrated the need to include the effects of variable dispersion in three-phase systems where the gas is being generated inside the reactor, as the dispersion coefficient varied by more than an order of magnitude across the bed.

Enhancement of the volumetric rate of oxygen transfer in three-phase fluidized bed bioreactors was studied by Kang et al. (1991). The rates of oxygen transfer from air bubbles to viscous liquid media were promoted by floating bubble breakers in three-phase fluidized beds operated in the bubble coalescing regime. The liquid-phase volumetric oxygen transfer coefficient was recovered by fitting the axial dispersion model to the resultant data. The dependence of the mass transfer coefficient on the experimental variables, such as the gas and liquid flow rates, particle size, concentration of bubble breakers, and liquid viscosity was examined. The liquid-phase volumetric oxygen transfer coefficient was enhanced by 20-25% through the use of bubble breakers. The coefficient has a maximum with respect to the volume ratio of the floating bubble breakers to the fluidized solid particles; it increases with increase in either the gas and liquid flow rates or size of fluidized particles and decreases with an increase in the liquid viscosity. Similarly, Sun and Furusaki (1990) describe the continuous production of acetic acid using immobilized *Acetobacter aceti* in a three-phase fluidized bed bioreactor. Experimental results and/or theoretical calculations based on the kinetic models showed that suspended cells were important in the production of acetic acid if the solid holdup was small or if gel radius was large. Theoretical calculations showed that an optimal solid holdup or gel size existed at higher dilution rates because of the volumetric oxygen transfer coefficient dependence on solid holdup and particle diameter.

Vos et al. (1990a; b; c) discuss aspects of the multistage fluidized bed reactor (MFBR) which are different from single stage fluidized beds operated in the batch mode with respect to the solids. Semicontinuous transport of the particles requires perfect mixing in the reactor compartments because particles are mainly transported from the bottom. A large spread in the physical properties of the biocatalyst, especially of both size and density, may cause the particles to segregate into layers with different diameter and/or density. This affects the efficient use of the biocatalyst.

In the MFBR, flow of biocatalyst is countercurrent to the substrate solution, achieving a plug flow mode of contacting. In an experimental study, particles withdrawn from the reactor compartments were investigated using an image analyzer. Histograms of particle size distribution did not indicate segregation and it is concluded that the particles used were mixed completely within the compartments. As a result, transport of the biocatalyst is nearly plug flow.

For reactions in the first order regime, enzyme requirements in this new reactor are slightly less than for fixed bed processes. Overall, the multistage fluidized bed appears to be an attractive reactor design to use biocatalyst to a low residual activity. However, uniformity of the particles is essential for plug flow transport of the biocatalyst.

In related studies, Kiesser and coauthors (Kiesser and Bauer, 1988; Kiesser et al., 1990a; b) have worked on the development and application of a solid-liquid fluidized bed reactor for the enzyme catalyzed reaction with immobilized glucose isomerase. Experiments concerning hydrodynamics, overall conversion and conversion profile as a function of reaction height were performed. In other experiments, a static-mixer was used to reduce the expansion ratio of the fluidized bed. The results indicate that conversion for fixed and fluidized bed reactors is comparable. For the calculation of the concentration profile inside the reactor, a semianalytical solution is derived which takes into account the height dependence of the effectiveness factor. Calculated and experimentally determined values for the axial dispersed plug flow model were in agreement. It was concluded that the overall model is well suited to predicting the effects of specific parameters on the effective kinetics of the biocatalyst and the expansion of the fluidized bed.

Asif et al. (1991) examined the effect of flow nonuniformities existing near the distributor of a liquid fluidized bed containing very low density particles (p_s = 1.05 g/ml) (e.g., as immobilized enzyme or cell bioreactors). With poorly designed multihole distributors, dead zones can be created which seriously undermine the performance of the bioreactor. Design guidelines are provided for distributors to minimize the inlet flow disturbances, and hence eliminate potential distributor region problems. As a result, the dispersion model, which is normally used for design purposes, can then be applied.

3.14 NOMENCLATURE

a_m	Surface area for mass transfer, cm^2/g catalyst.
D_e	Effective diffusivity, cm^2/sec.
E	Enzyme concentration in the reactor, g/L.
J	Mass flux, g/cm^2 min.

K'	Combined mass transfer-kinetic coefficient, cc of reactor fluid/cm^2 (catalyst surface) min.
K_m	Briggs-Haldane/Michaelis-Menten constant, g/L.
K'_m	Apparent Michaelis-Menten constant, g/L.
k	Membrane mass transfer coefficient, m/hr.
k_2	Reaction rate constant, min^{-1}.
k_d	Enzyme decay constant, min^{-1}.
k_f	Pseudo first order rate constant, min^{-1}.
k'_f	Modified pseudo first order rate constant, min^{-1}.
k_L	Mass transfer coefficient, cm/sec.
ℓ	Characteristic dimension of catalyst particle, cm.
m	Order of reaction.
$\bar{P}$	Catalyst packing density, cc catalyst/cc reactor fluid.
r	Reaction rate, g/L/min.
r_m	Rate of mass transfer, g/min/g catalyst.
S	Substrate concentration, g/L.
S_F	Bulk substrate concentration, g/L.
S_0	Inlet substrate concentration, g/L.
S_S	Substrate concentration at the catalyst surface, g/L.
t	Time, min.
V	Volume, ml.
V_m	Maximum reaction rate, g/L/min.
V'_m	Apparent maximum reaction rate, g/L/min.
V_R	Reactor volume, L.
χ	Fractional conversion of substrate.
χ_o	Conversion at time t = 0.
χ_t	Conversion at time t.
ϵ	Fractional void volume of packed-bed reactor.
η	Effectiveness factor.
ν	Macrodiffusional efficiency factor.
τ	Reactor space-time = V_R/Q, min.
τ'	Reactor space-time based on reactor fluid volume = $V_R \epsilon/Q$, min.
ϕ	Thiele modulus = $\ell(k_{true} S_S^{\ m-1})/D_e)^{0.5}$.
ϕ_m	Modified Thiele modulus = $\ell(V'_m/[K'_m D_e])^{0.5}$.

3.15 REFERENCES

Abbott, B. J., In "Annual Reports on Fermentation Processes," D. Perlman, Ed., p. 205. Academic Press: New York (1977).

Adams, J. M., L. A. Ash, A. J. Brown, R. James, D. B. Kell, G. J. Salter and R. P. Walter, "A Range of Ceramic Biosupports," Am. Biotech. Lab. Oct. (1988).
Arbeloa, M., R. J. Neufeld and T. M. S. Chang, *J. Mem. Sci.*, **29**, 321 (1986).
Aris, R., *Chem. Eng. Sci.*, **6**, 262 (1957).
Asif, M., N. Kalogerakis and L. A. Behie, *AIChE J.*, **37**, 1825-1832 (1991).
Atkinson, B., "Biochemical Reactors," p. 153. Pion Limited: London (1974).
Bajpai, R., J. E. Thompson and B. H. Davison, *Appl. Biochem. Biotechnol.*, **24-25**, 485-496 (1990).
Bakken, A. P., C. G. Hill and C. H. Amundson, *Appl. Biochem. Biotechnol.*, **28-29**, 741-756 (1991).
Bakken, A. P., C. G. Hill and C. H. Amundson, *Biotechnol. Bioeng.*, **39**, 408-417 (1992).
Barrer, R. M., *Trans. Farad. Soc.*, **35**, 628 (1939).
Bly, M. J. and R. M. Worden, *Appl. Biochem. Biotechnol.*, **24-25**, 553-564 (1990).
Chithra, N. and A. Baradarajan, *J. Chem. Technol. Biotechnol.*, **49**, 115-127 (1990).
Constantinides, A. and B. Adu-Amankwa, *Biotechnol. Bioeng.*, **22**, 1543 (1980).
Coulet, P. R. and D. C. Gautheron, In "Analysis and Control of Immobilized Enzyme Systems," D. Thomas and J. P. Kernevey, Eds., p. 165. North Holland: Amsterdam (1976).
Crank, J. and G. S. Park, "Diffusion in Polymers," Academic Press: New York (1968).
Crank, J., "The Mathematics of Diffusion," 2nd Ed., Clarendon Press: Oxford (1975).
Davison, B. H., In *"Biochemical Engineering," Ann. N. Y. Adad. Sci.*, **589**, 670-677 (1990).
Davison, B. H. and C. D. Scott, *Appl. Biochem. Biotechnol. Symp.*, **18**, 19-34 (1988).
Daynes, H. A., *Proc. Roy. Soc. Lond.*, Ser. A **97**, 286 (1920).
Doran, P. M. and J. E. Bailey, *Biotechnol. Bioeng.*, **28**, 73-87 (1986).
Doran, P. M. and J. E. Bailey, In "Enzyme Engineering 8," A. I. Laskin, K. Mosbach, D. Thomas and L. B. Wingard, Eds. *Ann. N. Y. Acad. Sci.*, **501**, 330-335 (1987).
Enfors, S. O. and B. Mattiasson, In "Immobilized Cells and Organelles," B. Mattiasson, Ed. CRC Press: Boca Raton, FL (1983).
Eskamani, A., Ph. D. Thesis, Dept. of Chemical and Biochemical Engineering, Rutgers University (1972).
Ganetsos, G., P. E. Barker and A. Akintoye, *Inst. Chem. Eng. Symp. Ser.*, **118** (Adv. Sep. Processes), 17-24 (1990).
Goldberg, B. S., "A Novel Immobilized Enzyme Reactor System," paper presented at the 24th Annu. Spring Symp., AIChE, East Brunswich, NJ, May 10 (1984).
Goldgerg, B. S. and R. Y. Chen, US Patent Appl. 595954, Apr. 2 (1984); Patent Appl. 85/302224, Mar. 29 (1985).
Goldstein, H., P. Barry, K. Venkatasubramanian and W. R. Vieth, *J. Ferm. Technol.*, **55**, 516 (1977).
Grant, R. A., R. W. Cox and R. W. Horne, *J. Royal Microscopical Soc.*, **87**, Part 1. 143 (1967).

Hamilton, B. K., L. J. Stockmeyer and C. K. Colton, *J. Theor. Biol.*, **41**, 547 (1973).
Howell, J. M., Ph. D. Thesis, Dept. of Chemical and Biochemical Engineering, Rutgers University (1981).
Ichijo, H., T. Suehiro, J. Nagasawa, H. Uedaira, A. Yamauchi and N. Aisaka, *Kenkyu Hokoku - Sen'i Kobunshi Zairyo Kenkyusho*, (163), 165-171 (1990).
Ikemi, M., N. Koizumi and Y. Ishimatsu, *Biotechnol. Bioeng.*, **36**, 149-154 (1990).
Ikemi, M., Y. Ishimatsu and S. Kise, *Biotechnol. Bioeng.*, **36**, 155-165 (1990).
Inushima, T., Jpn. Patent JP 03072879 A2, Manufacture of Polymer Fibers Containing Immobilized Microorganisms as Bioreactors (1991).
Kalevi, V., Eur. Patent, EP 341503 A2, Manufacture of Crosslinked Water-Insoluble Glucose Isomerase with Dialdehydes and Amines (1989).
Kang, Y., L. T. Fan, B. T. Min and S. D. Kim, *Biotechnol. Bioeng.*, **37**, 580-586 (1991).
Kiesser, T. and W. Bauer, *DECHEMA Biotechnol. Conf.*, Volume Date 1987, **1** (Technol. Biol. Processes - Saf. Biotechnol. - Appl. Genet. Eng.), 173-180 (1988).
Kiesser, T., G. A. Oertzen and W. Bauer, *Chem. Eng. Technol.*, **13**, 20-26 (1990).
Kiesser, T., G. A. Oertzen and W. Bauer, *Chem. Eng. Technol.*, **13**, 80-85 (1990).
King, J., *C&EN*, April, 32-54 (1989).
Laskin, A. I., Ed., "Enzymes and Immobilized Cells in Biotechnology," Benjamin Cummings: London (1985).
Levenspiel, O., "Chemical Reaction Engineering," 2nd Ed. John Wiley and Sons Inc.: New York (1973).
Leypoldt, J. K. and D. A. Gough, *J. Phys. Chem.*, **84**, 1058 (1980).
Ludolph, R. A., W. R. Vieth and H. L. Frisch, *J. Phys. Chem.*, **83**, 2795 (1979).
Luo, K. M., J. R. Giacin, S. G. Gilbert and E. R. Lieberman, *J. Mol. Catalysis*, **5**, 15-26 (1979).
Marconi, W., F. Bartoli, F. Morisi and A. Mariani, in "Enzyme Engineering, Vol. 5," H. H. Weetall and G. P. Royer, Eds., p.269-278. Plenum Press: New York (1978).
Marconi, W. and F. Bartoli, In "Advances in Biotechnology, Fermentation Products," Vol. III, pp. 349-354. Pergamon Press: Toronto (1981).
Mogensen, A. O. and W. R. Vieth, *Biotechnol. Bioeng.*, **15**, 467 (1973).
Montana, M., M. Sc. Thesis, Dept. of Chemical and Biochemical Engineering, Rutgers University, New Brunswick, NJ (1977).
Nagashima, M., M. Azuma, S. Noguchi, K. Inuzuka and H. Samejima, *Biotechnol. Bioeng.*, **26**, 992-997 (1984).
Nillson, K., In "Large Cell Culture Technology," Hanser Publishers: New York (1987).
Pastore, M., F. Morisi and L. Leali, *Milchwiss*, **31**, 362 (1976).
Petersen, J. N. and B. H. Davison, *Appl. Biochem. Biotechnol.*, **28-29**, 685-698 (1991).
Peterson, R. S., Ph. D. Thesis, Univ. of Wisconsin-Madison (1987).
Radovich, J. M., *Biotech. Adv.*, **3**, 1-12 (1985).

Rai, V. R., U.S. - Japan Biotechnology Conference, Honolulu, Hawaii (December 1984).
Ram, K. A. and K. Venkatasubramanian, *Biotechnol. Bioeng.*, **24**, 355 (1982).
Ramachandran, G. N. and Ramakrishnan, C., In "Biochemistry of Collagen," G. N. Ramachandran and A. H. Reddi, Ed., Chap. 2. Plenum Press: New York (1976).
Saini, R. and W. R. Vieth, *J. Appl. Chem. Biotechnol.*, **25**, 115 (1975).
Scott, T. C., C. G. Hill and C. H. Amundson, *Biotechnol. Bioeng. Symp.*, **15**, 431-445 (1985).
Seko, H., S. Takeuchi, K. Yajima, M. Senuma and T. Tosa, In "Biochemical Engineering," *Ann. N. Y. Adad. Sci.*, **589**, 540-552 (1990).
Siegel, R. D. and R. W. Coughlin, *Recent Advances in Separation Techniques*, AIChE Symp. Series, **68**, 58 (1972).
Siegel, R. A., *J. Mem. Sci.*, **26**, 251-62 (1986).
Sueki, M., N. Kobayashi and A. Suzuki, *Biotechnol. Lett.*, **13**, 185-190 (1991).
Sun, Y. and S. Furusaki, *J. Ferment. Bioeng.*, **69**, 102-110 (1990).
Suzuki, S., I. Karube and K. Namba, In "Analysis and Control of Immobilized Enzyme Systems," D. Thomas and J. P. Kernevey, Eds., p. 151. North Holland: Amsterdam (1976).
Tzeng, J. W., L. S. Fan, Y. R. Gan and T. T. Hu, *Biotechnol. Bioeng.*, **38**, 1253-1258 (1991).
Van't Riet, K. and J. Tramper, "Basic Bioreactor Design," Marcel Dekker, Inc.: New York (1991).
Venkatasubramanian, K., W. R. Vieth and A. Constantinides, In "Enzyme Engineering, Vol. 3," E. K. Pye and H. H. Weetall, Eds., p.29. Plenum Press: New York (1977).
Venkatasubramanian, K., Ed., "Immobilized Microbial Cells," ACS Symp. Ser., **106**. Academic Press: New York, NY (1979).
Vieth, W. R., S. G. Gilbert and S. S. Wang, *Biotechnol. Bioeng.*, Symp., **3**, 285 (1972).
Vieth, W. R., K. Venkatasubramanian, A. Constantinides and B. Davidson, In "Applied Biochemistry and Bioengineering," L. B. Wingard, E. Katchalski-Katzir and L. Goldstein, Eds., Vol. 1, p. 221. Academic Press: New York (1976).
Vieth, W. R. and K. Venkatasubramanian, *Adv. Enzymol.*, **34**, 243 (1976).
Vos, H. J., D. J. Groen, J. J. Potters and K. C. Luyben, *Biotechnol. Bioeng.*, **36**, 367-376 (1990).
Vos, H. J., M. Zomerdijk, D. J. Groen and K. C. Luyben, *Biotechnol. Bioeng.*, **36**, 377-386 (1990).
Vos, H. J., C. Van Houwelingen, M. Zomerdijk and K. C. Luyben, *Biotechnol. Bioeng.*, **36**, 387-396 (1990).
Webb, C., G. M. Black and B. Atkinson, Eds., "Process Engineering Aspects of Immobilized Cell Systems," Pergamon Press: Elmsford (1986).
Wrotnowski, C., *Genetic Eng. News*, **Oct.**, 9 (1990).
Yang, S. T. and M. R. Okos, *Biotechnol. Bioeng.*, **34**, 763-773 (1989).

4

BIOPROCESS KINETICS

4.0 INTRODUCTION

The motivations for studying Bioprocess Kinetics include the gaining of a deeper understanding of the biochemical and biological mechanisms which are operative in the living organism, expressible in a concise, quantitative form, to allow prediction of performance under a variable range of conditions. Process optimization and control are corollary entities.

This is a very challenging task, due to the wide variety of microorganisms which are encountered, encompassing procaryotes (including bacteria), eucaryotes (including yeasts and molds), as well as mammalian cells and plant cells. The associated cellular machinery consists of about 2,000 enzymes and extensive compartmentation (for example, nucleus, mitochondria, rhibosomes, etc.).

Shuler and Kargi (1992) have attempted to deal with this complexity in a simplified fashion in the structured unicellular model for growth of *E. coli* B/r A, shown in Fig. 4.1. The cell is shown, growing in a glucose-ammonium salts medium with glucose or ammonia as the limiting nutrient. At the time shown, the cell has just completed a round of DNA replication and initiated cross-wall formation and a new round of DNA replication. Solid lines indicate the flow of material, while dashed lines indicate flow of information. (In this model, the term E refers to key enzymes; P, amino acids and ribonucleotides; M, protein and RNA; W, waste products, etc.) The number of equations in the simulation is reduced to a manageable set (about 35); predictive quality of the model pertaining to transient response of cell cultures to perturbations in limiting substrate concentrations seems quite good for the specific case referred to earlier (Ataai and Shuler, 1985).

Gu and Chang (1990) recently prepared "artificial cells," each containing leucine dehydrogenase, urease, soluble dextran-NAD$^+$ and one of the following coenzyme regenerating dehydrogenases: glucose dehydrogenase, yeast alcohol dehydrogenase, malate dehydrogenase or lactate dehydrogenase. The artificial cells were packed in small columns.

L-Leucine, L-valine and L-isoleucine were continuously produced with simultaneous dextran-NADH regeneration. The maximum production ratios depended on the coenzyme regenerating systems used, 83-93% for the D-glucose and glucose dehydrogenase system, 90% for ethanol and yeast alcohol dehydrogenase, 45-55% for L-malate and malate dehydrogenase, and 64-78% for L-lactate and lactate dehydrogenase.

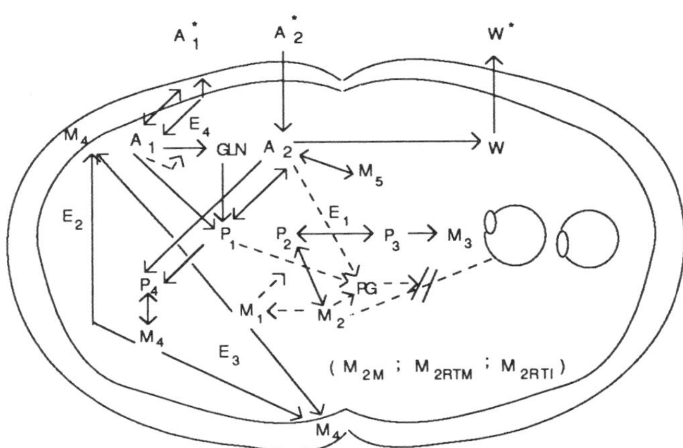

Figure 4.1 Schematic diagram of a highly structured single-cell model for growth of *E. coli* B/r A on glucose-ammonium salts medium (after Shuler and Kargi, 1992).

Common to all structured models is their consideration of certain key elements: enzyme kinetics, compartmentation/diffusion and *local* effects, induction/repression and cell age considerations. In contrast, unstructured models employ an empirical, *global* approach. The Monod model of cell growth on a limiting substrate (eqn. [4.3]), introduced a little later in this chapter, is a good example.

Further on the subject of substrate utilization, the concepts of biomass yield and product yield are conveniently incorporated (Pirt, 1975) as:

$$\frac{dS}{dt} = -\frac{1}{Y_g}\left(\frac{dX}{dt}\right) - \frac{1}{Y_p}\left(\frac{dP}{dt}\right) - mX \qquad [4.1]$$

where the respective yield coefficients are defined as Y_g and Y_p, respectively; m = maintenance energy (for respiration) and S, P refer to substrate and product concentrations. (dX/dt) represents the rate of cell proliferation while (dP/dt) represents the rate of product formation. A pertinent example of application of the above is afforded by the work of Kiss and Stephanopoulos (1991), which addresses the use of microbial environmental control (via reactor operating strategy) as a means of shifting the metabolism to favor the desired pathway. Batch and continuous cultures of the L-lysine producer *Corynebacterium glutamicum* were employed to establish the relationships between the culture growth rate and the instantaneous process yield and productivity. This suggested that significant improvements in process yield, productivity and product titer could be obtained through fed-batch cultivation of the organism under conditions of restrained growth. Constant feed rate fed-batch fermentations confirmed this hypothesis and further suggest that simple control strategies based on available on-line measurements can be used to manipulate the microbial metabolism to further improve process performance. Defined minimal media conditions were used to assess and subsequently enhance the production of subtilisin by genetically characterized *B. subtilis* strains. Subtilisin production was initiated by the exhaustion or limitation of ammonium in batch and fed-batch cultures (Pierce et al., 1992).

Chen et al. (1990) studied minimum nutritional requirements for immobilized yeast (*Kluyveromyces fragilis*). Experiments on reduced ammonium sulfate in the defined medium, and reduced yeast extract and peptone in YEP medium (3 g/L yeast extract and 3.5 g/L peptone) indicated that stable ethanol productivity could be maintained for extended periods (80 h) in the complete absence of any nutrients other than a few salts (potassium phosphate and magnesium sulfate). Productivity rates dropped by 35-65% from maximal values as nitrogenous nutrients were eliminated from the test media, while growth rates (as determined by released cell density) dropped by 75-95%. Thus, nutritional deficiencies largely decoupled growth and productivity of the immobilized yeast which suggests productivity is both growth- and nongrowth-associated for the immobilized cells. A yeast extract concentration of 0.375 g/L with or without 1 g/L ammonium sulfate was determined to be the minimum level which gave a sustained increase in productivity rates as compared to the nutritionally deficient salt medium. This represents a 94% reduction in complex nitrogenous nutrient levels compared to standard YEP batch medium.

4.1 OXYGEN AS THE LIMITING SUBSTRATE

As described by Singh et al. (1987), the approach to fermentor scale-up is based on the fact that most fermentations are oxygen limited, and scale-up is conducted on the basis of providing equal oxygen transfer at various scales of operation. For industrial-scale fermentors, heat removal may also be a design limitation. With regard to the latter, Hamer and Heitzer (1990) examined fluctuating environmental conditions in scaled-up bioreactors with regard to heating and cooling effects. In large-scale bioreactors, the proportional surface area available for heat transfer decreases with increasing volume and ultimately becomes insufficient to maintain the temperature within the optimum range if high rates are to be maintained. In many large-scale bioreactors, external cooling loops are employed for temperature control. When the process culture passes through such loops, it is subjected to a significant temperature gradient that could have both kinetic and physiological consequences. The effects of subjecting a model process culture to superoptimal and suboptimal growth temperatures were studied. Short-term cooling shocks of a growing culture of *Klebsiella pneumoniae* had negligible impact on overall culture performance; recovery appeared to be virtually instantaneous. In contrast, short-term heating of *K. pneumoniae* to temperatures above the superoptimal range for growth caused significant delays with respect to recovery.

Most correlations to estimate oxygen transfer capacity are based on power input and gas superficial velocity. The validity of this type of correlation, at least for local mass transfer, has been well justified by Van't Riet (1979). In general:

$$k_L a = k_1 \times \left(\frac{P_g}{V}\right)^{k_2} \times (v_s)^{k_3} \qquad [4.2a]$$

The constant k_1 is characteristic of fermentor geometry, and Bartholomew (1960) showed that exponents k_2 and k_3 are dependent on scale. Usually, the values of the constants k_1, k_2 and k_3 must be experimentally determined for a given configuration.

Moderate shear transport augmentation is described in papers by White et al. (1991) and Voit et al. (1989). In the former, the SGI Ultrafermentor is described, it being an external loop bioreactor with circulation through an ultrafiltration module allowing removal of medium and solution products during fermentation. The contents are continually circulated during operation and the vessel is also equipped with a stirring turbine. The interaction of these two mixing agents on gas transfer was investigated. Both mechanisms produced similar increments in $k_L a$ over their working ranges, with values ranging 0-700/h at 200 dm^3/h airflow. The effects of these two mechanisms on $k_L a$ were approximately additive at low values but the combined mixing produced a maximum k_La value also in the region of 700/h. The power draw of mixing using the two agents was calculated, and stirring was 10-20 times more efficient than circulation. In the latter study alluded to above, growth and overproduction of lipase with *S. carnosus* were investigated under increased acceleration in a centrifugal field bioreactor as an example of a typical fermentation at low viscosity. According to the maximum growth rate of *S. carnosus* and lipase yield, the maximum lipase productivity with strain pLipPS1, which may be obtained in a reactor at maximum cell density of about 100 g/L, is 6600 units/L-h. This requires an oxygen rate of about 60 g/L-h and can be achieved in the centrifugal field bioreactor. Lipase activities could be obtained which were 10-fold higher than those of shaken cultures. The maximum productivity of lipase, obtained in batch fermentations, was 578 units/L-h. This value is not limited by the available oxygen transfer in the reactor, but by substrate solubility and substrate inhibition at high initial concentrations.

Though oxygen transfer-limited scale-up is simplified by designing for equal mass transfer coefficient k_La, it is really based on providing equal oxygen transfer rate (OTR). k_La and OTR are related by the following equation:

$$OTR = k_La\,(C^* - C) \qquad [4.2b]$$

where C* is the dissolved oxygen concentration in equilibrium with the gas and C is the actual dissolved oxygen concentration in the liquid.

The difficulty with this technique is that it assumes that the fermentor is sufficiently well mixed so that a local mass transfer rate can provide an estimate of the overall oxygen consumption. In viscous, non-Newtonian fluids, or in very large fermentors, large oxygen gradients are known to exist and fluid mixing may be a major problem (Oosterhuis and Kossen, 1984; Van't Riet, 1979). Fermentors appropriate for the biological production of fuels and chemicals are of necessity large. A mismatch frequently occurs between the scale of turbulence in the large fermentor, the scale of turbulence in the laboratory fermentor on which growth and yield data were generated, and the size of microorganism (Dunlop, 1990). Typically, the scale of turbulence (size of the smallest energy-containing eddy) in a laboratory fermentor is 20-50 μ. In a large fermentor it is usually difficult to get below 200-500 μ. This is in contrast to the typical dimensions of the cell of 1-3 μ. This means that the cell spends substantial portions of its time in a stagnant eddy depleted of nutrients and gives correspondingly poor yields that may destroy the economics of the process. In these cases the conventional k_La-based technique may significantly underestimate the power input required for adequate performance. Similar considerations apply to mass transfer at a free surface in stirred tank bioreactors (Kawase and Moo-Young, 1990). A theoretical model for liquid-phase mass transfer coefficients was developed on the basis of a periodic transitional sublayer model and the Kolmogoroff theory of isotropic turbulence. The capability of the proposed theoretical model was discussed in light of experimental data for Newtonian and non-Newtonian media. Satisfactory agreement was found between the model and the experimental data over a wide range of conditions.

Thus, in large fermentors, due to incomplete mixing, microorganisms are exposed to a continuously fluctuating environment with respect to physical and chemical parameters such as pressure, dissolved oxygen, pH and substrate concentrations. A study of the effects of such fluctuations on growth and product formation is important for scale-up. Antibiotic production in *Streptomyces clavuligerus* is affected by dissolved oxygen, and the enzymes responsible for biosynthesis of antibiotics are derepressed by higher dissolved oxygen levels in complex media. This suggests that DO fluctuations in large fermentors would have an effect on antibiotic yields (Yegneswaran and Gray, 1991).

Oxygen transfer in bioreactors with slurries having a yield stress was investigated by Kawase and Moo-Young (1991). The volumetric mass transfer coefficients in a 40-L bubble column with simulated fermentation broths, the rheological properties of which were represented by the Casson model, were measured. Experimental data were compared with a theoretical correlation developed on the basis of a combination of Higbie's penetration theory and Kolmogoroff's theory of isotropic turbulence. Comparisons between the proposed correlation and data for the simulated broths show good agreement. The mass transfer data for actual mycelial fermentation broths reported previously by the authors were then reexamined. Their rheological data were correlated by the Bingham plastic model. The oxygen transfer rate data in the mycelial fermentation broths fit the predictions of the proposed theoretical correlation.

4.2 LOW SHEAR ENVIRONMENTS

Using the method of equi-inocular synchronized comparative fermentation, the cultivation of *Sorangium cellulosum* and production of the polyketide antibiotic ambruticin S were compared in stirred tank and airlift reactors of different geometry (Hopf et al., 1990). This method requires that inocula originate from the same preculture, and cultivation parameters are synchronized to similar values. Similar ambruticin yields were obtained from both reactor systems, provided that the concentration of dissolved oxygen was maintained above a certain value (ca. 40% sat.). For cultivation of *S. cellulosum* it is the dissolved oxygen level rather than the oxygen transfer rate that presents the proper criterion for scale-up and comparative reactor studies. Using the same equi-inocular technique, Yonsel et al. (1991) found that the growth of *Trichosporon cutaneum* under oxygen-limitation is directly proportional to oxygen transfer rate (OTR). Stirred tank reactors allow a higher energy input per volume; thus, higher OTR-values can be reached than in airlift reactors. However, airlift reactors have a significantly higher efficiency (E_{02} = OTR/P/V_L); i.e., OTR-values higher than in stirred tanks can be reached at the same power input.

As a means of low-shear transport augmentation, the capability of perfluorocarbon (PFC) emulsions in enhancing oxygen transfer in bioreactors was investigated by Ju et al. (1991). Based on the penetration theory, a model was derived for evaluating the potential oxygen transfer enhancement effects of PFC emulsions with very fine PFC particles. Defined as the ratio between the maximum oxygen transfer rates in systems with and without the emulsions, the enhancement factor achievable with dilute PFC emulsions can be estimated as the square root of the product of ratios of oxygen "permeability" and solubility in media with and without the emulsions. The effect of PFC emulsions on oxygen transfer in low-shear cultivation systems was further studied experimentally with direct measurements of volumetric oxygen transfer coefficients, k_La, in bioreactors and with microbial fermentations conducted under low-shear conditions. Significantly higher cell populations could be maintained in an aerobic state by using systems supplemented with the subject emulsions. The feasibility of enhancing oxygen transfer in bioreactors by introduction of adequate amounts of PFC emulsions was clearly demonstrated.

In the airlift system, the gas motion carries fluid and cells up a draft tube, at the top of which gas disengages from the liquid. The denser spent liquid recontacts the entrant gas stream after flowing back down an annulus outside the draft tube. Recently, Chisti et al. (1990; Chisti and Moo-Young, 1989) carried out studies in such systems. The former examined the effect of static mixers on the overall gas-liquid volumetric mass transfer coefficient (k_La), as examined in an external-loop airlift bioreactor. Over a broad range of power law parameters, the presence of static mixers in the riser was found to enhance k_La relative to the mixer-free mode of operation. The latter study is concerned with a mathematical model, comprising six equations, for estimation of the shear rate and apparent viscosity of non-Newtonian fluids in airlift and bubble column bioreactors. In the context of correlating hydrodynamic parameters in non-Newtonian fluids, a more common approach involves assuming that there exists an average shear rate in the column that is proportional to the superficial gas velocity. This average shear rate is then used to evaluate an effective viscosity of the non-Newtonian fluid that is subsequently used to quantify the fluid's rheological behavior in the correlation. A recent paper illustrates that this approach, which has mainly been applied to bubble columns, can also be applied to external loop airlift contactors, replacing the superficial gas velocity in the column by the superficial gas velocity supplied to the riser of the contactor. This extension is based upon consideration of the relevant characteristic velocity in the active zone (i.e., the riser section) of the reactor (Allen and Robinson, 1991).

A simple model for prediction of liquid velocity in external-loop airlift bioreactors has been developed (Kawase, 1990). Theoretical correlations for the friction factor of two-phase flows and for liquid velocity in the riser were derived, using the concept of an eddy diffusivity. The predictions of the proposed model were compared with the available experimental data for the friction factor and the liquid velocity in the riser of external-loop airlift contactors, and satisfactory agreement was obtained. Similarly, in a rather extensive study, local two-phase flow measurements were obtained by Young et al. (1991) in a pilot-scale, external-loop airlift bioreactor by using hot-film anemometry and resistivity-probe techniques. The radial dependence of both gas and liquid velocities and of the void fraction was substantial. Developing flow effects were pronounced, as evidenced by distinct changes in the radial profiles of flow properties with changing axial position. For high gas flow rates, liquid acceleration effects near the sparger resulted in greatly reduced slip velocities in a substantial portion of the riser. A significant reduction in mass transfer may occur under such conditions. The point equations of continuity and motion were used to develop a differential, two-fluid model for two-phase flow in airlift risers. The only empirical parameters in the model represented frictional effects. The developing two-phase flow characteristic of airlift risers created significantly higher frictional effects at the wall than are routinely observed for a fully developed flow. Model predictions were compared with the experimental results; agreement between the predicted and measured values was within 10% for both cases.

In a parallel study, the local properties of the dispersed gas phase (gas holdup, bubble diameter and bubble velocity) were measured and evaluated at different positions in the riser and downcomer of a pilot plant reactor and, for comparison, in a laboratory reactor (Froehlich et al., 1991c). These were previously described during yeast cultivation and with model media (Froehlich et al., 1991a; b). In the riser of the pilot plant reactor, the local gas holdup and bubble velocities varied only slightly in the axial direction. The gas holdup increased considerably with the aeration rate, while the bubble velocity increased only slightly. The bubble size diminished with increasing distance from the aerator in the riser, since the primary bubble size was larger than the equilibrium bubble size. In the downcomer, the mean bubble size was smaller than in the riser. The mean bubble size varied only slightly, the bubble velocity was accelerated, and the gas holdup decreased from top to bottom in the downcomer.

In the pilot plant at constant aeration rate, the properties of the dispersed phase were nearly constant during the batch cultivation, i.e., they depended only slightly on the cell concentration. In the laboratory reactor, the bubble sizes were much larger than in the pilot plant reactor. Also, the bubble velocities in the riser and downcomer increased, and the mean gas holdup and bubble diameter in the downcomer remained constant as the aeration rate was increased.

4.3 MEMBRANE-ASSISTED BIOPROCESSES

Kang et al. (1990) describe a process which employs chopped hydrophilic hollow fibers as supports for yeast cells in a conventional tubular bioreactor for ethanol production. The high productivity and cell density achieved demonstrate the usefulness of this technique.

Watanabe et al. (1990) carried out the production of ethanol in repeated-batch fermentation with a membrane-type bioreactor. At a yeast concentration of 110 g/L, an EtOH productivity of 3.5 g/L-h was attained, which is nearly an order of magnitude higher than that in conventional batch fermentation. Cho and Shuler (1990) disclose a multimembrane bioreactor which allows product to be extracted with a solvent toxic to the cell layer. The reactor cycles gas pressure to force a convective flux of nutrient into the cell chamber and the product out. With this system, the rate and extent of fermentation are significantly increased relative to prior art processes. In a related paper, Steinmeyer and Shuler (1990a) provide mathematical modeling and simulations of membrane bioreactor extractive fermentations. In other variations, Steinmeyer and Shuler (1990b) analyze continuous operation of a pressure-cycled membrane bioreactor, while Shuler et al. (1991) perform an economic evaluation.

Numura et al. (1989) studied the continuous production of acetic acid by an electrodialysis bioprocess with computerized control of fed batch culture. The maximum rate of production under the improved electrodialysis conditions was 5.5 gh^{-1}, which was about 1.7 times higher than the rates obtained by bioprocesses without the electrodialysis system. The improvement in the rate of production and yield of acetic acid resulted from the alleviation of the inhibitory effect of the acetic acid that is produced; a favorable pH and a low level of acetic acid in the reaction mixture were maintained by electrodialysis. In a different sort of contacting arrangement, Lotong et al. (1989) describe the production of vinegar by *Acetobacter* cells fixed in a film on a rotating disk reactor. Based on the capability of tolerating high temperatures and of forming very thin films, *Acetobacter* 249-1 was used; cotton towel cloth was employed as the support. Cells were densely fixed to the fibrils of the support material. Batch fermentation with a fixed film rotating disk reactor resulted in an acetic acid yield (Yp/s) of 0.95.

In addition to conventional membrane recycle bioreactors (e.g., Melzoch et al., 1991; Mehaia and Cheryan, 1990) units coupled with membrane pervaporation elements are now coming into play for ethanol production (Shabtai et al., 1991; Mori and Inaba, 1990; Strathmann and Gudernatsch, 1991). Maerkl (1989) describes a new fermentor type, with an outer wall consisting of a thin but strong polyamide film. This new fermentor is claimed to be superior to conventional transparent glass laboratory reactors because it can be safely sterilized *in situ* with no problems. The same technology allows the integration of cylindrical membrane sleeves in the fermentor, to make a compartmented bioreactor. Qureshi and Cheryan (1989) studied production of 2,3-butanediol in a membrane recycle bioreactor. They found that higher dilution rates resulted in higher productivity but lower utilization of glucose and low butanediol concentration.

4.4 GENERAL FRAMEWORK

As a means of integrating the previous concepts with others to follow, Roels and Kossen (1978) have attempted a systematic description of the architecture of bioprocess modeling, shown in Fig. 4.2, that unites the cellular behavior with the overall bioprocess response. This approach is quite useful in modeling anaerobic digestion, as will become apparent a little later in this chapter. Moser's recent book (1988) contains a very thorough and timely exposition of bioprocess kinetics; Schuegerl's (1989), bioreactors and their characterization. Moser's special topics include water activity, enthalpy-entropy compensation, and microkinetic equations derived from the kinetics of chemical and enzymatic reactions. Other topics include kinetic models for microbial product formation, multisubstrate and mixed population kinetics, dynamic models for transient operation techniques (non-stationary kinetics), etc.

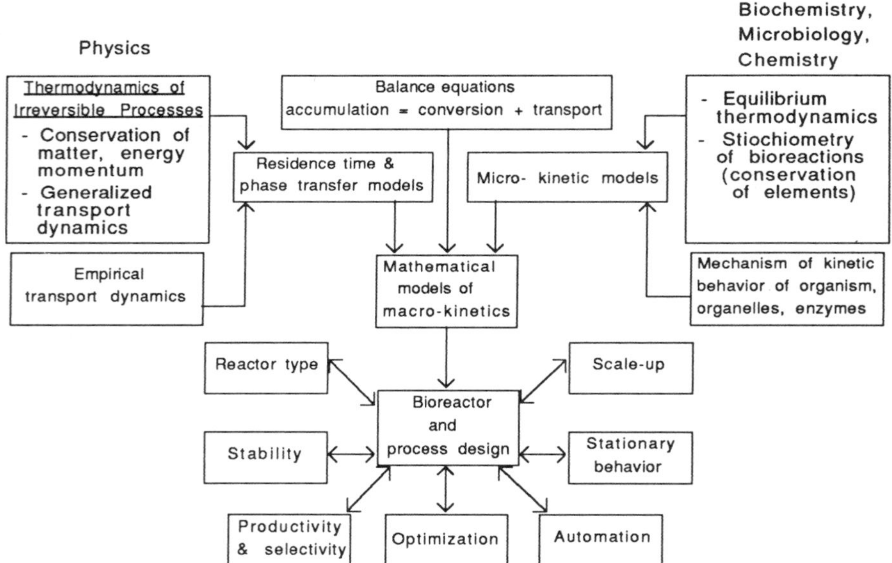

Figure 4.2 Architecture of bioprocess modeling (after Roels and Kossens, 1978).

4.5 IMMOBILIZED LIVING CELL SYSTEMS

Immobilization of microorganisms on the surface of solid supports has been practiced for centuries in the manufacture of vinegar. It is only within the past fifteen years that live cell immobilization has been studied for the purpose of developing more productive bioreactor systems (Hattori, 1972). Initial cell entrapment studies used synthetic polymers such as polyacrylamide; however, the immobilization conditions were harsh, and natural hydrocolloid gels were soon utilized, giving much more satisfactory results (Chibata, 1979). These materials, such as agarose, agar, collagen, calcium alginate and κ-carageenan, are, with the exception of collagen, all long chain polysaccharide polymers derived from seaweed. Agarose and agar undergo gelation when the solution temperature falls below approximately 40-45°C. A solution containing sodium alginate gels in the presence of calcium ions, while carageenan gels in the presence of potassium. Immobilization using alginate is simpler than with carageenan because in the liquid state alginate hydrocolloid suspensions are much less viscous, making for easier handling.

The history of laboratory experimentation with immobilized cells (IMC) is nearly as brief as that of recombinant cell technology. The potentials which have been uncovered in the past ten to fifteen years make it an attractive technique for widespread adoption in large scale processing. There is a definite technological driving force that will lead to the implementation of IMC reactor technology, to take its place alongside classical fermentation processing. Some of the characteristics of live-cell immobilization are listed below.

i. Very high cell loadings can be obtained in the immobilized phase. In some cases, immobilized cell concentrations may be as much as one hundred-fold those of free-cell concentrations (Karkare et al., 1985).

ii. IMC reactors can be operated at dilution rates well beyond the maximum specific growth rate of the microorganism without incurring cell washout.

iii. Significantly increased productivities of biomass and product formation can be achieved.

iv. Increased yield ratio of product formation to cell growth can provide higher carbon conversion efficiencies (Karkare et al., 1986).

v. Short and long-term reactor stability to system perturbations can be improved.

vi. In comparison to batch fermentors, IMC reactors have lower working volumes. Substantial capital savings, especially for new plants, can be realized from reduced size requirements.

vii. IMC bioreactors can achieve a physiological uncoupling between cell growth and secondary metabolite formation by feedstock alternation; i.e., one feed for cell growth and the other for metabolite production.

4.6 ANALYSIS OF LIVE CELL REACTORS IN AEROBIC PROCESSES

We have decided to introduce certain simple fermentation process concepts with those pertaining to immobilized cell behavior, *together*, in order to unite these entities into the discussion of a single bioprocess application. We do this in the context of the continuous flow stirred tank reactor, (CSTR), used for continuous culture, one of the reactors shown in Fig. 4.3. (The packed bed reactor has already been discussed in the context of enzyme reactions; the fluidized bed reactor will be discussed in Chapters 7 and 8; the hollow fiber reactor, Chapter 8.)

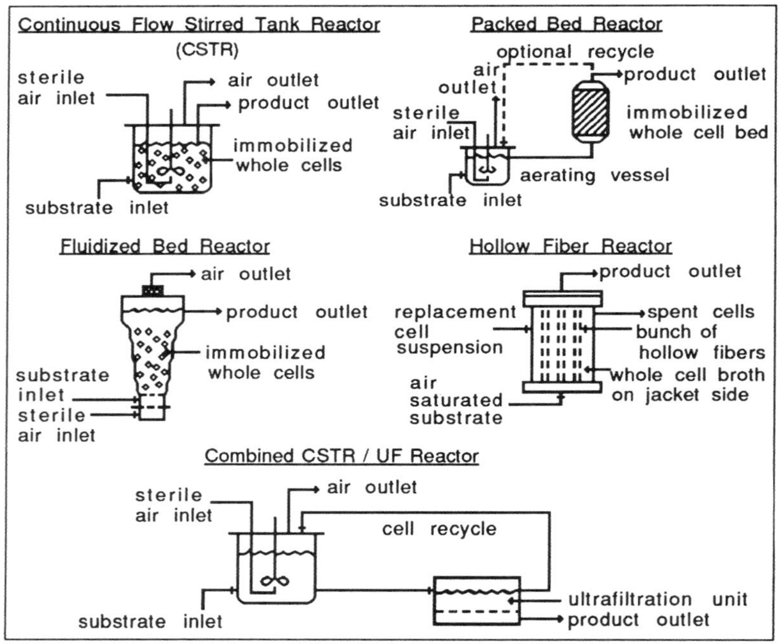

Figure 4.3 Bioreactor types.

In the following analysis (Venkatasubramanian et al., 1983), immobilized cell replication is considered to occur on the surface of the catalyst only. Oxygen limitation usually regulates maximum growth near the catalyst surface of available area, A (m^2/L), a characteristic of the supporting material used. A maximum biomass loading capacity, X_s^* (g dry cell wt/m^2), can be defined; it is a function of the support as well as the microorganism used. The cell surface concentration is defined as

X_S (g/m^2) and the bulk concentration is given by X (g/L). On the basis of reactor volume, cell concentration becomes $X_{im} = X_s \cdot A$ (g/L).

As usual, the reactor volume is V_R (liters) and substrate flow rate is Q (L/min). S is the concentration of limiting substrate in the reactor; S_0 at the inlet. A modification of the standard approach applied to analysis of a classical continuous culture system suffices; it was suggested in part by the work of Topiwala and Hamer (1971). A CSTR type of configuration is considered; the analysis also applies to relatively shallow fluidized beds.

Regarding the surface as the cell "hatchery," with a specific growth rate μ_S, and employing the Monod model of cell growth,

$$\mu_S = \frac{1}{X_S}\frac{dX_S}{dt} = \frac{\mu_{ms}S}{K_S + S} \qquad [4.3]$$

The next logical step is for surface growth to continue until catalyst loading capacity is reached. The data presented by Jirku et al. (1981) suggest that μ_S is different (lower) from the bulk growth rate, μ_b.

Once the catalyst is completely loaded, the cells *release* into the bulk with specific growth rate, μ_b, which depends on the substrate concentration:

$$\mu_b = \frac{1}{X_{total}}\frac{dX_{total}}{dt} = \mu_m \cdot \frac{S}{K_S + S} \qquad [4.4]$$

where $X_{total} = X_{im} + X$.

Live/growing cells remain attached to the catalyst while the dead cells wash away; this is supported by some recent experimental evidence which shows the preferential leaching of dead cells (Karkare et al., 1981).

4.7 IDEALIZED REACTOR PERFORMANCE EQUATIONS

For a steady state mass balance, the catalyst is assumed to be fully loaded, i.e., $X_{im} = X_S*A$. Taking inlet cell concentration to be zero, we have for a CSTR:

$$\mu_b X V_R + \mu_b X_S * A V_R = Q X \qquad [4.5]$$

Let the *dilution rate* $D = \frac{Q}{V_R} = \frac{1}{\tau}$

$$D \cdot X = \mu_b [X + X_S * A] \tag{4.6}$$

or

$$D \cdot X = \mu_b [X + X_{im}] \tag{4.7}$$

A balance on the limiting substrate provides:

$$D[S_0 - S] = \frac{\mu_b}{Y} [X + X_{im}] \tag{4.8}$$

where Y is the *biomass yield coefficient.*

Combine eqns. [4.7] and [4.8] to obtain:

$$X = Y [S_0 - S] \tag{4.9}$$

Equation [4.9] is identical to the analogous continuous culture equation. However, the value of S is always lower in the case of IMC reactors as shown below.

Substituting eqn. [4.9] and [4.6] in [4.8], one obtains:

$$\frac{\mu_{ms}}{K_S + S} = \frac{DY [S_0 - S]}{Y [S_0 - S] + X_{im}} \tag{4.10}$$

Equation [4.10] is quadratic in S and has *one* meaningful root between zero and S_0. In the limit as $X_{im} \to 0$, we regain the familiar continuous culture relationship:

$$\mu_b = D \tag{4.11}$$

The important result of this analysis is that $D > \mu_b$ at all values of D, implying that the exit substrate concentration is always lower for the IMC reactor at any given dilution rate. As a result, the exit cell concentration for the immobilized cell process must always be higher and, consequently, the IMC reactor is superior to the continuous culture system in terms of conversion efficiencies. Figure 4.4 shows the effect of increasing X_{im} on the exit biomass concentration. The calculations are for growth of *A. suboxydans*. The relevant constants were obtained from the data of Vera-Solis (1976). It is apparent that the washout condition in submerged continuous cultures is eliminated by using immobilized cells.

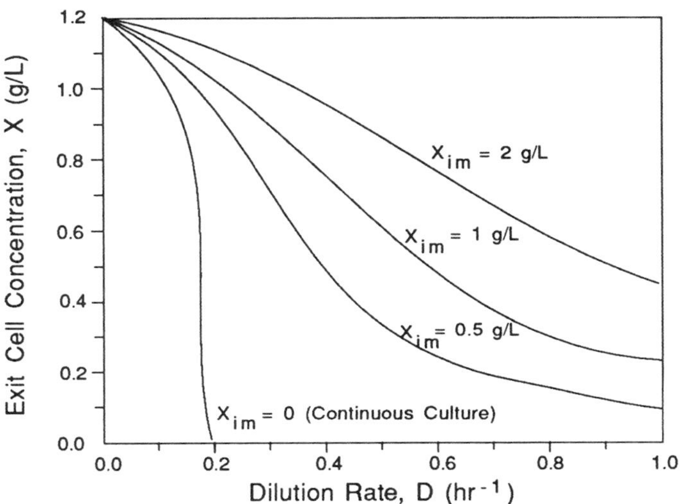

Figure 4.4 Effect of dilution rate on effluent cell concentration in immobilized live cell CSTR with varying amounts of immobilized cells. Calculations were based on data of Vera-Solis (1976) for growth of *Acetobacter suboxydans* on ethanol. μ_m = 0.23 hr^{-1}, K_S = 6.5 g/L, Y = 0.0265 g cell/g ethanol and S_0= 45 g/L ethanol.

For the special case of $D = \mu_m$ we have a unique solution for eqn. [4.10].

$$S = \frac{K_S\, Y\, S_0}{K_S\, Y + X_{im}} \qquad [4.12]$$

Again, the equation reduces to $S = S_0$ as $X_{im} \rightarrow 0$. Rearranging eqn. [4.10] we obtain:

$$D = \frac{\mu_m\, S\, Y\, [S_0 - S] + \mu_m\, S\, X_{im}}{Y\, [K_S + S]\, [S_0 - S]} \qquad [4.13]$$

Multiplying this by $X = Y\, (S_0 - S)$,

$$DX = Pr = \frac{\mu_m\, S\, Y\, [S_0 - S] + \mu_m\, S\, X_{im}}{K_S + S} \qquad [4.14]$$

For maximum productivity w.r.t. substrate concentration:

$$dPr / dS = 0 \quad \text{and} \quad d^2Pr / dS^2 \big|_{S_{opt}} < 0 \qquad [4.15]$$

Differentiation and simplification lead to:

$$Y\, S_{opt}^2 + 2K_S\, Y\, S_{opt} - K_S\, [Y\, S_0 + X_{im}] = 0 \qquad [4.16]$$

Using the meaningful positive root,

$$S_{opt} = \frac{\sqrt{K_S^2\, Y^2 + K_S\, Y\, [Y\, S_0 + X_{im}]} - K_S\, Y}{Y} \qquad [4.17]$$

and

$$D_{opt} = \frac{\mu_m\, S_{opt}\, Y\, [S_0 - S_{opt}] + \mu_m\, S_{opt}\, X_{im}}{Y\, [K_S + S_{opt}]\, [S_0 - S_{opt}]} \qquad [4.18]$$

It is readily verified that as $X_{im} \rightarrow 0$, equations [4.17] and 4.18] reduce to their counterparts in continuous culture systems.

For the example in Fig. 4.4, considering X_{im} = 2g/L,

$$D_{opt} = 0.77\ hr^{-1}$$

This is much greater than μ_m; i.e., the productivity at this dilution rate is 0.463 g/L·hr, or about 3.5 times the maximum productivity of a similar continuous culture. The exit cell concentrations for the two cases are quite comparable (X = 0.6 g/L for IMC reactor and X = 0.8 g/L for continuous culture). It is obvious that the cell concentration can be increased for the IMC reactor at the expense of a little productivity and the productivity levels will still be much higher than the continuous culture. In another study, Okita and Kirwan (1986) have reached much the same conclusions.

Fig. 4.5 shows the change in biomass productivity with dilution rates for various immobilized cell concentrations.

When whole living cells are immobilized, high cell densities are possible, as well as reactor operation at high dilution rates without washout, leading to higher productivities and simplified product purification. Thus, volumetric productivity of ethanol can sometimes be increased by one or two orders of magnitude (Hahn-Hägerdal, 1990).

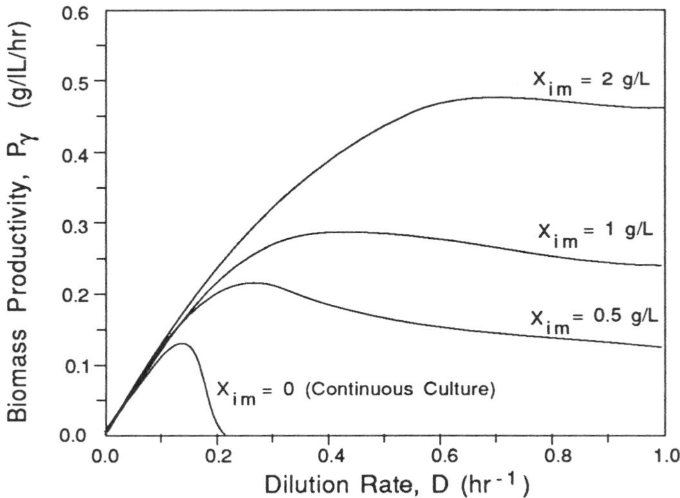

Figure 4.5 Effect of dilution rate on biomass productivity of immobilized live cell CSTR with varying amounts of immobilized cells. Calculations based on data of Vera-Solis (1976) for growth of *Acetobacter suboxydans* on ethanol. Parameter values are the same as in Figure 4.4.

By contrast, on a per cell basis, immobilized cell physiology exerts its influence on *specific productivities* and product yields. Bunch et al. (1990) report that the physiological environment of cells in hollow fiber membrane reactors can be controlled to allow limitations associated with excess cell growth and/or oxygen supply to be overcome.

4.8 CONCEPT OF A DUAL COLONY OR HYBRID REACTOR

The foregoing considerations lead one to the concept of a highly productive biomass generator (Messing et al., 1981). If a large amount of cells can be immobilized in a small volume (by using highly porous supports), biomass may be generable at much higher rates than otherwise possible. This is the argument for use of extended surfaces (Topiwala and Hamer, 1971) in fermentation. Since, for the production of biomass at enhanced rates we are utilizing cell growth both in surface culture and submerged culture, it is appropriate to term these reactors, in particular, as *hybrid reactors*. This term takes on added significance in the case of immobilized recombinant cells (described in a later chapter) where the overall gene pool is divided between the two colonies. Thus, hybrid reactors are a combination of the concepts of cell immobilization and continuous culturing of microorganisms, with a type of controlled or regulated release providing the communication between the two cell populations. In general, wherever live immobilized cells are used for continuous fermentation (whether for growth associated or nongrowth associated products), the term structured bed fermentation (Vieth, 1979) seems appropriate in the design sense.

In this connection, Gil et al. (1991) describe continuous ethanol production in an immobilized/suspended-cell bioreactor from a feedstock containing glucose. The reactor has now been in continuous operation for two years. The operational conditions required to maintain an ethanol product level of 98 gL^{-1} and total sugar consumption of 95.8% were an adequate feedstock (synthetic sugarcane juice with 200 gL^{-1} sucrose, mineral salts, and 5 gL^{-1} yeast extract), a nutrient supplement of yeast extract, peptone, and mineral salts to the second reactor stage, continuous aeration to both stages and a liquid recycle ratio of 20:1 within each reactor stage. The controlled aeration was necessary to maintain an active and stable yeast population. The dilution rates studied varied from 0.1 to 0.25 h^{-1}. The greatest ethanol productivity was achieved at the highest dilution rate tested (0.25 h^{-1}), but the most complete conversion of sugar to ethanol occurred at the lowest dilution rate. The most effective nutrient supplement for supporting maximal ethanol production was obtained with a combined addition of yeast extract, peptone and mineral salts. The effectiveness of the nutrient supplement could not be equalled by substituting vitamins, inositol, acetic acid, or trace minerals either alone or in combinations. High test molasses (commercially concentrated, natural sugarcane juice) was a suitable feedstock only when it was fortified with ammonium salts. The combination of the immobilized cells and high concentration of suspended free cells within the air-supplemented reactor stabilized the continuous ethanol fermentation for long-term performance, eliminated the necessity for cell conditioning and recycle, and effectively suppressed infection from extrinsic organisms.

In a related study, Gil et al. (1990) report that *Saccharomyces cerevisiae* was cultivated in a controlled aerated, dual-stage (column), continuous flow bioreactor in a hybrid free-cell/immobilized-cell state. The yeast cells maintained an ethanol concentration of 58-64 and 91-98 g/L in stages I and II, respectively. The lipid composition of the cells cultivated under these conditions was correlated to the effects of aeration by interrupting the aeration on days 113 and 266 of continuous operation. Under conditions of aeration or nonaeration, an alternating increase and decrease in the contents of squalene, sterols and fatty acids of the respiratory-competent and -deficient unattached free cells was observed. The cellular free lipid compositions of the immobilized cells in the aerated and nonaerated conditions were characteristic of respiratory-deficient cells, with the exception of the immobilized cells exposed to higher ethanol concentration (stage II). These cells contained a broader range of sterol components and increased levels of unsaturated fatty acids than immobilized cells at a lower ethanol concentration (stage I). The neutral lipid to phospholipid ratio decreased for respiratory-deficient cells, with phosphatidylethanolamine and phosphatidylinositol being the principal phospholipids. The essentiality of the hybrid bioreactor design for continuous long term performance and the importance of maintaining specific yeast lipid constituents for continuous high alcohol productivity were demonstrated.

4.9 INDUCER/SUBSTRATE TRANSPORT INTO THE IMMOBILIZED CELLS

Until recently, rather little information was available in the literature on the analysis of mass transfer effects in immobilized living cell systems. However, the treatment of bulk and pore diffusional effects up to the surface of the immobilized cell would proceed in the manner somewhat similar to the case of immobilized single enzyme type reactor systems. Next we have to consider the diffusion of the substrate and the product through the barrier imposed by the cell envelope (i.e., cell wall and cell membrane) itself.

It can safely be assumed that substrate transport into the microbial cell is characterized by passive diffusion in the case of monoenzyme-type immobilized cell systems. For such reactions the permeability of the cell envelope can sometimes be increased by specific treatments; e.g., heat treatment of cells containing glucose isomerase activity prior to immobilization (Vieth and Venkatasubramanian, 1976). With more complex reaction sequences and pathways, the total cell structure needs to be retained intact. Thus, the role of transport resistances through the cell envelope in the overall reaction assumes greater importance in the case of immobilized living cell systems.

The type of diffusion mechanism would depend on the organism and the limiting substrate itself. Active transport is the common mode of transport of sugars into cells. Hence, models based on this concept should be used to take into account the cell envelope resistance in the reactor performance calculation. Vieth et al. (1982) have described lactose transport through the cell membrane by an equation of the type:

$$J_L / V_C = \frac{A\,L_0}{B + L_0} \qquad [4.19]$$

where L_{in} is the intracellular lactose concentration, L_0 is the extracellular concentration, and A and B are concentration-independent parameters in continuous culture. J_L is the molar flux. The topic of active inducer transport is developed in detail a little later, in Chapter 7.

4.10 GENERAL MASS TRANSFER CONSIDERATIONS

A further challenge in development of immobilized live cell systems (for aerobic organisms) is the problem of oxygen transfer. Due to the increased cell densities in these systems, the oxygen demand is much higher compared to conventional submerged fermentation processes. Conversely, the productivity of such systems would be limited by the oxygen transfer efficiency of the system. Conventional oxygen transfer methods are neither always appropriate nor sufficient to provide the necessary oxygenation. Hence, novel methods for oxygenation need to be developed to further embellish the performance of immobilized cell reactors. Possibilities include membrane oxygenation, external oxygenation with rapid recycle, and use of high pressures in the reactors.

Focusing on the other end of the pathways, Tosa (1991) describes the formation of L-alanine from D-aspartic acid using immobilized *Pseudomonas dacunhae* by the pressurized bioreactor method, which suppressed the liberation of CO_2 formed as a byproduct. Productivity levels increased by ≥50% compared to the conventional method. L-aspartic acid and L-alanine were produced simultaneously from fumaric acid and ammonium by connecting a pressurized bioreactor using immobilized *P. dacunhae* to one using immobilized *Escherichia coli.* Simultaneous production of D-aspartic acid and L-alanine from DL-aspartic acid were also carried out by using a pressurized column reactor containing immobilized *P. dacunhae* cells.

Membrane oxygenator systems can be designed with predictable k_La values which can even exceed those which are achievable by conventional high shear transfer techniques. Jandel (1991) has demonstrated that silicone membranes allow bubble-free aeration of immobilized cell systems in fluidized bed bioreactors. Oxygen is transferred by diffusion into the culture medium and delivered directly to the cells. Limitation of growth and productivity by an insufficient oxygen supply is obviated. The use of hyperbaric oxygen and/or recycle stream oxygenation furnish two other possibilities for enhancement of oxygen transfer to growing cultures which are oxygen-limited. Yamada and Iwata (1989) report the development of an air-separation composite membrane for use in bioreactors. The bioreactor contains a composite membrane for selectively supplying oxygen to immobilized biocatalysts such as enzymes or microorganisms. The composite membrane is prepared by coating an air-separation membrane on one side of a porous supporting membrane. Meanwhile, the biocatalyst is immobilized on the other side and at the inner space. The technique was successfully applied to the production of acetic acid.

Hannoun and Stephanopoulos (1986) and Ruggeri et al. (1991) have reported on diffusion coefficients of glucose and ethanol in cell-free and cell-occupied calcium alginate membranes, while Tanaka et al. (1984) describe diffusion characteristics of substrates in Ca-alginate beads, and Itamunoala (1988) and Petersen et al. (1991) discuss limitations of methods of determining effective diffusion coefficients in cell immobilization matrices.

Effects of gelling cation variations have been considered by Simon (1989) and Bales et al. (1989). Chamy et al. (1989) produced gel particles hardened by treatment with aluminum nitrate, which gave them good operational stability. Teixeira and Mota (1990; Teixeira et al., 1990) describe a membrane microreactor for the experimental assessment of internal diffusion limitations in yeast flocs. It proved useful when comparing the effects of flocculation additives on the performance of a strain of *Kluyveromyces marxianus* in continuous ethanol production. The influence of recycling rate in the performance of a bioreactor of the external loop type was determined. An increase in recycling rate led to a decrease in floc size that simultaneously caused an increase in specific kinetic parameters as well as their stability (Mota and Teixeira, 1991).

Karel et al. (1990) point out that in IMC systems cell concentration varies with the distance from the site of nutrient provision; that the resulting local regions vary in local effective diffusivity (and possible convective channeling effects) and that hindered growth rates and concomitant prolongation of reaction/diffusion transients cast doubt on the validity of pseudo steady state assumptions.

Webb et al. (1990) point out that viable cells are apt to be distributed nonuniformly throughout the catalyst particle, creating intraparticle spatial variations of the diffusivity with consequent implications for the intrinsic rate of reaction. Furthermore, the importance of high product concentrations in most biochemical processes necessitates the use of complex rate laws incorporating substrate and product inhibitions; this, in turn, leads to the need to consider more than one diffusing species inside the particle. Moreover, the open pore structure of some carriers used to immobilize cells may mean that diffusion is not the dominant mechanism for nutrient transport throughout the aggregate. In this connection, Fowler and Robertson (1990), employing a hollow fiber system, found that mass transfer limitations are present throughout most of the cell aggregate in diffusionally supplied reactors. These limitations can be overcome by employing convective transport to supply nutrients and to remove products. However, continuing work is necessary to further the understanding of the resistance to convection in a convectively supplied reactor. On a practical side, it should be noted that continued cell growth in either diffusively or convectively supplied cell aggregates will inevitably lead to increased mass transfer resistances and to degradation of system performance. A general conclusion to be drawn from this work is that immobilized cell systems are amenable to quantitative description and that, for *E. coli*, the same metabolic parameters can be employed as for cells in suspension cultures.

The foregoing discussion points to the important fact that immobilized cell systems often colonize into thin layers near carrier surfaces, where they operate at the verge of starvation with respect to a limiting substrate. Considering the interplay of increasing total biocatalyst loading (i.e., total cell loading) and decreasing local substrate concentration with increasing layer thickness, it seems clear that an optimal shell size for maximizing metabolite production would exist. If the shell were too thin, insufficient catalyst is present; if too thick, substrate depletion is too severe and the inner portion of catalyst is ineffective.

Chotani (1984) studied ethanol production via yeast cell culture immobilized in particles of calcium alginate in a specially baffled plug-flow type reactor (see Table 4.1). Allowing for both substrate and product inhibition, the equation for the rate of change of substrate along the reactor length becomes:

$$-\frac{dS}{dt} = \frac{\eta \nu_{max} S}{K_{S0} + S + S^2 / K_{S2}} [1 - P / P_m] \qquad [4.20]$$

where P_m = 95 g/L, K_{S0} = 1 g/L, K_{S2} = 450 g/L and υ_{max} is the specific growth rate limit for the yeast cells (Chotani, 1984). The diffusion-limited conditions imposed by the membrane-like carrier envelope limit fermentation activity to the outer shell of the particles, resulting in effectiveness factors in the range $0.1 \leq \eta \leq 0.2$ ("biofilm particles").

Table 4.1 Comparisons of Immobilized Cell Bioreactors (Chotani, 1984)

Slurry-Type Fixed Bed	Fluid-Agitated Bed	Cross-Current
No mixing	Good mixing	Intermediate mixing
Low fermentation rate	High fermentation rate	High fermentation rate
Unstable operation	Stable operation	Stable operation
Low effectiveness factor	Higher effectiveness factor	Higher effectiveness factor
Radial temperature	Good temperature control	Good temperature gradients control
Solids-free feed	Suspensions processable	solids-free feed
Low axial dispersion	High axial dispersion	Intermediate axial dispersion

Mass transfer limitations commonly lead to strong spatial gradients in cell loading and activity within immobilized cell supports. Kuhn et al. (1991) have very recently developed a broadly applicable experimental technique which reveals, at high resolution, DNA synthesis rates as a function of position in the immobilization support, enabling determination of *local* specific growth rates. The method involves bromouracil (a thymine analog) pulse-labeling of immobilized cells undergoing active DNA synthesis, followed by immunofluorescent detection of incorporated label with scanning microfluorimetry.

The entrapment of yeast cells described just above was found to exercise better control on the growth nutrients supply to the immobilized colony, which could be maintained at levels in the range of 5 x 10^9 cells/ml within a typical bead. As a consequence, close to theoretical glucose to ethanol conversion efficiency and large doubling time resulted for such cells. So, while the cells did not replicate efficiently, carbon was channeled to the desired product, ethanol, quite effectively. Thus, we may consider a biofilm particle to consist of an inert core which is used as a carrier for a layer that contains the enzymes or microorganisms. Vos et al. (1990) present the modeling and effectiveness factor calculations for such a biofilm particle and a general model for an immobilized, non-growing biocatalyst. The model includes internal and external mass transfer resistance, the partitioning effect, and inhibition or reversible reaction kinetics.

In order to model product formation by the immobilized cell particles in the cross-current flow reactor, the effect of axial liquid dispersion was studied first. The tracer experimental study for the residence time distribution (RTD) was carried out by injecting one hundred microliters of isobutanol into the feed port of the reactor. The impulse response was analyzed over the next two space times by sampling at regular time intervals. For the prescribed conditions of inlet glucose concentration (185 g/L), 30°C, 40% packing density, 7-h residence time and 180-ml operating volume distributed equally among three chambers, RTD response has been plotted in Fig. 4.6. When the direction of the liquid flow through the main chambers was downward, this type of response corresponded to a train of seven stirred tanks (CSTR) in series. On reversal of the flow through the reactor, the distribution corresponded to eight CSTRs in series. Although there is no exact way to compare the tanks-in-series and dispersion models, the variances for the two models can be equated and the overall Peclet number comes out approximately as 14, showing an intermediate amount of axial dispersion.

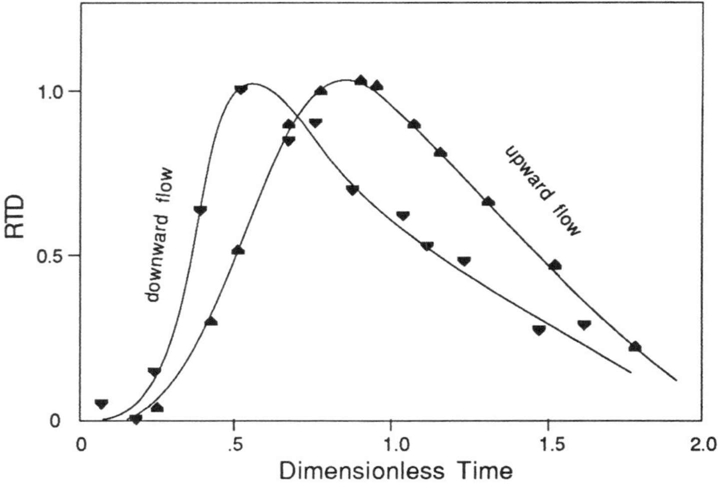

Figure 4.6 Experimentally measured RTD for the cross-flow reactor system.

Recently, Muslu (1990) studied dispersion in suspended growth systems. Correspondence was established between a bioreactor having a given dispersion number and the equivalent ideal reactors-in-series system. Design tables are provided to determine the required hydraulic residence time for a reactor with a given dispersion number, when the inlet and required outlet conditions and the Monod kinetic coefficients are known.

Ethanol production is, in practice, a very large scale bioprocess, and reactor design/operation methodologies can exhibit substantial differences, as tabulated below (Table 4.2).

Table 4.2 Ethanol by Free and Immobilized Systems (After Flannery & Steinschneider, 1983)

		Average Volumetric Productivity (gm/L/hr)
CONTINUOUS	Tower	12 - 14
	Vacuum + Recycle	65
	Immobilized Yeast	120 - 150
	Xymomonas	400 +
BATCH	Conventional	2.5 - 3.5
	Recycled Yeast	5.0 - 6.0

Seeking to optimize the process of ethanol production, a number of recent papers have appeared; e.g., Petersen and Whyatt (1989) have studied maximization of ethanol production by manipulating the feed rate to a bioreactor. Gupta and Chand (1990) and Ramakrishna et al. (1991a, b) studied bioconversion of sugars to ethanol in an immobilized cell packed bed bioreactor. The former also analyzed the dynamic response of the process to parameter perturbations. Sueki et al. (1991) employed a cascade feed technique to a fixed bed bioreactor. A process of ethanol fermentation in a stripping bioreactor was simulated, and the calculated data agreed with the experimental data (Sato et al., 1990a; b). The ethanol recovery rate increased with increases in the flow rate of stripping gas or in the fermentation temperature, and the ethanol productivity was improved by raising the temperature in spite of increasing the extinction rate of yeasts.

The central role of model bioreactors in bench scale applications is illustrated as a contribution to the optimal design of large scale bioreactors. Pilot plant data of a tubular reactor used for the production of ethanol with *Zymomonas* (and biopesticides with *Bacillus thuringiensis*) are presented (Moser, 1990; 1991a). Similarly, Moser (1991b) employs a systematic approach based on a holistic view called "integrating strategy" for the case of bakers' yeast processing, where the comparability of kinetic models including the concept of "dynamic flux" is critically discussed, showing that macro- and microkinetics have similar potential in describing bioprocesses according to the formal macroapproach.

4.11 ANAEROBIC PROCESSES

A subtle form of transport regulation of reaction pathways appears to be practiced by certain mixed cultures of anaerobic bacteria (Bhatia, 1983). Because relatively little attention has previously been paid to the subject of diffusion-reaction coupling in anaerobic systems, we go into this subject in some detail. Insight into this phenomenon is gained through process modeling and verification, as described below.

Until recently, proposed models of substrate consumption in anaerobic systems contained either simple Monod types of substrate consumption equations or slightly more complex equations, viz., substrate inhibition. Common to these approaches were two basic assumptions:

i. The substrate volatile fatty acid (VFA) consumption process is growth-associated, and

ii. The different substrates present in the feed (acetic, propionic and butyric acids) can be lumped together and reported as a composite substrate by measures such as COD (chemical oxygen demand) or BOD (biological oxygen demand).

A thorough experimental study demonstrated that feature i. above is not valid (Bhatia, 1983). This stems mainly from the fact that in all the transient studies performed, the production rate of methane stabilized at a new steady state in less than 0.5 hours when the influent conditions were changed. Such a rapid response of the culture to the environment would be possible only if the rate controlling step in the mechanism of methane production were very close to the cell's external environment. That part of the cell which maintains its internal environment and addresses for its maintenance energy requirement, i.e., the "ATP pumps," is known to be situated in the cell membrane.

Furthermore, the individual acids present in the feed interact; i.e., inhibit each other's consumption, so feature ii. also becomes invalid. Thus, models based on COD or BOD would not be able to account for these interactions without resorting to non-mechanistic model equations.

The above makes it obvious that a kinetic model incorporating equations for the consumption of each substrate is required. Such a model should not have to assume, implicitly or explicitly, that VFA consumption is a growth-associated process.

4.12 CARBON FLOW INSIDE THE CELL

The cell machinery is envisaged to function in the manner suggested by Ramakrishna (1966), Pirt (1975) and Van der Meer (1975). Carbon in the substrate is utilized to produce *energy* (primarily in the form of adenosine triphosphate or ATP) or is used to build the infra-structure of the cell for reproduction (see Fig. 4.7). Since methane is known to be a byproduct of the *energy*-consuming metabolic pathway, the higher the energy consumption, the higher is the production of methane. It has always seemed to the author that the potential exists for a membrane-moderated compartmentalized reaction system wherein an ATP-producing aerobic process is used to drive an ATP-consuming anaerobic process, the medium of exchange, ATP, being transported through the membrane.

In a somewhat similar vein, carbon monoxide, hydrogen and carbon dioxide in synthesis gas are converted to methane by employing a triculture of *Rhodospirillum rubrum*, *Methanosarcina barkeri* and *Methanobacterium formicium*, by coupling the photosynthetic reducing power of *R. rubrum* with the methane-producing pathways of the latter two microbes. Trickle-bed reactors were found to be effective for this conversion because of their high mass transfer coefficients. Mass transfer and scale-up parameters are defined by Kimmel et al. (1991), and light requirements for *R. rubrum* are considered in bioreactor design.

The energy produced is utilized for biochemical transformations that keep the cell's internal environment viable and for biochemical transformations necessary for the growth and replication of the cell. It is important to note the fine distinctions between various kinds of energy flow and how they lead to an unconventional definition of yield coefficient:

$$E_T = E_G + E_M \tag{4.21}$$

where:

E_T = Total energy production (units could be number of moles of ATP),

E_G = Energy required in cellular reproduction mechanism,
E_M = Energy required by the cell to maintain its viability.

From Fig. 4.7 it is shown that methane production is proportional to E_T and anything that increases it will cause methane production rate to increase. Since most of the carbon input to the cell goes into gas production, most of the carbon would be accounted for by E_T. Whatever carbon is left unaccounted for can be construed to go into waste products or new cell mass (this is usually 5 - 10%). Thus, the biomass yield coefficient can be defined as follows:

$$Y = \frac{\text{[Total amount of carbon consumed]} - \text{[Amount of carbon consumed for gas production]}}{\text{Amount of carbon in the new cell mass}}$$

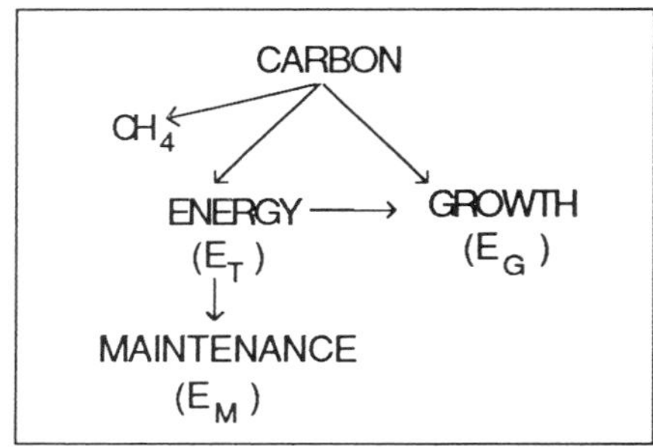

Figure 4.7 Intracellular carbon flow.

A similar definition of Y was developed by Van der Meer (1975), but since he accounted for "waste" products with a separate term, Y in his work became equal to unity.

4.13 MAINTENANCE ENERGY

Any abrupt change in the cellular environment would change the growth rate and/or the energy required to maintain the cell's viability (Ramakrishna, 1966). Thus, it is not surprising that past modeling efforts have assumed methane to be a growth-associated product. It was reasoned that, since an increase in substrate concentration increased growth rate which resulted in increased energy consumption (hence increased gas production), methane was growth-associated. Since anaerobic cultures grow very slowly (doubling time is of the order of a few days) it is difficult to conceive that they respond to substrate changes as rapidly as methane does. On the other hand, if methane is assumed to be a nongrowth-associated product then this enhanced gas production could be attributed to an increase in the maintenance energy coefficient (m); i.e., an increase in the maintenance energy requirement (E_M) of the culture. This is entirely possible since m is a very strong function of ionic strength (Pirt, 1975; Stouthamer and Bettenhaussen, 1973); i.e., as the ionic strength of the external environment increases, the cell machinery has to expend greater amounts of cellular energy to maintain its internal environment. In anaerobic systems this greater expenditure of energy will result in greater production of methane.

In laboratory experiments where fatty acid concentration increases, an equivalent amount of base has to be added to maintain the pH. This increases the ionic strength of the medium, causing the chain of events which leads to enhanced methane production. Thus the following expression, appropriately modified to take into account the existence of any inhibition, can be used for maintenance energy (Ramakrishna, 1966):

$$m = \frac{M \cdot S}{K_m + S} \qquad [4.22]$$

The above expression has tacitly been used by Van der Meer (1975) and Bastin and Wandrey (1981) in their analysis of the kinetics of VFA consumption.

Whether methane is growth- or nongrowth-associated, the model proposed in the following sections is unaffected, because most of the substrate carbon goes into gas production (90-95%), making the substrate consumption rate directly proportional to gas production rate and not to a weighted sum of growth rate and product generation rate. The model presented is, therefore, generally applicable.

4.14 SCHEMATIC DESCRIPTION OF THE MODEL

The model being proposed is designed to simulate the action of "methane-formers" in as much detail as possible. Thus, as a starting point, the block diagram proposed by Cohen et al. (1979; 1980) has been adopted.

Since most higher fatty acids are almost insoluble in water, most soured soluble organic wastes would contain mainly lower fatty acids (Van der Meer, 1975; Mahr, 1969). Hence, it is assumed that the feed input to the reactor consists of acetic, propionic and butyric acids only. Fig. 4.8 shows the flow of carbon from the three acids into and out of the different bacterial cultures. Since the oxidation of VFA mixtures requires a common enzyme for activation, it is assumed that propionic acid-consuming cultures can also consume acetic acid.

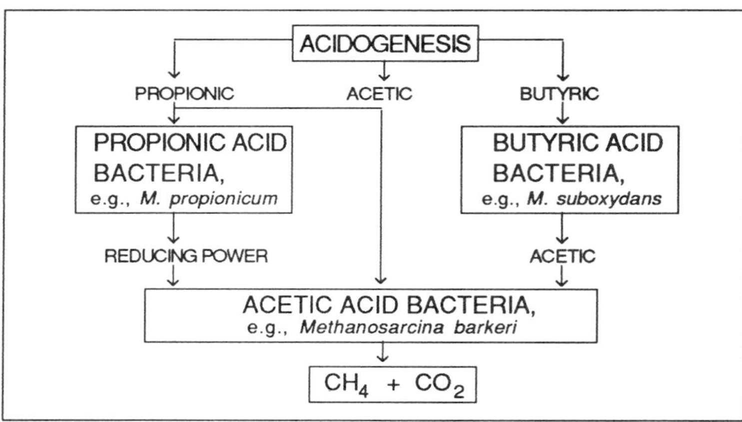

Figure 4.8 Extracellular carbon flow.

Strict anaerobic (methanogenic) cultures are incapable of consuming molecules larger than C_2 (Stadtman, 1967) but they readily consume acetic acid and reduce CO_2. They are also known to exist in a symbiotic relationship with cultures that preferentially consume hydrogen (Ghose et al., 1975). Thus the presence of an intermediate (I in Fig. 4.9) is postulated. The overall scheme of consumption is represented in Fig. 4.9. The lines represent all the possible places that each

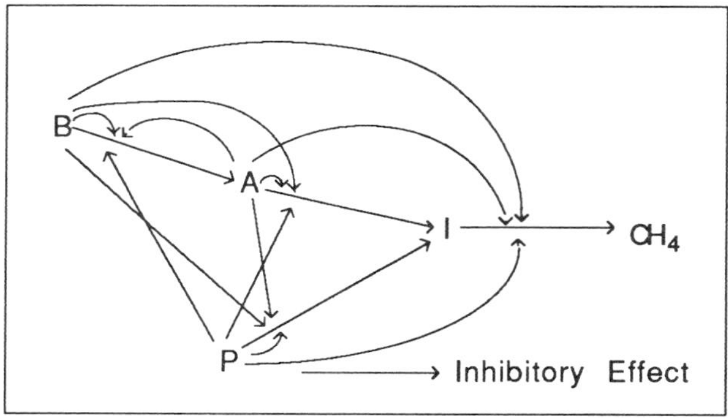

Figure 4.9 Model schema for substrate flow.

acid could exert its inhibitory influence. All the possible interactions, mapped out in the figure, do not necessarily occur but they have been included for completeness.

4.15 REACTOR SET UP

Since the upflow anaerobic sludge bed (UASB) reactor system responded like a CSTR to flow rate step-ups, the entire reactor has been modeled as a CSTR (Bhatia, 1983). Van der Meer in his doctoral dissertation (1975) found that the UASB system responds to tracer step and/or pulse inputs as two CSTRs in series. One CSTR represents the biomass bed (Fig. 4.10) and the other represents the disengaging space.

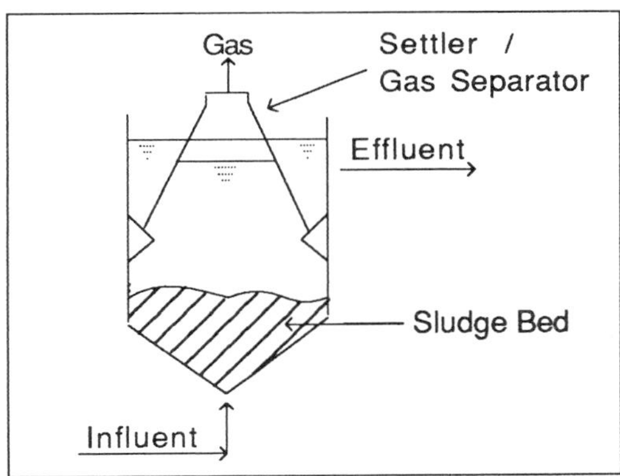

Figure 4.10 Upflow anaerobic sludge bed reactor.

Recently, Chen (1991) described what he terms an upflow attached bed (UFAB) continuous fermentor for ethanol production. When *Saccharomyces fragilis* was attached to cellulose acetate to ferment deproteinized and concentrated acid whey, ethanol productivity reached several multiples of batch reactor productivity, while for *Zymomonas mobilis* attached to vermiculite, another order magnitude improvement was reported. Sreekrishnan et al. (1991) studied the effect of various operating variables, such as initial inoculum circulation, dilution rate, COD, loading rate, and quantity and quality of inoculum on the process of film formation on a sand carrier surface, as well as reactor performance, using synthetic glucose-based wastewater. Film formation was favored by a high dilution rate, a large quantity of inoculum, and an inoculum having high methane-producing capacity. Biofilm formation apparently is initiated by methanogenic bacteria.

4.16 MODEL EQUATIONS

SUBSTRATE CONSUMPTION TERM

Using the notation developed by Roels and Kossen (1978) one can write the following general equations necessary for conducting a mass balance on component i of the liquid stream.

$$r_i = \frac{R_i \cdot S_i}{[1 + K_{Si} \cdot S_i] \cdot \left[1 + \sum_{j=1}^{3} K_{Iij} \cdot S_i \right]} \qquad [4.23]$$

where:

r_i = rate of consumption of the i-th substrate,

R_i = maximum rate of consumption of the i-th substrate,

K_{Si} = substrate affinity coefficient of the i-th substrate (similar to the reciprocal of the K_S in a Michaelis-Menten equation),

K_{Iij} = coefficient of inhibition depicting the inhibitory effect of the j-th substrate on the consumption of the i-th substrate.

It should be noted that the amount of active cell mass does not appear explicitly in eqn. [4.23] but does so implicitly in the term R_i; i.e., this term is a composite of the maximum specific consumption rate and the active biomass fraction consuming the i-th substrate. Since the experimental strategy was to perform the study of UASB in a manner designed to avoid growth effects, the fraction of biomass consuming a given substrate can be taken to be a constant.

Additionally, in case substrate inhibition, as suspected, exists, one can depict this situation using eqn. [4.23] wherein the inhibition coefficients for i = j are nonzero. Thus, eqn. [4.23] is a very general representation of various particular substrate consumption situations (including interaction of various substrates).

MECHANISMS CAUSING MULTIPLE STEADY STATES

The manifestation of multiple steady states is an indication that at low substrate concentrations the rate controlling step is different from that observed at higher concentrations. The two possible mechanisms that can explain such a phenomenon are discussed in the following two sections.

SUBSTRATE INHIBITION

In the case of simple substrate inhibition of an enzyme or a rate controlling sequence of reactions inside a cell, the following is the mechanistic representation:

$$E + S \leftrightarrow ES \rightarrow E + P \tag{4.24}$$

$$ES + S \leftrightarrow ES2 \tag{4.25}$$

where: E = enzyme or cell,
S = substrate,
ES, ES2 = enzyme substrate intermediates,
P = product.

In the above case, step [4.24] is rate controlling at low substrate concentrations and the product generation rate exhibits a Monod or Michaelis-Menten type of functionality. As the substrate concentration increases, the reaction represented in eqn. [4.25] moves more and more towards the right of the equilibrium. This increasingly inhibits the ES-complex and causes the production rate, which is proportional to the concentration of the complex, to decrease. Thus, at very high substrate concentrations, a switchover of rate controlling mechanisms occurs; i.e., the product generation rate now becomes a monotonically decreasing function of the substrate concentration. This reaction scheme can lead to the manifestation of multiple steady states (Bhatia, 1983).

4.17 HYSTERESIS DUE TO STRUCTURE OF THE BEADS

The UASB system is not set up like a conventional chemostat; i.e., the cultures present therein are not well mixed but rather are flocculated. The granular nature of the catalyst can lead to the manifestation of multiple steady states due to the way the various bacterial colonies are structured inside the beads.

Butyric acid-consuming bacteria are *aerobic* and/or facultative. Thus, in the UASB system, these bacteria are predominantly found on the outer layers of the biomass particles. Since the propionic acid bacteria are *anaerobic*, and methanogenic bacteria are strictly *anaerobic*, the model envisages such bacteria to form layers as shown in Fig. 4.11. In view of the fact that the biomass studied had been adapted to an undeaerated feed, it is highly plausible that over the protracted process of adaptation the aerobic bacteria clustered on the outside of the particle while the anaerobes accumulated inside.

Shinmyo et al. (1982) have shown that carageenan-immobilized aerobic bacteria growing inside the matrix could only penetrate 50 microns of the catalyst particle diameter. Their study was conducted in an aerated reactor in which oxygen transfer would have been less a problem than in the UASB reactor system. Additionally, Wada et al. (1980) found that immobilized yeast grown under anaerobic conditions could penetrate the beads up to 400 microns. These studies indicate that the less severe the oxygen requirement of the culture, the easier it is to penetrate deeper into the bead. Thus, in the case of a strict anaerobe, the entire population would have to reside in the core of the particle (Fig. 4.11).

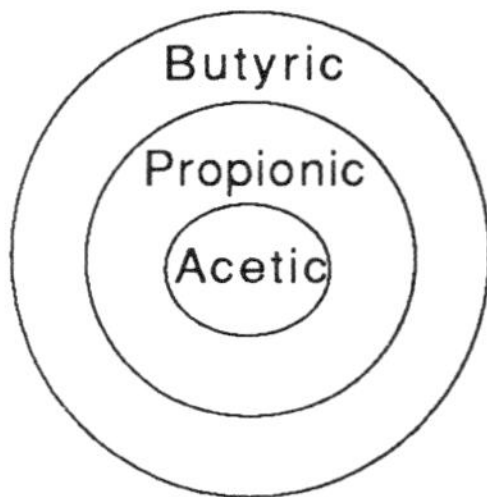

Figure 4.11 Schematic diagram of the proposed arrangement of the cultures inside the biomass particle.

The above layering of cultures inside the particle, combined with the stoichiometry of Fig. 4.9, can explain the manifestations of multiple steady states as follows:

At low concentration all the propionic acid substrate entering the bead would be readily taken up by the propionic acid-consuming bacteria. Furthermore, propionic acid is a known inhibitor of anaerobic cultures (i.e., the cultures that oxidize the reducing power generated [Fig. 4.9]). Thus the layered structure of the beads would effectively protect the innermost layer of strict anaerobes from propionic acid's inhibitory effect. Beyond the threshold concentration, propionic acid cannot be consumed completely by the outer layers of bacteria, and the strict anaerobes which convert the intermediate to methane are inhibited by the excess propionic acid. At this point the conversion of intermediate (I in Fig. 4.9) to methane becomes the rate controlling step; i.e., switchover occurs.

Regarding process instabilities encountered elsewhere, Garcia et al. demonstrate that cheese whey can be anaerobically treated (1991). The main operational limitation for single phase reactors is the onset of instability which necessitates the addition of alkalinity. A two-phase

UASB reactor was used and a new operational strategy is proposed. The recirculation of the effluent of the methanization reactor produces a dilution of the influent that allows a good stability of the system without the necessity of adding alkalinity. Greater than 99% COD removal can be obtained.

Similarly, the fermentation kinetics of methane production from whey permeate in a packed bed immobilized cell bioreactor at mesophilic temperatures and pHs around neutral was studied by Yang and Guo (1990). Propionate and acetate were the only two major organic intermediates found in the methanogenic fermentation of lactose. Based on this finding, a three-step reaction mechanism was proposed: lactose was first degraded to propionate, acetate, carbon dioxide and hydrogen by fermentative bacteria; propionate was then converted to acetate by propionate-degrading bacteria; and finally, methane and carbon dioxide were produced by methanogenic bacteria. The second reaction step was the rate-limiting step in the overall methanogenic fermentation of lactose. Monod-type mathematical equations were used to model these three-step reactions. The kinetic constants in the models were sequentially determined by fitting the mathematical equations with the experimental data on acetate, propionate and lactose concentrations. A mixed-culture fermentation model was also developed. This model simulates the methanogenic fermentation of whey permeate very well.

In adaptive optimal control, conditions are adjusted to obtain the best performance, as determined by a performance index (PI), which is often a function of more than one output variable. An empirical, linear input-output model is used, whose parameters are updated at each sampling time using online measurements. In one example, for anaerobic wastewater treatment, two relations were used. These are linear models relating feed flow rate to methane production rate and organic acid concentration. The PI was taken as a compromise between high methane and low organic acid concentration; it was maximized by the method of steepest ascent. This dynamic optimization method has considerable advantage for slow systems, in that a steady state is not required (Ryhiner et al., 1992).

4.18 ENZYME KINETICS

The conversion of VFA to methane is (like any other biochemical transformation) a multistep process involving various enzymes. Generally, one of the enzymes involved is appreciably slower than the others; this results in that particular enzymatic step becoming the rate controlling step. The overall kinetics of the biochemical transformation then become a manifestation of this key enzyme's kinetic character. In such a situation the other enzymatic reaction steps can be construed to be at equilibrium. In relation to one such mechanistic point of view the following sequence of reactions is proposed:

$$E + I \xrightarrow{k_1} CH_4 \qquad [4.26]$$

$$E + A \underset{k_{-2}}{\overset{k_2}{\leftrightarrow}} EA \qquad [4.27]$$

$$E + P \underset{k_{-3}}{\overset{k_3}{\leftrightarrow}} EP \qquad [4.28]$$

$$EA + I \xrightarrow{k_4} CH_4 \qquad [4.29]$$

where: [E] = concentration of the key enzyme,
[I] = concentration of the intermediate,
[A] = concentration of acetic acid,
[P] = concentration of propionic acid.

This is only a representation of the actual reactions taking place, assuming that the other reaction steps involved are at equilibrium. All the ks shown in the above sequence of reactions are composite; i.e., they incorporate ks of all the individual steps at equilibrium, as mentioned above. Eqns. [4.26] and [4.29] are represented as irreversible reactions because the standard free energy of formation of methane is highly negative (-12.1 Kcal/g-mole). In addition, the gas produced is essentially insoluble in water. Thus, the reaction depicting the production of methane is essentially driven to the product side; i.e., methane.

From the above sequence of reactions it is obvious that the rate of conversion of the intermediate would be proportional to the sum of the concentrations of [E] and [EA]; i.e., assuming that propionic acid inhibits the conversion completely.

$$\text{rate} \propto \{ [E] + [EA] \} \qquad [4.30]$$

$$[E_O] = [E] + [EA] + [EP] \qquad [4.31]$$

where $[E_O]$ = total amount of reactive sites available, and if reaction eqns. [4.27] and [4.28] are assumed to be at equilibrium, then:

$$\text{rate} = k_1 \cdot [E] + k_4 \cdot [EA] \qquad [4.32]$$

From eqn. [4.32]:

$$[EA] = K_2 \cdot [E] \cdot [A] \qquad [4.33]$$

where $K_2 = k_2 / k_{-2}$, and from eqn. [4.28]:

$$[EP] = K_3 \cdot [E] \cdot [P] \qquad [4.34]$$

where $K_3 = k_3 / k_{-3}$.
Thus, by substituting eqns. [4.33] and [4.34] into [4.31], one obtains:

$$[E] = \frac{[E_O]}{1 + K_2 \cdot [A] + K_3 \cdot [P]} \qquad [4.35]$$

Substituting eqns. [4.33] and [4.35] into eqn. [4.32] we have:

$$\text{rate} = \frac{[k_1 + k_4 \cdot K_2 \cdot (A)] \cdot E_O}{1 + K_2 \cdot [A] + K_3 \cdot [P]} \qquad [4.36]$$

Generally, the amount of free or unattached [E] available is negligible; i.e.,

$$[E_O] \approx [EA] + [EP]$$

so eqn. [4.36] becomes:

$$\text{rate} = \frac{[k_4 \cdot K_2 \cdot E_O] \cdot [A]}{K_2 \cdot [A] + K_3 \cdot [P]}$$

$$= \frac{k_4 \cdot E_O}{1 + \dfrac{K_3}{K_2} \cdot \dfrac{[P]}{[A]}} \qquad [4.37]$$

The above equation system was used to model the right limb of Fig. 4.12.

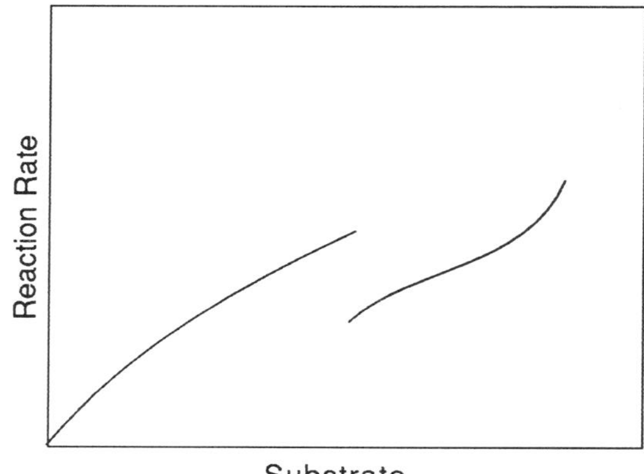

Figure 4.12 Pattern of reaction rate versus substrate from actual data.

Figures 4.13 to 4.15 show superimposed plots of actual and predicted response of the effluent concentrations when butyric acid is stepped up and down. The predicted response of the UASB fit the actual response for all three acids very well. It should, however, be noted that

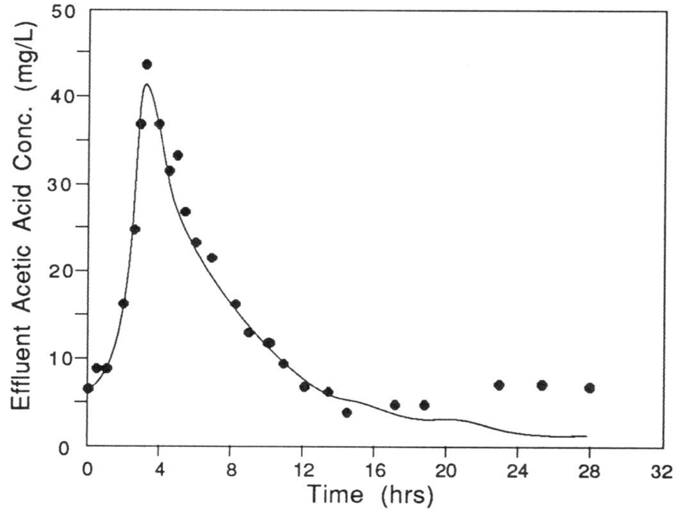

Figure 4.13 Response to 50% butyric acid step-up/down.

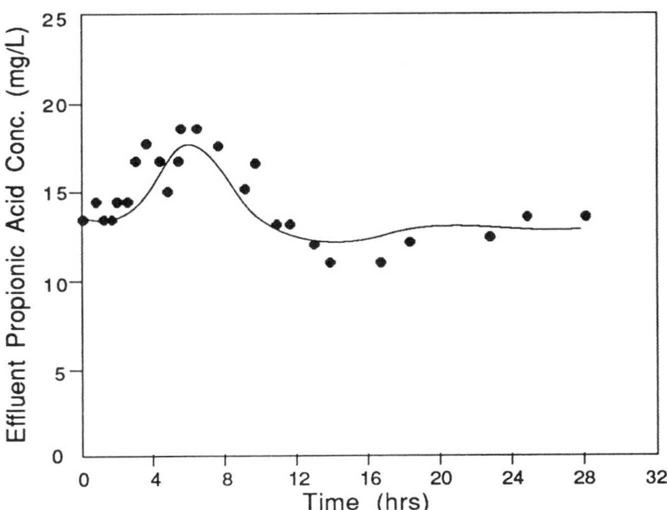

Figure 4.14 Response to 50% butyric acid step-up/down.

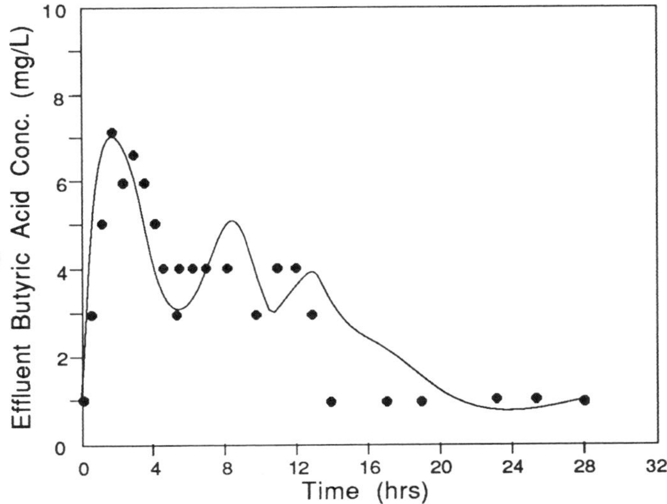

Figure 4.15 Response to 50% butyric acid step-up/down.

the predicted curves do not match the data as well when the *step down* occurs; i.e., the predicted curve dips more than does the actual data. This is in agreement with chemostat data reported in the literature; i.e., the response of a biological reactor (CSTR) peaks when inlet concentrations are stepped up but the effluent concentration decreases monotonically when influent concentration is stepped down.

Most bioreactor modeling efforts have characterized biocatalysts, whether free or immobilized, as uniform unsegregated colonies of cells with spatially constant properties. In the case of disgregated free cell systems operating in a chemostat, such a characterization results in relatively simple system equations and very stable transient responses to step inputs (i.e., the eigenvalues of the Jacobian matrix of the linearized system equations are real).

On the other hand, if a single microbial specie is immobilized (either on a carrier or in a network of auto-excreted polysaccharides), then the well-mixed chemostatic conditions cease to apply and diffusional aspects have to be considered as an important part of the system equations. Explicit incorporation of a diffusion equation for a limiting substrate could lead to more complicated system equations. The eigenvalues generated by such a system might be complex, leading to the possibility of oscillatory response and manifestation of multiple steady states. (The standard technique for locating such bifurcations is to look for model parameters which cause the eigenvalues (of the Jacobian matrix of the linearized equations) with largest real part to make a transition from negative to positive real root.)

The system equations become more nonlinear still when multiple-specie diffusional effects are being considered in structurally anisotropic systems; e.g., anaerobic sludge reactors. It has been shown (Bhatia et al., 1985) that analysis of the system equations in such a situation makes it necessary to take into account kinetic, biological and, at least implicitly, diffusional interactions. A ternary VFA substrate, combined with a layering of biocatalytic colonies, led to a cubic equation with, of course, multiple roots.

Explicitly, detailed modeling would lead to a set of nonlinear differential equations which invariably have complex eigenvalues. Nonetheless, it is this sort of analysis that must ultimately be used to analyze complicated systems like anaerobic digestors. The analysis in turn would help in elucidating the nature of spatial segregation of the various biological populations, resulting in manifestation of diffusional time delays, which invariably lead to erractic reactor response unless the reactor system is closely controlled.

4.19 NOMENCLATURE

A	Surface area available for immobilization, m^2/L.
C	Dissolved oxygen concentration, mMO_2/L or %sat.
C*	Dissolved oxygen concentration at saturation, mMO_2/L or %sat.
D	Dilution rate, hr^{-1}.
D_{opt}	Optimum dilution rate for maximum volumetric productivity, hr^{-1}.
K_S	Monod equation constant - "Half growth velocity constant", g/L.
k_La	Mass transfer coefficient.
L_{in}	Intracellular lactose concentration, moles/L.
L_o	Extracellular lactose concentration, moles/L.
P	Product concentration, g/L.
P_g	Gassed power, kw.
Pr	Reactor volumetric productivity, g/L/hr.
Q	Flow rate through the reactor, L/min.
S	Substrate concentration, g/L.
S_{opt}	Optimum substrate concentration for maximum productivity, g/L.
S_0	Inlet substrate concentration, g/L.
V_c	Specific volume of cells, L/g cell dry weight.
v_s	Superficial gas velocity, cm/sec.
X_{im}	Immobilized cell concentration expressed in terms of reactor volume, g/L.
X_S	Cell concentration on the surface of support, g/m^2.
X_S*	Maximum possible cell concentration on the surface of support, g/m^2.
X_{total}	Total cell concentration in immobilized cell reactor including free and immobilized cells, g/L.
Y	Product yield coefficient, g product/g substrate.
μ_b	Specific growth rate of cells in bulk, hr^{-1}.
μ_m	Maximum specific growth rate of cells in bulk, hr^{-1}.
μ_s	Specific growth rate of cells on the surface of the support, hr^{-1}.
μ_{ms}	Maximum specific growth rate of cells on the surface of the support, hr^{-1}.

4.20 REFERENCES

Allen, D. G. and C. W. Robinson, *Biotechnol. Bioeng.*, **38**, 212-216 (1991).

Ataai, M. M. and M. L. Shuler, *Biotechnol. Bioeng.*, **27**, 1051 (1985).

Bales, V., A. Bukovska, L. Pach, J. Herain and I. Langfelder, *Chem. Pap.*, **43**, 733-742 (1989).

Bartholomew, W. H., In "Adv. Appl. Microbiol," W. W. Umbreit, Ed., **2**, 289. Academic Press: New York (1960).

Bastin, K. and C. Wandrey, *Ann. N.Y. Acad. Sci.*, Biochemical Engineering II, **369**, 135 (1981).

Bhatia, D., Ph.D. Thesis, Dept. of Chemical and Biochemical Engineering, Rutgers University (1983).

Bhatia, D., W.R. Vieth and K. Venkatasubramanian, *Biotechnol. Bioeng.*, **27**, 1199 (1985).

Bunch, A. W., G. Higton, J. Ball and C. J. Knowles, In "Physiology of Immobilized Cells," J. A. M. deBont, J. Visser, B. Mattiasson and J. Tramper, Eds. Elsevier Science Publishers: Amsterdam (1990).

Chamy, R., M. J. Nunez and J. M. Lema, *Meded. Fac. Landbouwwet., Rijksuniv. Gent*, **54**, 1489-1499 (1989).

Chen, C., M. C. Dale and M. R. Okos, *Biotechnol. Bioeng.*, **36**, 993-1001 (1990).

Chen, H. C., *Biotechnol. Prog.*, **7**, 311-314 (1991).

Chibata, I., In " Immobilized Microbial Cells," ACS Symp. Series (1979).

Chisti, Y. and M. Moo-Young, *Biotechnol. Bioeng.*, **34**, 1391-1392 (1989).

Chisti, Y., M. Kasper and M. Moo-Young, *Can. J. Chem. Eng.*, **68**, 45-50 (1990).

Cho, T. and M. L. Shuler, U. S. Patent US 4940547 A, Use of Inhibitory Solvents in Multi-Membrane Bioreactor (1990).

Chotani, G. K., Ph. D. Thesis, Dept. of Chemical and Biochemical Engineering, Rutgers University (1984).

Dunlop, E. H., *Proc. Intersoc. Energy Convers. Eng. Conf.*, 25th, **3**, 500-504 (1990).

Flannery, R. J. and A. Steinschneider, *Biotechnology*, **Nov.**, 773-776 (1983).

Garcia, P. A., J. L. Rico and F. Fernandez-Polanco, *Environ. Technol.*, **12**, 355-362 (1991).

Ghose, S., J. R. Conrad and D. L. Klass, *J. W. P. C. F.*, **47**, 30 (1975).

Gil, G. H., W. J. Jones and T. G. Tornabene, *Biochem. Cell Biol.*, **68**, 661-668 (1990).

Gil, G. H., W. J. Jones and T. G. Tornabene, *Enzyme Microb. Technol.*, **13**, 390-399 (1991).

Gu, K. F. and T. M. S. Chang, *Biotechnol. Bioeng.*, **36**, 263-269 (1990).

Gupta, S. K. and S. Chand, *Chem. Eng. J.* (Lausanne), **43**, B1-B12 (1990).

Hahn-Hägerdal, B., In "Physiology of Immobilized Cells," J. A. M. deBont, J. Visser, B. Mattiasson and J. Tramper, Eds. Elsevier Science Publishers: Amsterdam (1990).

Hamer, G. and A. Heitzer, *Ann. N. Y. Acad. Sci.*, **589** (Biochem. Eng. 6), 650-664 (1990).

Hannoun, B. J. M. and G. Stephanopoulos, *Biotechnol. Bioeng.*, **28**, 829-835 (1986).
Hattori, R., *J. Gen. Appl. Microbiol.*, **18**, 319 (1972).
Hopf, N. W., S. Yonsel and W. D. Deckwer, *Appl. Microbiol. Biotechnol.*, **34**, 350-353 (1990).
Itamunoala, G. F., *Biotechnol. Bioeng.*, **31**, 714-717 (1988).
Jandel, A. S., *BioEngineering* (Graefelfing, Ger.) **7**, 35-36 (1991).
Jirku, V., J. Turkova and V. Krumphanzl, *Biotechnol. Lett.*, **3**, 509 (1981).
Ju, L. K., J. F. Lee and W. B. Armiger, *Biotechnol. Prog.*, **7**, 323-329 (1991).
Kang, W. K., R. Shukla and K. K. Sirkar, *Ann. N. Y. Acad. Sci.*, **589** (Biochem. Eng. 6), 192-202 (1990).
Karel, S. F., P. M. Salmon, P. S. Stewart and C. R. Robertson, In "Physiology of Immobilized Cells," J. A. M. deBont, J. Visser, B. Mattiasson and J. Tramper, Eds. Elsevier Science Publishers: Amsterdam (1990).
Karkare, S. B., G. K. Chotani and K. Venkatasubramanian, Unpublished results, Rutgers Univ. (1981).
Karkare, S. B., R. C. Dean and K. Venkatasubramanian , *Biotechnology*, **3**, 247 (1985).
Karkare, S. B., K. Venkatasubramanian and W. R. Vieth, *Ann. N. Y. Acad. Sci.*, Biochem. Eng. IV, **469**, 83 (1986).
Kawase, Y., *Biotechnol. Bioeng.*, **35**, 540-546 (1990).
Kawase, Y. and M. Moo-Young, *Chem. Eng. Res. Des.*, **68**, 189-194 (1990).
Kawase, Y. and M. Moo-Young, *Biotechnol. Bioeng.*, **37**, 960-966 (1991).
Kimmel, D. E., K. T. Klasson, E. C. Clausen and J. L. Gaddy, *Appl. Biochem. Biotechnol.*, **28-29**, 457-469 (1991).
Kiss, R. D. and G. Stephanopoulos, Abstract, Biochemical Technology Div., American Chemical Society Mtg., New York, NY, Dec. (1991).
Kuhn, R. H., S. W. Peretti and D. F. Ollis, Abstract, Biochemical Technology Div., American Chemical Society Mtg., New York, NY, Dec. (1991).
Lotong, N., W. Malapan, A. Boongorsrang and W. Yongmanitchai, *Appl. Microbiol. Biotechnol.*, **32**, 27-31 (1989).
Maerkl, H., *Forum Mikrobiol.*, **12**, 234-237 (1989).
Mahr, I., *Water Res.*, **3**, 507 (1969).
Mehaia, M. A. and M. Cheryan, *Bioprocess Eng.*, **5**, 57-61 (1990).
Melzoch, K., M. Rychtera, N. S. Markvichev, V. Pospichalova, G. Basarova and M. N. Manakov, *Appl. Microbiol. Biotechnol.*, **34**, 469-472 (1991).
Messing, R. A., R. A. Opperman, L. B. Simpson and M. Takeguchi, U. S. Patent 4,286,061 (August 25, 1981).
Mori, Y. and T. Inaba, *Biotechnol. Bioeng.*, **36**, 849-853 (1990).
Moser, A., *Biotech Forum Eur.*, **7**, 321-327 (1990).
Moser, A., *Biotechnol. Bioeng.*, **37**, 1054-1065 (1991).
Moser, A., *Bioprocess Eng.*, **6**, 205-211 (1991).
Mota, M. and J. A. Teixeira, *Chem. Biochem. Eng. Q.*, **5**, 63-65 (1991).
Muslu, Y., *Chem. Eng. J.* (Lausanne), **44**, B15-B23 (1990).
Okita, B. and D. J. Kirwan, *Biotechnol. Progress*, **2, No. 2**, 83 (1986).

Oosterhuis, N. M. G. and N. W. F. Kossen, *Biotechnol. Bioeng.*, **26**, 546-550 (1984).

Petersen, J. N. and G. A. Whyatt, *Biotekhnol. Khim.*, **1**, 16-17 (1989).

Petersen, J. N., B. H. Davison and C. D. Scott, *Biotechnol. Bioeng.*, **37**, 386-388 (1991).

Pierce, J. A., C. R. Robertson and T. J. Leighton, *Biotechnol. Prog.*, **8**, 211-218 (1992).

Pirt, S.J., *Proc. R. Soc., London, Ser. B*, **163**, 224 (1975).

Qureshi, N. and M. Cheryan, *Process Biochem.*, **24**, 172-175 (1989).

Ramakrishna, D., A.G. Fredrickson and H.M. Tsuchiya, *J. Gen. Appl. Microbiol.*, **12**, 4 (1966).

Ramakrishna, S. V., P. Prema and P. S. T. Sai, *Bioprocess Eng.*, **6**, 117-21 (1991).

Ramakrishna, S. V., T. Ratnakumari and P. S. T. Sai, *Bioprocess Eng.*, **6**, 141-144 (1991).

Roels, J. A. and N. W. F. Kossen, *Prog. Ind. Microbiol.*, **14**, 95 (1978).

Ruggeri, B., A. Gianetto, S. Sicardi and V. Speechia, *Chem. Eng. J.* (Lausanne), **46**, B21-B29 (1991).

Ryhiner, G., I. J. Dunn, E. Heinzle and S. Rohani, *J. Biotechnol.*, **22**, 89-105 (1992).

Sato, K., K. Yoshizawa and T. Nishiya, *Nippon Jozo Kyokaishi*, **85**, 645-650 (1990).

Sato, K., K. Yoshizawa and T. Nishiya, *Nippon Jozo Kyokaishi*, **85**, 745-749 (1990).

Shabtai, Y., S. Chaimovitz, A. Freeman, E. Katchalski-Katzir, C. Linder, M. Nemas, M. Perry and O. Kedem, *Biotechnol. Bioeng.*, **38**, 869-876 (1991).

Shinmyo, A., H. Kimura and H. Okada, *Eur. J. Appl. Microbial Biotechnol.*, **19**, 7 (1982).

Shuler, M. L. and F. Kargi, "Bioprocess Engineering," p. 179. Prentice Hall: Englewood Cliffs, NJ (1992).

Shuler, M. L., D. E. Steinmeyer, A. P. Togna, S. Gordon, P. Cheng and S. J. Letai, *Appl. Biochem. Biotechnol.*, **28-29**, 571-588 (1991).

Simon, J. P., *Belg. J. Food Chem. Biotechnol.*, **44**, 183-191 (1989).

Singh, V., R. Fuchs and A. Constantinides, In " Biotechnology Processes: Scale-up and Mixing," C. S. Ho and J. Y. Oldshue, Eds., p. 200. American Institute of Chemical Engineers: New York (1987).

Sreekrishnan, T. R., K. B. Ramachandran and P. Ghosh, *Biotechnol. Bioeng.*, **37**, 557-566 (1991).

Stadtman, T. C., *Ann. Rev. Microbiol.*, **21**, 121 (1967).

Steinmeyer, D. E. and M. L. Shuler, *Biotechnol. Prog.*, **6**, 362-369 (1990).

Steinmeyer, D. E. and M. L. Shuler, *Biotechnol. Prog.*, **6**, 286-291 (1990).

Stouthamer, A. H. and C. Bettenhaussen, *Biochimica et Biophysica Acta*, **301**, 53 (1973).

Strathmann, H. and W. Gudernatsch, *Bioprocess Technol.*, **11** (Extr. Bioconvers.), 67-89 (1991).

Sueki, M., K. Arikawa and A. Suzuki, *Hakko Kogaku Kaishi*, **69**, 7-13 (1991).

Tanaka, H., M. Matsumura and I. A. Veliky, *Biotechnol. Bioeng.*, **26**, 52-58 (1984).

Teixeira, J. A. and M. Mota, *Chem. Eng. J.* (Lausanne), **43**, B13-B17 (1990).
Teixeira, J. A., M. Mota and G. Goma, *Bioprocess Eng.*, **5**, 123-127 (1990).
Topiwala, H. H. and G. Hamer, *Biotechnol. Bioeng.*, **23**, 1683 (1971).
Tosa, T., *Nippon Nogei Kagaku Kaishi*, **65**, 185-188 (1991).
Van der Meer, R. R., Ph. D. Thesis, Delft Univ. Press, The Netherlands (1975).
Van't Riet, K., *Ind. Eng. Chem. Process Des. Dev.*, **18**, 357 (1979).
Venkatasubramanian, K., S. B. Karkare and W. R. Vieth, In "Applied Biochemistry and Bioengineering," **4**, p. 312. Academic Press: New York (1983).
Vera-Solis, F., M. S. Thesis, Dept of Food Science and Nutrition, M.I.T (1976).
Vieth, W. R. and K. Venkatasubramanian, In "Methods in Enzymology," K. Mosbach, Ed., Vol. 44, pp. 243 and 768. Academic Press: New York (1976).
Vieth, W. R., In "Biochemical Engineering," *Ann. N. Y. Adad. Sci.*, **326**, 1-6 (1979).
Vieth, W. R., K. Kaushik and K. Venkatasubramanian, *Biotechnol. Bioeng.*, **24**, 1455 (1982).
Voit, H., F. Goetz and A. B. Mersmann, *Chem. Eng. Technol.*, **12**, 364-373 (1989).
Vos, H. J., P. J. Heederik, J. J. Potters and K. C. Luyben, *Bioprocess Eng.*, **5**, 63-72 (1990).
Wada, M., J. Kato and I. Chibata, *Eur. J. Appl. Microbial Biotechnol.*, **10**, 275 (1980).
Watanabe, T., T. Aoki, H. Honda, M. Taya and T. Kobayashi, *J. Ferment. Bioeng.*, **69**, 33-38 (1990).
Webb, C., G. A. Dervakos and J. F. Dean, In "Biochemical Engineering," *Ann. N. Y. Adad. Sci.*, **589**, 593-598 (1990).
White, C., N. L. Morgan and E. G. Killick, *J. Chem. Technol. Biotechnol.*, **52**, 49-56 (1991).
Yamada, K. and K. Iwata, Jpn. Patent JP 01312993 A2, Air-Separation Composite Membrane for Bioreactor (1989).
Yang, S. T. and M. Guo, *Biotechnol. Bioeng.*, **36**, 427-436 (1990).
Yegneswaran, P. K. and M. R. Gray, Abstract, Biochemical Technology Div., American Chemical Society Mtg., New York, NY, Dec. (1991).
Yonsel, S., B. Langer and W. D. Deckwer, In "Biochem. Eng." - Stuttgart (Proc. Int. Symp.) 2nd, Mtg. date, 1990, 192-5, M. Reuss, Ed. Fisher: Stuttgart (1991).
Young, M. A., R. G. Carbonell and D. F. Ollis, *AIChE J.*, **37**, 403-428 (1991).

5

BIOPROCESS OPTIMIZATION AND CONTROL: ANTIBIOTIC CASE STUDY

5.0 INTRODUCTION TO ECONOMICS OF THE BIOTECHNOLOGY INDUSTRY

The biotechnology industry is expected to grow quite rapidly over the next decade. However, in common with other high technology industries, the pattern may be one of spurts and pauses rather than a smoothly continuous one. Taking into account these factors, Table 5.1 summarizes the projected situation for an important market segment in the years 1995 and 2000.

Table 5.1 Summary of U. S. Biotechnology Sales Forecasts by Key Market Segments: 1995 & 2000[1] (after Shamel and Chow, 1989)

Key Sectors	Base Year (1989)	Forecast Years 1995	Forecast Years 2000	1989/2000 Growth (PPA)
Contaminant monitoring	5	80	250	43
Agriculture	50	560	2,100	41
Specialties	55	240	900	29
Human diagnostics	360	1,000	2,200	18
Human therapeutics	640	2,410	5,950	23
TOTAL	1,110	4,290	11,400	24

[1]Millions of 1989 dollars. Source: Consulting Resources Corp.

The table shown above pertains specifically to genetically engineered products, which are considered to be synonymous with biotechnology. But, as was pointed out in Chapter 1, biotechnology in the larger sense encompasses the entire range of products produced from enzymatic reactions carried out in bioprocessing. A number of such products correspond to large volume applications. Factors which must be considered include complex reaction pathways and relatively low product concentrations with attendant high isolation and purification costs. Feed stocks may have to be complex (balanced nutrients) in order to obtain desired yields. Energy requirements and pollution abatement may pose other constraints.

To obtain some further perspective, consider the major components of production costs: bioreaction, isolation and finishing/packaging. Within those broad categories, the key factors include volumetric productivity, efficiency of aeration and mixing, product yield and product concentration in the broth ("titer value"). With regard to the latter, Table 5.2 illustrates typical product concentrations obtained in fermentation.

Table 5.2 Typical Product Concentrations in Fermentations

	Concentration %, w/v, dry
Vitamin B_{12}	0.002
Riboflavin	1
Benzyl penicillin	3
Bakers yeast	5
Ethanol	8
Glutamic acid	10
Citric acid	12
Lactic acid	13

Based in part on the data presented in this table, it is possible to determine that the fermentation/recovery cost ratios follow the trend 4.0, 2.0 and 0.16 for penicillin, a typical enzyme and ethanol, respectively (Datar, 1986). For newer antibiotics the situation can demonstrate a reversal (i.e., to 0.25).

Major antibiotic products generate very substantial revenues, as shown in Table 5.3. In general, antibiotic production involves fermentation, cell separation, extraction, further purification and crystallization/spray drying. Given the relatively high cost factors which are involved and their considerable range, it is clear that there is ample room for process optimization in the production of antibiotics.

Table 5.3 Major Antibiotic Products (Datar, 1986)

	1985 Revenue ($MM)
Amoxicillin	59
Erythromycins	144
Tetracyclins	136
Penicillins	193
Aminoglycosides	121
Cephalosporins[1]	896
Cefoxitin	315
Other	549
Total	2,413

[1]includes Keflex (Cephalexin) @ $300MM

5.1 INTRODUCTION TO PROCESS OPTIMIZATION

In order to achieve process optimization, detailed data acquisition is often necessary, implying the need for precision biosensing. As an example, Li and Humphrey (1991) disclose a new fluorescent bioreactor monitoring probe, which incorporates a multiple excitation fluorometric system (MEFS). It was found that the most effective way of monitoring a bioreactor by fluorometry may be to monitor several fluorophors in the whole culture broth simultaneously and to relate these fluorescence signals to various biological parameters.

Pesl and Seichter (1990) discuss a recent relevant case involving parameter selection for industrial bioreactors. The optimization procedure is based on an analysis of production costs and their possible effects on the overall process economics. A new method of finding the optimum split of energy for mechanical agitation and that supplied by pressurized air is also presented. Another recent example appears in the work of Nishiwaki et al. (1990). The optimal extent of longitudinal mixing with respect to substrate conversion for a steady state bioreactor in which substrate limited cell growth occurs was investigated numerically. The tanks-in-series and the dispersion mixing models with Monod kinetics were employed. The optimal numbers of tanks and Peclet numbers to yield the maximum substrate conversion were determined and presented. In a related study, Eisenmann (1989) reported on characterization and operating regimes of reactors with regard to optimization of product quality and yield. Achievement of ideal mixing in relation to reactor design, e.g., a double-axial turbulence mixing reactor, was considered. More generally, Hardman (1989) provides a review of strategies in bioprocess design involving enzyme and microbial biotechnology with extensive reference to biosensors, while Yoshida and Kishimoto (1989) describe the optimization of bioprocesses by application of an artificial intelligence technique using a "fuzzy set" theory approach. The authors claim that the construction of the knowledge data base from the experimental data and semi-quantitative information found in the literature proceeded more rapidly than that achieved by the more traditional approach of mathematical modeling. A neural network program with efficient learning ability for bioprocess variable estimation and state prediction has been developed (Linko and Zhu, 1991). The well trained neural network accurately and rapidly estimated the state variables for the system with or without noise, even under varying process dynamics.

Lin and Lim (1990) reveal a kinetic model for *M. mucosa* which includes specific rates of growth and polysaccharide formation as functions of methanol concentration, that show a maximum and a minimum, respectively. Because of the nonmonotonicity of the specific growth rate, there are two nontrivial steady states, one stable and the other unstable, for a single continuous stirred-tank bioreactor (CSTBR). Cell growth is favored at stable steady states while polysaccharide yields are favored at unstable steady states. These phenomena favor a system of two CSTBRs in series, in which the first bioreactor is operated at a stable steady state to feed a constant supply of cells to the second bioreactor for polysaccharide production. It is shown that such a system can be made naturally stable and that the polysaccharide yield can be maintained high at dilution rates much above the maximum specific growth rate for a single bioreactor.

Pertaining more specifically to the work to be reported here shortly, Kim et al. (1989) have investigated the effects of the metabolic regulator, methionine, on cephalosporin C production in a fluidized bed bioreactor using particles of *Cephalosporium acremonium*. The authors found that the initial concentration of methionine was a key factor for higher cephalosporin C production.

Carbon consumption rate increased significantly in the presence of methionine while the rate of product formation was greatest near the point of exhaustion of the regulator. It was therefore considered important to feed an optimal methionine concentration early for effective antibiotic production over the entire process interval.

5.2 MODELS FOR SECONDARY METABOLITE SYNTHESIS

METABOLITE PRODUCTION

In general, metabolite production in fermentation systems can be described by the Leudeking-Piret model:

$$\frac{dP}{dt} = K_1 X + K_2 \frac{dX}{dt} \qquad [5.1]$$

where P is the metabolite concentration; if $K_2 >> K_1$ we have a growth-associated product (or primary metabolite), and when $K_1 >> K_2$ we have a secondary metabolite.

a. Primary (growth-associated) metabolites.

In this case the rate of product formation is given by:

$$\frac{dP}{dt} = K_2 \frac{dX}{dt} \qquad [5.2]$$

and the metabolite productivity is described by:

$$Pr = \frac{dP}{dt} = K_2 DX \qquad [5.3]$$

Hence, maximizing the metabolite productivity is the same as maximizing biomass productivity. Therefore, the calculation of optimum dilution rate is the same as before.

b. Secondary metabolites.

Rate of biosynthesis of secondary metabolites is dependent primarily on the cell concentration. Hence we can write this rate as:

$$\frac{dP}{dt} = K_1 X \qquad [5.4]$$

For an immobilized-cell process, this would become:

$$\frac{dP}{dt} = K_1 [X + X_{im}] \qquad [5.5]$$

Hence, maximizing the productivity in this case involves maximizing X_{im}. This can be done by using a catalyst support with high loading capacities. It is also advantageous to keep X as high as possible. From Fig. 4.4 (on page 109) one can see that X does not change drastically until D becomes greater than μ_m. Thus any value of D would yield substantially the same productivity. Therefore, the choice of D would depend on the yield requirement of the process. The product mass balance in this case is given by:

$$DP = K_1 \, [X + X_{im}]$$

or

$$P = \frac{K_1}{D} \, [X + X_{im}] \qquad [5.6]$$

Because X is relatively constant, D can be chosen to suit the requirement of P (often dictated by the recovery process).

c. Metabolites with mixed growth model.

When K_1 and K_2 are both significant, the product mass balance becomes:

$$DP = K_1[X + X_{im}] + K_2\mu_b[X + X_{im}] = [K_1 + K_2\mu_b] \, [X + X_{im}] \qquad [5.7]$$

Again, using the same techniques as in biomass productivity, we can calculate optimum dilution rate for maximum productivity. In this case,

$$S_{opt} = \frac{\sqrt{Y^2K_S^2[K_1+\mu_mK_2]^2 - YK_S[K_1+\mu_mK_2]\,[YS_0+X_{im}]} - YK_S[K_1+\mu_mK_2]}{Y[K_1 + \mu_m \, K_2]} \qquad [5.8]$$

and, once again,

$$D_{opt} = \frac{\mu_m S_{opt} Y[S_0 - S_{opt}] + \mu_m S_{opt} X_{im}}{Y[K_S + S_{opt}]\,[S_0 - S_{opt}]} \qquad [5.9]$$

Again, we can verify that as $K_1 \rightarrow 0$, one obtains S_{opt} identical to that of growth-associated products.

5.3 BIOSYNTHESIS OF CANDICIDIN

Up to the present, relatively few studies have been directed at production of secondary metabolites by immobilized living cell systems. Ideally, an immobilized cell system should lend itself more readily to the production of secondary metabolites because of its ability to maintain a high concentration of cells in a non-growing condition. However, the complexity of the organisms and the rather elaborate pathways involved in their biosynthesis, as well as the problem of oxygen transfer, appear to have hindered progress in this direction. Morikawa et al. (1979) studied the production of penicillin by immobilized *Penicillium chrysogenum* and found the half life to be about six days, but no attempt was made at rejuvenating the activity of the cells by inducing cell growth.

As elaborated by Constantinides and Mehta (1991), in the case of microbial cells, secondary metabolism often occurs in a narrow range of specific growth rates. Furthermore, the optimal expression of secondary metabolism takes place in a subset of this range, often at zero growth (Bu'Lock et al., 1975; Bunch and Harris, 1986; Martin and Demain, 1980). Consequently, batch fermentations for products such as antibiotics often follow a biphasic pattern. During the first phase, there is rapid growth of cell mass without significant production of secondary metabolites. The second phase involves the production of secondary metabolites accompanied by slow growth, or no growth, of the microorganism.

It should be pointed out, however, that while the process of cell growth is detrimental to the cellular specific productivity of secondary metabolites in these cases, the instantaneous volumetric productivity of the reactor is obtained by multiplying the cell concentration and the corresponding cellular productivity of the metabolite. Due to the bilinear nature of this relationship, higher cell concentrations are essential for increased volumetric productivity. Obtaining higher cell concentration in the reactor, however, involves microbial growth.

The foregoing suggests that by controlling the physiology of the cell, one might be able to optimize antibiotic productivity and also achieve significant economies of substrate consumption. In other words, it might be possible to divert the substrate consumption from cell growth to product formation, thus increasing the carbon conversion efficiency and overall productivity of the system. (The pathway for candicidin synthesis has already been introduced in Fig. 1.12.)

This approach had been anticipated earlier in our laboratory through a study of the carbon conversion efficiency of *Streptomyces griseus* under various growth conditions (Karkare et al., 1986) and it was indeed shown that, by proper administration of the nutrients, the substrate could be channeled into metabolite formation by suppressing growth. The growth versus product synthesis conditions can be effectively controlled by the phosphate level in the input medium (Martin, 1980). In this case, $PO_4^{\equiv}$ was used as a metabolic switch in the production of candicidin, as shown in Table 5.4.

Table 5.4 Selected Conditions

Day of Operation	Dilution Rate (h^{-1})	Mode of Operation	Phosphate Concentration in Medium (moles/L)
18	0.145	steady, growth	10^{-3}
21	0.49	steady, growth	10^{-3}
27	0.149	steady, nongrowth	0
39	0.149	transient, growth	5×10^{-4}

Phosphate promotes growth and suppresses antibiotic production (Liu et al., 1975). Thus, it provides a relatively simple means to increase productivity of the system, and at the same time rejuvenate its catalytic activity periodically. Phosphate plays a similar role in several secondary metabolite fermentations (Martin, 1980), hence, the method described herein may be applied to other antibiotic fermentations. Other appropriate growth factors may be employed in a similar manner in most secondary metabolite fermentations. Recently, Abulesz and Lyberatos (1987) simulated the periodic variation in dilution rate in a continuous fermentation. They showed that, in a chemostat, periodic operation of bioreactors can result in higher biomass and protein productivities. Sayles and Ollis (1989) simulated the performance of a single immobilized cell particle subject to periodic substrate cycling. Their simulation showed that cycling increased the average product yield in the case of nongrowth associated product formation by up to a factor of three.

On the whole, few attempts have been made to model the batch kinetics of antibiotic synthesis in general (Blanch and Rogers, 1971; Brown and Voss, 1973; Martin and McDaniel, 1975). Another approach is to invoke the concept of a "maturation time" to explain the delay in the synthesis of secondary metabolites. However, Martin and McDaniel (1976) have shown that when the cells of *Streptomyces griseus* are placed in a phosphate-free medium, they start producing candicidin at a constant rate regardless of whether the cells are "mature" or not. Hence, the "maturation" concept seems to be more a function of the phosphate content of the medium than the actual age of the cells. In this study, a synthetic medium was used to isolate the effect of phosphate on growth of cells and candicidin production. The resting cell system (phosphate-free) was found to produce candicidin at a constant rate for about 36 hours. These results suggest that during batch fermentation, antibiotic synthesis begins (or is promoted) when phosphate is depleted from the medium. Since phosphate is the growth factor in the case of many microbes producing secondary metabolites, it seems reasonable to deduce that secondary metabolite synthesis is controlled by the growth versus nongrowth condition of the cells.

The reactor used to investigate this hypothesis was a stirred tank fitted with a low-shear impeller and a sintered-steel (Pall Trinity grade G) sparger. The reactor volume was 120 ml. The low-shear impeller kept the 1-mm diameter beads of carageenan-immobilized *S. griseus* well stirred without causing significant bead attrition. The beads were contained within the reactor with a 100-mesh stainless steel screen that was kept clean by a periodic backflush of sterile air. Since the effluent stream has the same concentration as the bulk of the reactor, kinetic measurements are simplified. The entire continuous system is shown in Fig. 5.1.

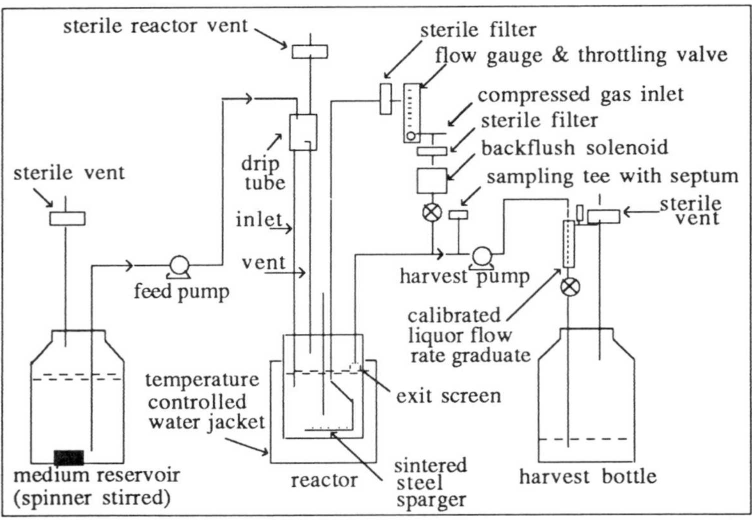

Figure 5.1 Continuous immobilized living cell reactor system.

Experiments were carried out under various nutrient feed conditions with the following purposes in mind:

i. To examine rejuvenation of catalytic activity by inducing brief periods of cell growth.

ii. To determine steady state productivity values for candicidin synthesis under various growth and nongrowth conditions.

iii. To test kinetic models capable of describing the productivities obtained experimentally.

iv. To examine the carbon conversion efficiency of the cells under growth and nongrowth conditions and to determine the apportioning of carbon into growth and production channels in each case.

Initially, the reactor was fed with soy peptone, cerelose (SPC) medium until the candicidin productivity appeared to stabilize. Mycelium growth could be observed in the broth. When the reactor had stabilized, the medium was changed to the synthetic one devoid of phosphate. After this change, the productivity first dropped about 20%, possibly due to washing away of the free mycelium in the reactor. The rate then increased to a higher level and remained there for several days. During this period, there was no visible growth occurring in the reactor. Clearly, the nongrowth medium appears to shift the cell from growth to antibiotic production.

Subsequently, the reactor was operated to obtain data for carbon balancing under various growth conditions; the reactor was operated for 40 days. The data is summarized here, but the full results are also readily obtainable (Karkare, 1983). For our purposes, the four different growth conditions appearing in Table 5.4 are selected. The first two represent growth continuing at steady state, the third represents nongrowth steady state, while the last one is under unsteady state conditions.

The transient state selected was obtained as a result of increasing the phosphate concentration in the feed from 50 mM to 500 mM. The cell and candicidin productivities increased dramatically to high levels and then declined very quickly. This seems to be due to synchronous cell growth induced by nutrient deprivation (Bailey et al., 1985).

To find out how the substrate carbon was apportioned between cell growth and candicidin production, a carbon balance was performed on the system for the conditions shown in Table 5.4. This was accomplished by measuring the glucose and asparagine concentrations at the inlet and outlet of the reactor. This determines the amount of carbon being consumed. The exit cell concentration, candicidin concentration, and carbon dioxide evolution rates were also measured. The dissolved carbon dioxide leaving the reactor was estimated using the equilibrium solubility value at 36°C and the neutralizing capacity of the NaOH added to the medium.

Table 5.5 gives the raw data necessary for the calculations. For the purpose of calculations, it was assumed that:

(1) 50% by weight of cell mass was carbon, and

(2) each milligram of candicidin uses up 0.04 g of glucose (based on data of Martin and McDaniel, 1976).

Table 5.6 gives the results of the carbon balance for the four cases mentioned. The high percentage of unaccounted carbon may be due to production of some unknown metabolites and partly due to the inaccuracies in glucose and asparagine measurements.

Table 5.6 indicates that the carbon conversion efficiency of the nongrowing cells is more than three times higher than that of growing cells. Even at two different growth rates, the conversion efficiency remains fairly constant. Thus, we can surmise that during cell growth, less than one-third of the carbon (that is normally used for product formation) is channeled to metabolite production, while the rest goes toward cell growth and CO_2 formation. It also means that on a "per cell" basis, the rate of production of candicidin will be less than one-third, during the growth phase, compared to the nongrowth phase.

The last result is also very interesting. In this case, the conversion efficiency was very high, while the cells were growing vigorously. There was also no unaccounted carbon. All the cell's energy seems to be focused on growth and product formation during this transient phase. It should be noted though, that this efficiency was obtained for only a short time. Clearly, further analysis is necessary to delineate the underlying reasons for the observed results.

It is apparent from the results of the carbon balance that under steady state conditions, the highest productivity and efficiency is obtained when the reactor is maintained in a nongrowth mode. However, a reactor cannot be constantly maintained in a nongrowth mode, because the catalytic activity starts to decline after about six days of operation. Fortunately, the activity can be completely rejuvenated by inducing cell growth for a brief period of a few hours. Thus, the appropriate reactor operation strategy should be to stimulate growth of cells to their maximum extent and then use a nongrowth medium to enhance reactor productivity and efficiency. The reactor should be periodically rejuvenated by a brief introduction of growth-supporting medium at regular intervals.

It became of obvious interest to follow up on the transient results obtained in the candicidin experiments. One of the possibilities was to

Table 5.5 Data for Carbon Balance

Day	Phosphate (M)	Glucose In (g/L)	Glucose Out (g/L)	Asparagine (g/L)	Cells Out (g/L)	Exit CO_2 (%)	Liquid Flow (L/h)	Gas Flow (L/min)	Candicidin Out (mg/L)
18	0.001	45.04	42.16	5.99	1.548	0.0815	0.0134	0.16	12.675
21	0.001	45.04	45.04	6.30	0.715	0.1075	0.04	0.16	6.375
27	0	43.42	42.52	7.0	0.117	0.01	0.01	0.08	24.0
39	5×10^{-4}	43.42	39.63	5.80	3.35	0.092	0.014	0.14	60.30

Table 5.6 Carbon Balance and Conversion Efficiencies for Candicidin Synthesis by Immobilized Cells

Mode	Glucose Carbon Consumed (g/h)	Asparagine Carbon Consumed (g/h)	Carbon to Candicidin (g/h)	Carbon to Cells (g/h)	Carbon to CO_2 (g/h)	Carbon Conv. Eff. (%)	Carbon Unaccounted (%)
Continuous growing cells	0.0154	0.0195	0.0027	0.0103	0.0115	7.73	32
Continuous growing cells	0.0041	0.0538	0.0041	0.0143	0.0276	7.04	22
Continuous nongrowing cells	0.0038	0.0109	0.0038	0.0006	0.0058	25.85	30.6
Continuous unsteady state growing cells	0.0211	0.0214	0.0134	0.0234	0.0057	31.52	---

operate the reactor in a semi-continuous transient mode in order to take advantage of the high productivity obtained, and this was carried out and recently reported (Constantinides and Mehta, 1991).

The system used for studies on periodic operation of immobilized live cell bioreactors is shown schematically in Fig. 5.2. The reactor is a three column system consisting of the fluidized bed-immobilized live cell column, a bubble column for gas exchange between the liquid phase and fresh sterile air, and a disengagement column for separating the gas and liquid phase. The volume of gel beads in the reactor was approximately 200 ml. The operating volume of the reactor was 650 to 700 ml.

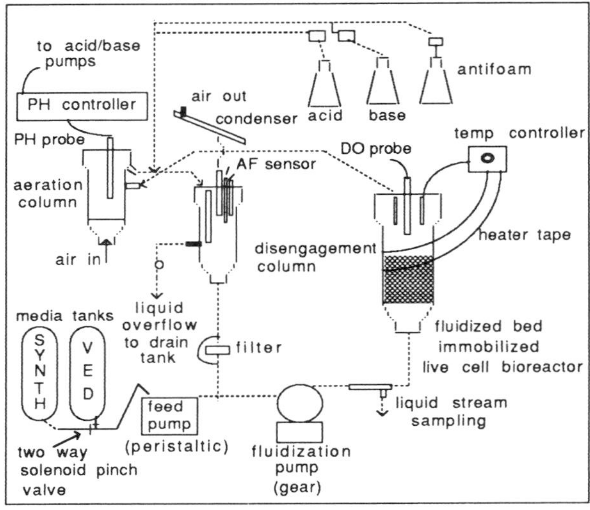

Figure 5.2 IMC bioreactor.

The kinetics of growth and candicidin production by *S. griseus* in complex and in synthetic media were investigated. On-line glucose analysis of the fermentation was used to identify nutrient limitation by multiple substrates during the fermentation. The growth kinetics of *S. griseus* under limitation by two nutrients were studied. The Monod type of model was used to analyze cell growth under multiple substrate limitation. Regions of phosphate limitation were identified and the nutrient medium for the forced periodic operation of the immobilized cell bioreactor was developed.

A lumped model for cell growth and candicidin production was developed. The model was used to perform a simulation study of the periodic operation of the bioreactor. An IMC bioreactor with forced periodic operation was used to study the effect of cycling frequency on reactor performance.

The model itself involved a modification of eqn. [5.4] in that the right hand side was multiplied by the term K_c / K_c + P^2, to allow explicitly for the effect of phosphate inhibition. Substrate uptake by the cells was modeled by the following equation:

$$r_{phosphate} = \frac{1}{Y}\frac{\mu_m PX}{(K_S + P)} \qquad [5.10]$$

where Y is the yield coefficient for the phosphate limited growth case. It should be noted that no maintenance term is required in the equation for substrate uptake since it was assumed that phosphate was not a source of available energy for the cells.

In order to evaluate the differences between steady state and periodic operation, comparable conditions had to be used. Periodic operation will be advantageous only when the conversion of reactant to the desired product in unit time and volume can be increased. In other words, for a given reactor volume, the amount produced per unit time, unit volume, or unit biomass must be the criterion for comparison. To achieve this, the amount of reactant reaching the reactor in unit time must be constant. The experimental and simulation studies, therefore, are based on reactor performance over an integral number of cycles, over a period of 192 hours, with equal amounts of reactants supplied during this interval. Likewise, in order to study the rate of candicidin production (from its concentration in the effluent stream), the amplitude of the phosphate pulse (as yeast extract) had to be chosen such that a substantial concentration of the product was maintained in the reactor under conditions of declining productivity, for at least two cycles of the largest period chosen. Single pulses of 2, 3 and 5 g of yeast extract corresponding to phosphate concentrations of 0.094, 0.141, 0.235 g/L were injected at 96 hours into the reactor in three separate experiments. It was found that the 5 g pulse would provide sustained operation for up to two 96 hour cycles.

In periodic systems, forced oscillation with very small amplitude often results when the input varies rapidly with response time, and the system can be approximated as being at steady state. The term "relaxed" has been used by Bailey and Ollis, to describe this approximate steady state (1977). This is the form of the behavior which is apparent in the 12 hour period experiment, as shown in Fig. 5.3. The experimentally determined candicidin *production rate* is flat and at a value in the vicinity of 0.8 mg/g-hr. [In batch fermentations in different complex media the highest *production rate* obtained was 0.92 mg/g-hr (Mehta, 1988)]. The "steady state" cell concentration in the 12 hour case is higher than that in the 16 hour pulsing case (Fig. 5.4). The response of the culture becomes increasingly oscillatory in terms of its amplitude, with a lower frequency of pulsing.

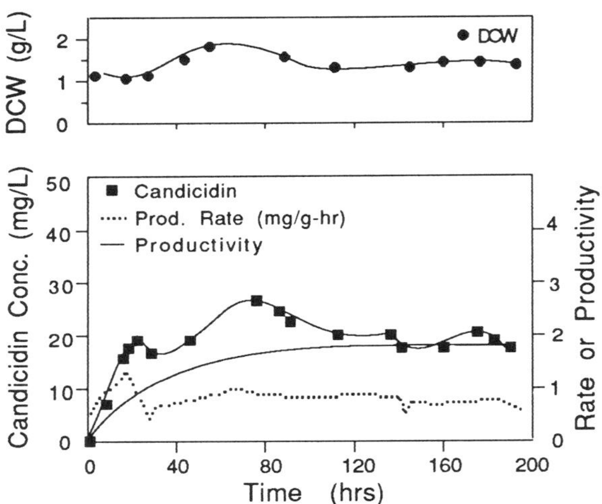

Figure 5.3 Bioreactor periodicity profiles (12 hour pulsing).

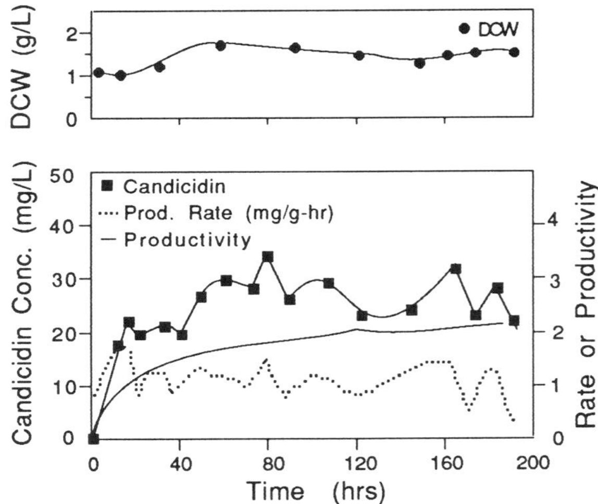

Figure 5.4 Bioreactor periodicity profiles (16 hour pulsing).

It was shown that rapid dosing of the growth medium resulted in a much more stable output of candicidin at a lower specific productivity than regeneration at longer times. At a pulsing frequency of 12 hours, an average reactor volumetric productivity of 1.63 mg candicidin/L-hr was obtained, as compared to an average productivity of 2.135 mg/L-hr obtained with 16 hour pulsing.

When the period is of the order of the system's dynamic response time, transient behavior assumes significant importance. Under these circumstances, performance shifts by transient behavior can be unexpected and occasionally quite large. The phenomenon of increased secondary metabolite productivity during transient operation was noticed by Karkare (1983), as already mentioned. In his study, the efficiency of converting the carbon source to candicidin increased very sharply in continuous unsteady state growing cells, as compared to both continuous steady state growing cells and continuous nongrowing cells (Table 5.6).

Figure 5.5 shows the volumetric productivities attained in each of the four bioreactor runs. The volumetric productivity in all cases of periodic operation compares favorably with that during steady state operation of the immobilized bioreactor. The 16 hour pulsing generated the highest levels of volumetric productivity: 2.135 g/L-hr as compared to 1.41 g/L-hr obtained during the steady state operation of the bioreactor. Additionally, the system oscillated with a much smaller amplitude than the 24 hour case.

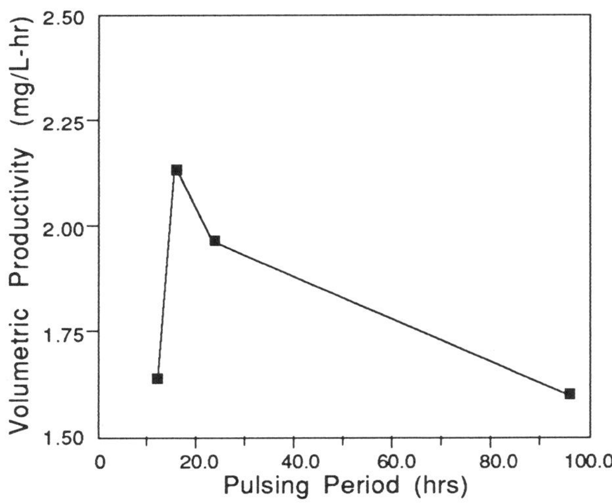

Figure 5.5 Volumetric productivity in periodic operation.

The results of the two studies just described suggest that periodically operated immobilized live cell bioreactors can provide a potent alternative for the production of nongrowth associated biochemicals, as compared to free cell fermentations, pulsed fermentations with process cycle regeneration, and non-regenerated bioreactors. This work has demonstrated that by frequent pulsing of the growth limiting nutrient, stable extended production can be obtained at high specific cellular productivities.

5.4 REFERENCES

Abulesz, E.-M. and G. Lyberatos, *Biotechnol. Bioeng.*, **29**, 1059-1065 (1987).

Bailey, J. E. and D. F. Ollis, "Biochemical Engineering Fundamentals." McGraw-Hill, Inc.: New York (1977).

Bailey, K. M., K. Venkatasubramanian and S. B. Karkare, *Biotechnol. Bioeng.*, **27**, 1208-1213 (1985).

Blanch, H. W. and P. L. Rogers, *Biotechnol. Bioeng.*, **13**, 843 (1971).

Brown, D. E. and R. C. Voss, *Biotechnol. Bioeng.*, **15**, 321 (1973).

Bu'Lock, J. D., D. Hamilton, M. A. Hulme, A. J. Powell, H. M. Smalley, D. Shepherd and G. N. Smith, *Canadian J. Microbiol.*, **11**, 765-778 (1975).

Bunch, A. W. and R. E. Harris, *Biotechnol. & Genetic Eng. Reviews*, **4**, 117-144 (1986).

Constantinides, A. and N. Mehta, *Biotechnol. Bioeng.*, **37**, 1010-1020 (1991).

Datar, R., *Process Biochem.*, **Feb.**, 19-25 (1986).

Eisenmann, A., *Chem.-Tech.* (Heidelberg), **18**, 84-88 (1989).

Hardman, N., *Adv. Biochem. Eng./Biotechnol.*, **40** (Bioprocesses Eng.), 1-18 (1989).

Karkare, S., Ph. D. Thesis, Dept. of Chemical and Biochemical Engineering, Rutgers University (1983).

Karkare, S., D. H. Burke, R. C. Dean, Jr., J. Lemontt, P. Souw and K. Venkatasubramanian, *Ann. N. Y. Acad. Sci.*, Biochem. Eng. IV, **469**, 91 (1986).

Kim, E. Y., Y. J. Yoo and Y. H. Park, *Sanop Misaengmul Hakhoechi*, **17**, 611-618 (1989).

Li, J. K. and A. E. Humphrey, *Biotechnol. Bioeng.*, **37**, 1043-1049 (1991).

Lin, C. S. and H. C. Lim, *J. Biotechnol.*, **16**, 137-151 (1990).

Linko, P. and Y. Zhu, *J. Biotechnol.*, **21**, 253-269 (1991).

Liu, C. M., L. E. McDaniel and C. P. Schaffner, *Antimicrob. Agents and Chemotherap.*, **7**, 196 (1975).

Martin, J. F., *Adv. Biochem. Eng.*, **6**, 105 (1980).

Martin, J. F. and L. E. McDaniel, *Biotechnol. Bioeng.*, **17**, 925 (1975).

Martin, J. F. and L. E. McDaniel, *Eur. J. Appl. Microbiol.*, **3**, 135 (1976).

Martin, J. F. and A. L. Demain, *Microbiol. Reviews*, **44**, 230-251 (1980).

Mehta, N., Ph. D. Thesis, Dept. of Chemical and Biochemical Engineering, Rutgers University (1988).

Morikawa, Y, I. Karube and S. Suzuki, *Biotechnol. Bioeng.*, **21**, 261 (1979).
Nishiwaki, A., I. J. Dunn and J. R. Bourne, *Chem. Eng. Res. Des.*, **68**, 387-390 (1990).
Pesl, L. and P. Seichter, *DECHEMA Biotechnol. Conf.*, **4** (Pt. B, Lect. DECHEMA Annu. Mtg. Biotechnol. 8th), 903-906 (1990).
Sayles, G. D. and D. F. Ollis, *Biotechnol. Bioeng.*, **34**, 160-170 (1989).
Shamel, R. E. and J. J. Chow, *Chem. Eng. Prog.*, **85**, 33-37 (1989).
Yoshida, T. and M. Kishimoto, *Microb. Util. Renewable Resour.*, **6**, 433-445 (1989).

6

THE NEW BIOTECHNOLOGY

6.0 RECOMBINANT GENES AND CLONING

The introduction of heterologous DNA into hardy, serviceable microorganisms such as *Escherichia coli, Bacillus subtilis* or yeast has by now become almost routine. The introduction of DNA into organisms of a higher order, such as mammalian cells, is proceeding rapidly, as well. Figure 6.1 is a simple schematic diagram of the process of gene cloning, accomplished through the agency of a plasmid vector.

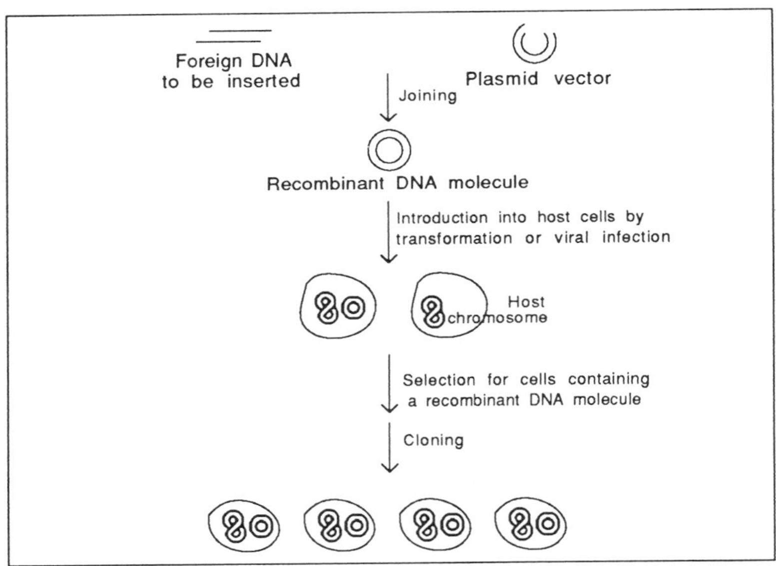

Figure 6.1 Construction and cloning of recombinant DNA molecules.

Plasmids are small circular, extrachromosomal DNA molecules that are found in most bacteria. Plasmids are independently replicative and may be present in high copy number; i.e., 30-50 molecules per chromosome. Naturally occurring plasmids usually carry genes that are not required for cell viability, but they may confer useful traits, such as resistance to antibiotics.

Plasmids make versatile vehicles to carry, maintain and amplify target genes, especially since molecular biologists have streamlined them by cutting and resecting them repeatedly to make them more suitable as cloning vectors. Small (2000-4000 DNA base pairs) and of known complete DNA sequence, plasmids contain several essential elements: an origin of replication, a genetically selectable marker (antibiotic resistance) and several unique restriction enzyme cleavage sites. The unique sites are used to insert a sequence containing the gene of interest.

To obtain a copy of a complete target gene lacking introns (intervening sequences), a cDNA (complementary DNA) clone must be provided, using the enzyme reverse transcriptase to make a complementary DNA copy of the mRNA of the target gene. The scheme for preparing cDNA is shown in Fig. 6.2. Strong base specifically degrades RNA, not DNA. S1 nuclease is needed to break the closed left end after polymerization. RNA from a cell producing the target gene product is purified and treated as outlined in the figure. The cDNA so prepared is then cloned into a suitable vector (Silhavy, 1985).

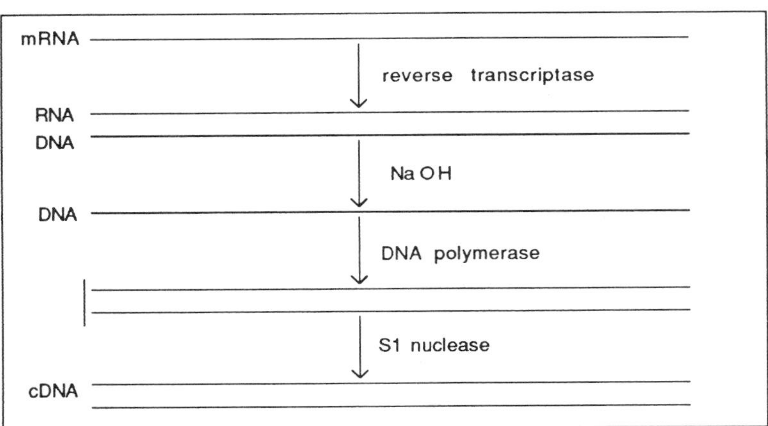

Figure 6.2 Preparation of cDNA.

The appropriate transcription and translation initiation signals must be attached before the cDNA clone in *E. coli* can be expressed. A vector displaying the required signals is shown in Fig. 6.3. The synthetic DNA, though containing a variety of cleavage sites, does not interrupt expression of *lacZ* (β-galactosidase). The plasmid is opened at one of the restriction enzyme cleavage sites and the appropriate cDNA is inserted. Such inserts can be recognized because they will separate *lacZ* from its expression signals and thus prevent production of β-galactosidase.

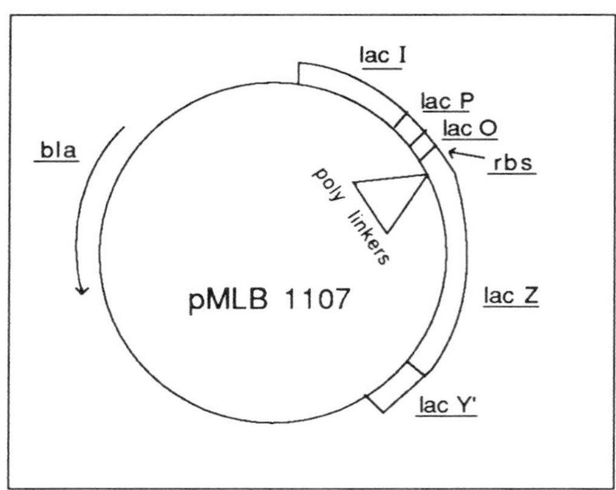

Figure 6.3 Genetic map of pMLB 1107.

These manipulations will position the target gene in approximately the correct location for expression driven by the expression signals of the lactose operon. Further and more subtle manipulations may be required for exact positioning. The resulting construct should allow expression of the target gene product in *E. coli.* Similar manipulations have been done with a variety of eucaryotic genes. In many cases, this allows very high levels of target gene expression; e.g., (20-30%) of total cell protein. Under such conditions, purification is occasionally a relatively simple task.

A general method has been developed for the introduction of any cloned sequence into the chromosome of *E. coli* (Resnik and LaPorte, 1991). This method employs an Hfr strain which carries a fragment of *bla* (the pBR322 gene imparting ampicillin resistance) between *lacI* and *lacZ*. Plasmid-borne inserts which are flanked by sequences from *bla* and *lacZ* can be introduced at this locus by homologous recombination. The isolation of recombinants is enhanced by selection for transfer of an integrated copy of the plasmid during conjugation. Once introduced into the chromosome, the inserted sequences can be transferred to other strains by conventional methods, such as P1 transduction or conjugation. This method is suitable for the transfer of any cloned sequence to the chromosome and is particularly well suited to the construction of chromosomal gene and operon fusions with *lacZ*.

The existence and development of this sophisticated technology has brought about the rise of a new industry commonly referred to as *biotechnology*, while the new contacting modes which are employed are being referred to under the heading *bioreactors*. It is convenient to use these terms for quick reference, but their connotation does a disservice to the evolutionary process by which the current state of biotechnology has been reached, as is already, or will become, apparent to the reader.

The *corps d'elite* of this new industry are the proteins, such as those found in blood (including antigens), as well as hormones, antibodies, etc. But even simple hydrolytic enzymes, such as the amylases, can also be important for large scale processing in the food industry.

The gram-negative procaryotic cell, *E. coli*, has an outer membrane with a peptidoglycan support; the inner cytoplasmic membrane is separated from it by the periplasm. The inner membrane is composed of about 50% protein, 30% lipids and 20% carbohydrates. The double envelope of the cell thus acts as a barrier both to intrusion from the outside and extrusion from the inside. When the envelope breaks, cell lysis and death take place. The eubacterium, *Bacillus subtilis*, is gram-positive, having only a multilayer peptidoglycan structure in the cell wall. With just a cytoplasmic membrane, protein excretion is facilitated.

Fungi (yeasts and molds), algae, protozoa and animal and plant cells constitute the eucaryotes. Eucaryotes are five to ten times larger than procaryotes in diameter. They have a true nucleus and a number of cellular organelles inside the cytoplasm. Animal cells themselves may be the organisms of choice for the production of certain medically important proteins. With such systems, secretion of a perfectly normal protein is possible. Expression vectors for these cells are available and procedures to amplify the target gene have been described. However, these cells can be somewhat difficult and expensive to grow.

Transcription and translation are closely coupled in procaryotes, whereas they are spatially and temporally separate in eucaryotes. In procaryotes, the primary transcript serves as mRNA and is used immediately as the template for protein synthesis (A in Fig. 6.4). In eucaryotes, mRNA precursors are processed and spliced in the nucleus before being transported to the cytosol (B in Fig. 6.4) (Darnell et al., 1986).

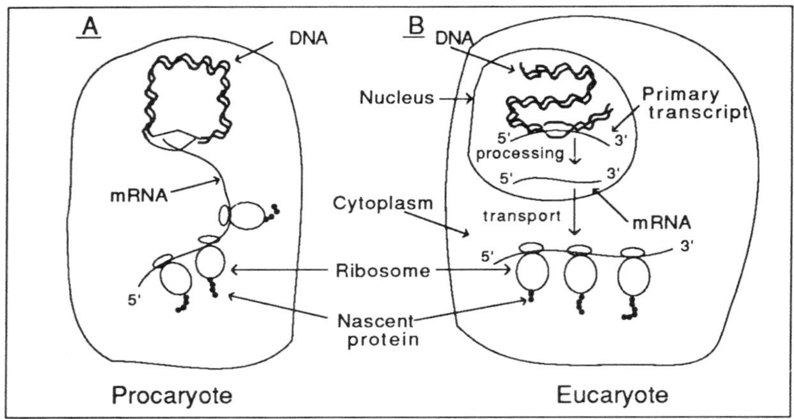

Figure 6.4 Messenger RNAs (After Darnell et al., 1986).

6.1 OPERONS

Expression of newly acquired genes is commonly exercised by control elements called operons, which exhibit both positive (induction) and negative (repression) elements. Thus, transcription can be amplified or suppressed virtually at will. In wild-type *Escherichia coli*, expression of the *gal* operon is negatively regulated by the Gal repressor and is induced ten- to fifteen-fold when the repressor is inactivated by an inducer. In strains completely deleted for *galR*, the gene which encodes the Gal repressor, the operon is derepressed by only ten-fold without an inducer. But this derepression is increased further by three-fold during cell growth in the presence on an inducer, D-galactose or D-fucose. This phenomenon of extreme induction in the absence of Gal repressor is termed ultrainduction, a manifestation of further inducibility in a constitutive setup (Tokeson et al., 1991). Construction and characterization of gene and operon fusion strains between *galE* and *lacZ*, encoding β-galactosidase as a reporter gene, show that ultrainduction occurs at the level of transcription and not translation. Transcription of the operon, from both the cAMP-dependent P1 and the cyclic nucleotide-independent P2 promoters, to be described further in a moment, is subject to ultrainduction.

Backing off for just a bit, among the simpler enzymes and better-understood operons is the combination of β-galactosidase and the lactose operon, the latter functioning in the rapidly-dividing *E. coli* (doubling time of approximately 20 min.), as shown in Figure 6.5.

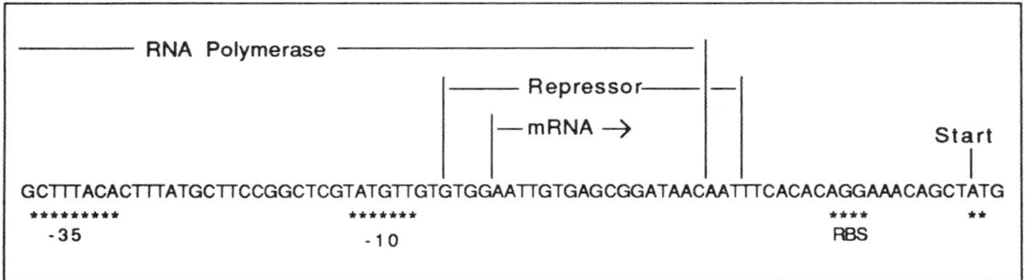

Figure 6.5 *lac* operon.

Shown is the sequence of a single strand of DNA encompassing the control region of the lactose operon (Silhavy, 1985). The process of gene expression is initiated by the binding of RNA polymerase to the designated region. The -35 and -10 regions are important elements of this binding reaction. Transcription is initiated by RNA polymerase at position +1 forming the messenger (m) RNA. Translation of the mRNA is accomplished by ribosomes that bind at the ribosome binding site (RBS or Shine-Dalgarno sequence) and start synthesis at the methionine codon ATG.

Expression of this operon is controlled by a repressor that interacts with an operator site slightly upstream of the -10 region. In the absence of inducer, repressor binds and this prevents recognition by RNA polymerase which, of course, prevents transcription. In the presence of inducer, repressor no longer binds to the operator and transcription occurs. In short, an operon may be summarized as a set of contiguous genes, encoding proteins with related functions, under the control of a single promoter-operator (Shuler and Kargi, 1992).

6.2 LAC OPERON (E. COLI)

INDUCTION AND REPRESSION

E. coli lac repressor is a tetrameric protein composed of 350 amino acid subunits. Considerable attention has focused on its N-terminal region, which is isolated by cleavage with proteases yielding N-terminal fragments of 51 to 59 amino acid residues. Because these short peptide fragments bind operator DNA, they have been extensively examined in NMR structural studies. Longer N-terminal peptide fragments that bind DNA cannot be obtained enzymically. To extend structural studies and simultaneously verify proper folding *in vivo*, the DNA sequence encoding longer N-terminal fragments was cloned into a vector system with the coliphage T7 RNA polymerase/promoter (Khoury et al., 1991). Both wild-type and mutant *lac* repressor N-termini were effective in binding operator DNA as judged by reduced β-galactosidase synthesis and methylation protection *in vivo*.

Pace et al. (1990) recently crystallized the intact *lac* repressor tetramer in the native form with an inducer, and in a ternary complex with operator DNA and an anti-inducer. Cocrystals were obtained with a number of different *lac* operator-related DNA fragments. These protein-DNA cocrystals crack upon exposure to the gratuitous inducer iso-Pr β-D-thiogalactoside (IPTG), suggesting a conformational change in the repressor-operator complex.

Computer graphics were used by Kisters-Woike, et al. (1991) to build a molecular model of the complex of Lac repressor and *lac* operator. The model is based on NMR data and on genetic and biochemical data including specificity changes. Effects of amino acid exchanges in the recognition helix could be predicted by the model and were subsequently tested and confirmed by genetic experiments. Comparison of the modeled *lac* complex with the known crystallographic structures of several helix-turn-helix DNA complexes reveals striking similarities and suggests rules which govern the recognition between particular amino acid side chains and particular base pairs in these systems. The results of a molecular dynamics simulation of a *lac* headpiece-operator complex in aqueous solution are reported by De Vlieg et al. (1989). The complex satisfies essentially all experimental distance information derived from two-dimensional studies. The interaction between *lac* repressor headpiece and its operator is based on many direct- and water-mediated hydrogen bonds and nonpolar contacts which allow the formation of a tight complex. No stable hydrogen bonds between side chains and bases are found, whereas specific contacts occur between both nonpolar groups and, to a lesser extent, through water-mediated hydrogen bonds.

Large sequencing projects require an efficient strategy to generate a series of overlapping clones. This can be accomplished by protecting one end of a linear DNA molecule while sequential deletions are introduced into the other end by exonuclease digestion. The *lac* repressor can protect the ends of linear nucleotide sequences from digestion by exonuclease, if these ends contain the *lac* operator sequence (Johnson et al., 1990).

Recently, the entire *E. coli* lactose operon was inserted into an *E. coli/Corynebacterium glutamicum* shuttle vector and introduced into the gram-positive host organism *C. glutamicum* R 163 (Brabetz et al., 1991). Recombinant *C. glutamicum* strains carrying the *lac* genes downstream of an efficient promoter displayed rapid growth with lactose as the sole carbon source. Two prerequisites were necessary to obtain good growth of *C. glutamicum* R 163 on lactose: presence of the *lacY* gene in addition to *lacZ* and an appropriate promoter for efficient transcription in *C. glutamicum*. The galactose moiety of lactose was not utilized but accumulated in the culture broth. *C. glutamicum* strains carrying the *lacZ* (β-galactosidase) gene but not *lacY* (lactose permease) were not able to grow in lactose minimal medium. The same lactose-utilizing host was prepared by transformation with a recombinant DNA containing the *E. coli* genes for β-D-galactosidase and β-galactoside permease for use in the manufacture of amino acids (Katsumata et al., 1989). *C. glutamicum* thus transformed was able to produce glutamic acid at a level of 15.9 g/L versus 0.6 by the parental strain, when cultured in a medium containing whey powder (containing 75% lactose).

CYCLIC AMP RECEPTOR PROTEIN

Transcription from the promoter region is under the positive control of catabolite activator protein (CAP)-cAMP and is also enhanced during growth in the presence of glucose, a favored energy source as well as a PTS substrate. (More about the latter in just a moment.) The lactose promoter-operator region of *Escherichia coli* contains two binding sites for cAMP receptor protein (CAP)-cAMP, abbreviated as CAP, two for the lactose repressor, and two for RNA polymerase. The high density of binding sites makes cooperative interactions between these proteins likely. In a recent study, the gel electrophoresis mobility shift assay and binding partition analytical techniques were used to determine whether the secondary CAP site influences the binding of CAP to the principal CAP site in the lactose promoter when both are present on a linear DNA molecule. Such an effect could occur through the formation of a bridged DNA-CAP-DNA structure, through the interaction of CAP molecules bound to each of the sites, or through allosteric effects caused by CAP-mediated DNA bending. It was found, however, that the interaction of CAP with these sites was not cooperative, indicating that CAP sites 1 and 2 bind CAP in an independent manner (Hudson and Fried, 1991).

Wilcox (1991) modified the *lac* operon to remove the CAP binding site and promoter/operator and fused it to the *lacI* gene (encoding a repressor) at its 3' end for constitutive expression of the heterologous gene from the *lacI* gene promoter. Using this modified *lac* operon, heterologous gene expression can be induced with lactose, rather than with iso-Pr β-D-thiogalactoside (IPTG). The induction of a lepidopteran insect toxin gene expression (from the constitutive *lacI* gene promoter) with lactose and IPTG was shown. The induction with lactose was comparable to that with IPTG. However, the concentration of lactose used was ten- to twenty-fold higher than IPTG.

PTS SYSTEM

The *pts* operon of *E. coli* is composed of the genes *ptsH*, *ptsI* and *crr*, which code for three proteins of the phosphoenolpyruvate-dependent phosphotransferase system (PTS): the HPr, enzyme I (EI), and $EIII^{Glc}$ proteins, respectively. These three genes are organized in a complex operon in which the major part of expression of the distal gene, *crr*, is initiated from a promoter region within *ptsI*. Expression from the promoter region of the *ptsH* and *ptsI* genes has been studied *in vivo* by using gene fusions with *lacZ* (De Reuse and Danchin, 1991). A genetic characterization of the mechanism by which growth on glucose causes transcriptional stimulation of the *pts* operon is now available. This regulation is dependent on transport through the glucose-specific permease of the PTS, EII^{Glc}. Apparently, transcriptional regulation of the *pts* operon is the consequence of an increase in the level of unphosphorylated EII^{Glc} which is produced during glucose transport. Furthermore, overproduction of EII^{Glc} in the absence of transport stimulated expression of the *pts* operon. Also, CAP-cAMP could cause stimulation independently of the EII^{Glc}. Thus, glucose acts like an environmental signal through a mechanism of signal transduction.

Overproduction and rapid purification of the phosphoenolpyruvate:sugar phosphotransferase system proteins Enzyme I, HPr, and protein III^{Glc} of *E. coli* are reported by Reddy et al. (1991). Methods for their rapid, simple purification are presented, using plasmids overproducing gene products. We conclude this section by noting that a series of isogenic strains harboring known deletions in the *pts* operon of *E. coli* have been constructed by reverse genetics (Levy et al., 1990). Strains bearing deletions for the whole *pts* operon failed to grow on maltose or on carbon sources of the same class. In these strains the total cAMP synthesis was significantly lower than in a strain deleted only for the *crr* gene. This indicated that Enzyme I or phosphorylated histidine-containing phosphotransferase protein, in addition to its role in phosphorylating Enzyme III^{Glc}, is involved in adenylate cyclase (AC) activation or cAMP excretion. It was further shown that deletions in the *pts* operon do not affect synthesis of AC.

Analogously, Fischer et al. (1991) call attention to the fact that the enzyme III^{Mtl} is part of the mannitol phosphotransferase system of *E. faecalis*. It is phosphorylated in a reaction sequence requiring enzyme I and heat-stable phosphocarrier protein (HPr). The phospho group is transferred from enzyme III^{Mtl} to enzyme II^{Mtl}, which then catalyzes the uptake and concomitant phosphorylation of mannitol. The internalized mannitol-1-phosphate is oxidized to fructose-6-phosphate by mannitol-1-phosphate dehydrogenase.

MESSENGER RNA

McCormick et al. (1991) analyzed processing of mRNA from the *lac* operon in an *E. coli* strain carrying the *lac* on a multicopy plasmid; the mRNA was analyzed by hybridization. Nuclease protection of pulse-labeled RNA and precursor-product relationships were determined by quantitating radioactivity in primary and processed transcripts at various times after induction of the *lac* promoter or inhibition of transcription with rifampicin. Results support the existence of two types of processed transcripts with endpoints in the *lacZ-lacY* intercistronic region.

The polycistronic *lacZY A* mRNA of *E. coli* is cleaved during decay at approximately intergenic sites: current research (Murakawa et al., 1991) indicates that six *lac* mRNA species are present in the following order of decreasing abundance: *lacZ, -A, -ZY A, -ZY, -Y A* and *-Y*. Direct examination of the kinetics of *lac* messenger synthesis revealed that after initiation, most transcription continued to the end of the operon. During normal growth, the operon is apparently transcribed in its entirety and the individual *lac* mRNAs are formed by cleavage. These results confirm earlier work implying that the *lac* operon is transcribed in its entirety, but are in conflict with several recent reports suggesting that internal termination occurs. The results indicate that the natural polarity of the operon (*lacZ* is expressed six-fold more strongly than *lacA*) is based on post-translational effects and not on polarity of transcription.

PRODUCT SECRETION

According to a plausible model (Silhavy, 1985), the translation of the mRNA for a secreted protein initiates in the cytoplasm. Subsequently, the translation complex diffuses to the membrane where a specific export signal, the amino-terminal signal sequence, interacts with the membrane (endoplasmic reticulum) and directs a vectorial transfer of the protein across the bilayer as translation proceeds. The signal sequence is removed by a specific protease, probably before synthesis of the protein is complete. When synthesis of the molecule is complete it is discharged into the lumen of this organelle.

This mechanism is particularly relevant because it suggests that proteins destined for secretion never exist in completed form in the cytoplasm. It seems likely that these molecules were not designed to fold correctly in this compartment. Rather, they were designed to fold in a noncytoplasmic environment. Available evidence supports this hypothesis. For example, the cytoplasm may be too reducing to allow formation of disulfide bridges, a structural feature characteristic of many secreted proteins. If true, then secretion may be a requirement for the correct folding of certain proteins.

Studies of protein export in *E. coli* suggest that the basic mechanisms of protein localization have been conserved throughout biology. Accordingly, a eucaryotic protein may be secreted from the cytoplasm of the bacteria if it is expressed with its signal sequence still intact. In at least certain cases (proinsulin, ovalbumin), experimental evidence supports this prediction. In other cases, the eucaryotic signal sequence appears not to be recognized efficiently. In these instances, it may be necessary to replace the signal with a functional procaryotic signal sequence. This could be done in much the same way as has been described for replacing expression signals.

In *E. coli*, with few exceptions, secreted proteins are not released into the growth media. Rather, they are localized to the periplasm, an aqueous compartment between the cytoplasmic membrane and the outer membrane. This compartment offers certain advantages. First, proteins that are normally secreted can fold correctly in this location. Second, periplasmic proteins are still cell-associated and thus can be concentrated by pelleting the cells. Simple procedures for their release have been described. Finally, because periplasmic proteins are segregated from the cytoplasm, protein degradation may be to some extent minimized. Dennis (1991) describes culture conditions for the manufacture of poly-β-hydroxybutyrate (I) with *Escherichia coli* harboring the genes for I biosynthesis of *Alcaligenes eutrophus*. When the genes are under control of the *lac* promoter, then I is efficiently manufactured in a minimal medium containing moderate concentrations of whey. The I granules are recovered from lysates by agglomeration with salts, especially calcium chloride at a concentration level circa 10 mM. The polymer has potential for widespread application as a biodegradable plastic.

The outer membrane structure of gram-negative bacteria acts as a barrier against penetration of exogenous molecules, as well as against proteins secreted from the cells through the inner cell membrane. Secretion of proteins through the outer membrane of gram-negative bacteria thus becomes an interesting problem for possible solution through recombinant DNA technology.

Serratia marcescens is a gram-negative bacterium belonging to the *Enterobacteriaceae*, the same family as *E. coli*. The organism has been selected as a producer of an extracellular metalloprotease, which is utilized as an anti-inflammatory drug. During the course of cloning the extracellular protease gene of *S. marcescens* in *E. coli*, a clone producing another extracellular protease of serine type was obtained, which was different from the known metalloprotease of *S. marcescens*. A hybrid plasmid containing a 5.8-kb DNA fragment of *S. marcescens* was recovered from the clone. *E. coli* carrying pSP11 elaborated the serine protease into the medium during the exponential growth phase. No concomitant release of the intracellular or periplasmic enzymes of the host was observed.

Subcloning and sequencing of the cloned fragment revealed a single open reading frame of 3,135 base pairs coding a protein of 1,045 amino acids. Comparison of the 5' nucleotide sequence with the N-terminal amino acid sequence of the protease indicated the presence of a typical signal sequence of 27 amino acids. The C-terminal amino acid of the enzyme was found at position 408, as deduced from the nucleotide sequence. Artificial frameshift mutations introduced into the coding sequence for the assumed distal polypeptide after the C terminus of the protease caused complete loss of the enzyme production. It was concluded that the *Serratia* serine protease is produced as a 112-kilodalton proenzyme and that its N-terminal signal peptide and a large C-terminal part are processed to cause excretion of the mature protease through the outer membrane of *E. coli* cells. The *Serratia* protease and its gene may provide an effective mechanism of protein secretion and possibly a new secretion vector in *E. coli* (Beppu, 1989).

In studies with interleukin, production and secretion of heterologous proteins has been achieved with *Escherichia coli* (UV) mutants with leaky cell walls, that are transformed with a chimeric gene encoding the *OmpA* signal peptide fused to the heterologous protein (Lunn et al., 1991). Plasmid pRGT857-11, encoding the *OmpA* signal peptide fused to interleukin 4, was prepared; *E. coli* transformed with this plasmid secreted interleukin 4 into the medium. Likewise, human interleukins have been prepared from recombinant strains. Arthur et al. (1990; Arthur and Duckworth, 1990) constructed a recombinant bacterial strain for the large scale production of human interleukin-1β (IL-1β); the λP_R and the tryptophan systems were compared for efficiency of transcription and regulation. Plasmid #595 containing the λP_R region and a *ClaI* site linked to an interleukin-1β gene was constructed. Plasmid #599, which is #595 with a *trp* leader sequence including a ribosome binding site inserted into the *ClaI* site, was then prepared; *E. coli* transformed with this plasmid produced 500-900 mg interleukin-1β/L culture medium.

A TGATG vector system was developed that allows for the construction of hybrid operons with partially overlapping genes, employing the effects of translational coupling to optimize expression of cloned cistrons in *E. coli.* In this vector system (plasmid pPR-TGATG-1), the coding region of a foreign gene is attached to the ATG codon situated on the vector, to form the hybrid operon transcribed from the phage λp_R promoter (Mashko et al., 1990). This system was successfully tested in experiments on the optimization of expression of the genes encoding human or animal interferons, human interleukin-1β and *E. coli* β-galactosidase in *E. coli* cells. More directly, plasmid pAA1213-23 containing a gene for mature human interleukin 2 (IL2) under the control of the regulatable *E. coli* tryptophan operon promoter was expressed in *E. coli* following induction by 3-indolylacrylic acid (Avots et al., 1990a; b). A method for manufacture of interleukin 2 is described.

Thus, except for quite small peptides, *E. coli* proteins do not pass beyond the outer membrane. The overexpression of proteins that are secreted into the periplasm (such as the phosphate-binding protein) leads to the continuous secretion of highly pure protein by *E. coli* mutants (Furlong and Sundstrom, 1989). The use of appropriate regulation allows for protein production under nongrowing conditions, thus facilitating the construction of stable immobilized cell bioreactors for continuous protein secretion. The phosphate-binding protein of high purity produced by these bioreactors was immobilized and used to scrub phosphate from a feed stream. The protein bioreactor and bioadsorber were designed for simple scale-up and automation. Likewise, the presence of a high copy number plasmid (pUC8) was found to affect integrity of the cell envelope of *E. coli JM103*, causing, in turn, significant release of the plasmid-encoded protein (β-lactamase) in suspension and immobilized cell cultures (Ryan and Parulekar, 1991).

In another recent study (Georgiou and Baneyx, 1990) the genes for protein A and β-lactamase were fused in-frame using recombinant DNA techniques. The gene fusion was expressed in *E. coli* to give a 63-kDa hybrid polypeptide with both protein A and β-lactamase activities. The hybrid protein is secreted through the inner membrane. However, a fraction of the protein remains associated with the inner membrane and is not readily released from the cell. It was shown that the production of protein A-β-lactamase is limited by proteolytic degradation within the cell. The yield can be increased about 100-fold by using a suitable protease-deficient *E. coli* strain and by optimizing the fermentation conditions. The purified A-β-lactamase hybrid protein exhibits functional properties that are nearly identical to the unfused protein A and β-lactamase. Whereas the fusion of the two polypeptides had no apparent effect on either function, the stability of the hybrid protein was markedly decreased.

6.3 α-AMYLASE (B. SUBTILIS)

SPORULATION

Gram-positive bacteria belonging to the genera Bacillus and Clostridium form endospores, which are resistant to high temperature and conditions of extreme dessication. They remain viable in their metabolically inactive state for long periods and then germinate in a few minutes under conditions favorable for vegetative growth. The formation of endospores has been divided into stages based on the marked morphological and biochemical changes occurring during the process.

Early in the process of spore formation in *B. subtilis*, a septum is formed that partitions the sporangium into daughter cells, called the forespore and the mother cell. The daughter cells each have their own chromosome but follow dissimilar programs of gene expression (Margolis et al., 1991). Feavers et al. (1990) discuss the regulation of transcription of the *gerA* spore germination operon of *B. subtilis*. A high proportion of the *gerA* driven β-galactosidase detected in sporulating cells is found in the mature spore; the *gerA* promoter is therefore active in the forespore compartment of the sporulating cell.

Unfavorable conditions, for example nutrient deficiencies, trigger the formation of endospores. The nucleoid stretches out, forming an axial filament. It divides and the cytoplasm undergoes an unequal division. The smaller product of division (called the forespore) becomes the endospore. The forespore is engulfed by the remainder of the mother cells, which make many components for the endospore.

("Forespore and seven years ago . . .")

Then a heavy wall (the cortex and spore coat) is laid down between the pair of membranes formed around the forespore. When the endospore is mature, it is usually released by lysis of the surrounding cell (Niedhardt et al., 1990).

The morphological changes and biochemical changes are described by Maruo and Yoshikawa (1989). One of the biochemical changes is the production of α-amylase. It is temporally coincident with the onset of sporulation and hence it is relevant to understand the conditions that trigger the onset of sporulation and the regulation of the same.

GENES INVOLVED IN SPORULATION

One approach to the study of the initiation of sporulation (Stage 0) is the identification of genes' elucidation of the effect of their mutants in the initiation. Sporulation mutants have been isolated and classified according to the stage of sporulation which they affect as *spo0*, *spoI*, *spoII*, etc.

spo0 mutants are arrested at the earliest stage of sporulation and hence involved in the initiation of sporulation. Among these, *spo0A* is thought to play a central role. The *spo0A* mutant exerts maximum effects on sporulation-associated phenotypes. For instance, a temperature sensitive mutant could be normally induced by a shift down of the culture temperature even in the late stationary phase. The *spo0B* stage 0 sporulation operon encodes an essential GTP-binding protein (Trach and Hoch, 1989).

Rudner et al. (1991) have shown that spore formation in *B. subtilis* is a dramatic response to environmental signals that is controlled in part by a two-component regulatory system composed of a histidine protein kinase (SpoIIJ) and a transcriptional regulatory (Spo0A). In addition, the *spo0K* locus plays an important but undefined role in the initiation of sporulation and in the development of genetic competence.

The *spo0K* locus was cloned and sequenced. *spo0K* proved to be an operon of five genes that is homologous to the oligopeptide permease (*opp*) operon of *Salmonella typhimurium* and related to a large family of membrane transport systems. The transport system encoded by *spo0K* may have a role in sensing extracellular peptide factors that are required for efficient sporulation, and perhaps in sensing similar factors that may be necessary for genetic competence.

Weicker et al. (1990) employed site-directed mutagenesis of the *Bacillus subtilis* amylase catabolite repression operator to identify bases critical for repression of amylase production. Negative control of amylase was achieved by a temporal event which occurred reliably at T_2, two hours after amylase production was activated at T_0. This temporal shut-off was abrogated in mutations in the *spoIIA* operon. These strains continued to accumulate amylase for 24 hours or more after the onset of transcription of the amylase gene. Reversion of the sporulation defect to spore-plus restored the amylase turn-off. Mutations in the *spoIIA* operon also allowed protease accumulation to levels two- to four-fold above wild-type in overnight cultures.

Subtilisin gene expression was temporally extended in sporulation-deficient strains (*spoIIG*), relative to co-genic sporogenous strains, resulting in enhanced subtilisin production. Ammonium exhaustion not only triggered subtilisin production in asporogenous *spoIIG* mutants but also shifted carbon metabolism from acetate production to acetate uptake and resulted in the formation of multiple septa in a significant fraction of the cell population (Pierce et al., 1992).

DEGRADATIVE ENZYMES

Msadek et al. (1990) describe the signal transduction pathway which controls synthesis of a class of degradative enzymes in *B. subtilis.* Honjo et al. (1990) describe a gene fragment involved in both negative control of sporulation and degradative-enzyme production. More specifically, a 2.5-kilobase fragment of the *B. subtilis* genomic DNA was cloned, which caused the reduction of extracellular and cell-associated protease levels when present in high copy number. This fragment, in multicopy, was also responsible for reduced levels of α-amylase, levansucrase, alkaline phosphatase, and sporulation inhibition. The gene relevant to this pleiotropic phenotype is referred to as *pai* .

The *Bacillus subtilis degS, degU*, and *degQ* genes control degradative enzyme synthesis and also affect transformability, sporulation efficiency in the presence of glucose, and the presence of flagella. The DegS/DegU proteins share significant amino acid similarities with procaryotic two-component modulator/effector pairs. By analogy with these systems it may be possible that DegS is a protein kinase which could catalyze the transfer of a phosphoryl moiety to DegU, which acts as a possible regulator of degradative enzyme synthesis. It is shown that *degS* and *degU* are organized in an operon (Msadek et al., 1990a; b).

CATABOLITE REPRESSION

Since *spo0A* seems to be a key factor in initiating sporulation, it is relevant to probe into the factors influencing the regulation of *spo0A* expression, as well as the mechanism by which it acts to initiate sporulation (Ramasubramanyan, 1992). Schaeffer et al. (1965) found that the probability of sporulation increases greatly when a slowly metabolized carbon or nitrogen source is used. They suggested that the probability for a cell in a growth medium to become committed to sporulate must be determined by the intracellular concentration of at least one substance which catabolically-represses, directly or indirectly, the expression of sporulation genes.

A recent study (Chibazakura et al., 1991) on the differential regulation of *spo0A* transcription by glucose seems to support the above view. It was shown that *spo0A* gene is regulated by switching promoters during initiation of sporulation. Two discrete promoters were identified, the one upstream (P_v) being used for expression of *spo0A* when the cell is in vegetative phase and the promoter (P_s), 150bp downstream of P_v, being used at the initiation of sporulation. The latter promoter was found to be repressed in the presence of glucose, while the former was not shut off in its presence. The sporulation specific promoter was induced at the onset of sporulation while the vegetative phase promoter was shut off.

Further, the catabolite resistant mutants were able to switch even in the presence of glucose, indicating that the catabolite repression mechanism plays an important role in *spo0A* expression, and hence on initiation.

Freese et al. (1978) found that if the synthesis of guanine nucleotides is partially reduced in *Bacillus subtilis*, sporulation is induced in the presence of excess glucose, ammonia and phosphate. They conclude that the decrease in GTP (and GDP) concentration is a trigger for sporulation.

The reduction of purine nucleotides can be achieved by addition of drugs, such as decoynine and hadeidine, that interfere with purine synthesis. Decoynine was extremely effective in stimulating *spo0H* directed β-gal activity. Decoynine also rapidly induces the expression of *spo0A* and *spoVG*, which is totally dependent on the *spo0H* gene. Hence it has been suggested that some *spo0* gene product may sense the concentration of GTP in the cell just like GTP-binding proteins involved in yeast sporulation (Maruo and Yoshikawa, 1989).

In a recent study (Nicholson and Chambliss, 1987), it was observed that the addition of decoynine to *B. subtilis* growing in the presence of glucose causes the initiation of sporulation, overcoming the catabolite repression. One possible explanation may be that the induction of *spo0H* by decoynine is able to bypass the need for *spo0A* function, as the latter is regulated by glucose.

In the same experiments it was observed that α-amylase synthesis is not relieved from catabolite repression on the addition of decoynine, in the presence of glucose. However, addition of decoynine (hence the reduction of GTP levels) did temporally activate α-amylase synthesis in the absence of glucose.

Overall, it is clear that environmental signals play a role in the initiation of sporulation. How these signals are translated into regulation of genes has been partly elucidated. A signal transduction pathway, which uses sequential phosphorylation of specific proteins, has been proposed to be the mechanism involved. Burbulys et al. (1991) have observed that the end result of the environmental stimuli or signals is the phosphorylation of the *spo0A* protein which plays a regulatory role in transcription of genes with intimate roles in the initiation process.

Le Grice (1990) has clearly recognized that the ability to regulate heterologous genes in *B. subtilis* tightly has been hampered until recently by the lack of well-characterized gram-positive repressible systems, and have sought to rectify this situation. It was discovered that promoters of the *E. coli* bacteriophage T5 were efficiently recognized by the gram-positive transcriptional machinery. One such promoter, P_{N25}, had previously been fused to a synthetic *lac* operator fragment, generating the regulatable promoter $P_{N25/0}$, and successfully utilized in an *E. coli* expression system. It was assumed that insertion of this regulatable promoter and a functional *lacI* gene into a gram-positive vector would generate an inducible expression system for *B. subtilis*. Two expression systems displaying these features are described. The first of these is a single plasmid pREP9 which contains both a regulatable promoter and a *lacI* gene modified for expression in *B. subtilis*. The second system comprises a *B. subtilis* strain which constitutively expresses the *E. coli lac* repressor from a derivative of the *Staphylococcus aureus* plasmid pE194. This latter strain (designated BR:BL1) can be transformed with a compatible expression vector containing the regulatable promoter. In either the single- or dual-plasmid system, gene expression is tightly controlled and can be rapidly derepressed by addition of the inducer isopropyl-β-D-thiogalactopyranoside (IPTG). The usefulness of these systems in high-level production of heterologous proteins is illustrated. Furthermore, the utilization of the dual-plasmid system to produce an enzymically active heterologous protein is exemplified by the production of human immunodeficiency virus, type 1 (HIV-1) reverse transcriptase in *B. subtilis*.

6.4 YEASTS

The advent of recombinant DNA technology provides the means for manufacturing a wide spectrum of valuable proteins by fermentation, using either procaryotic or eucaryotic cell cultures. Due to the ease of genetic manipulation and bulk fermentation, the yeast *Saccharomyces cerevisiae* may provide a suitable host in which many of the eucaryotic proteins are produced as soluble, biologically active proteins. For instance, vaccines which were traditionally made from plasma, such as the hepatitis B. vaccines, are now being produced by recombinant yeast cells (Hong, 1992).

Hartley et al. (1989), followed by Kumar et al. (1992), describe the construction of a lactose-utilizing *S. cerevisiae* that expresses the cDNA for a secreted, thermostable β-galactosidase (*lacA*) from *Aspergillus niger*. Yeast cells expressing the *lacA* gene from the yeast *ADH1* promoter on a multicopy plasmid secrete up to 40% of the total β-galactosidase activity into the growth medium. The secreted product is extensively N-glycosylated, and cells expressing the *lacA* gene grow on whey permeate (4% w/v lactose), with a doubling time of 1.6 hours. Such strains may offer a solution to the increasing problem of waste whey disposal.

Jeong et al. (1989; 1990; 1991a) describe a membrane bioreactor in which living recombinant yeast cells are sandwiched between an ultrafiltration membrane and a reverse osmosis membrane. The concept has been successfully applied to the fermentation of lactose to ethanol (Jeong et al., 1991b). (The mass transfer coefficient in the yeast cell layer was found to be unexpectedly large, indicating the presence of convective effects.)

To prepare the strain, plasmid pSH096 was constructed by isolating lactose utilization genes from *Kluyveromyces lactis*, and then inserting that DNA into a yeast integrating vector, pRY296 (Yocum et al., 1984). pSH096 contains the lactose utilization genes *Lac4*, *Lac12* and *Lac13* from *K. lactis*; a region of the *HO* (homothallism) gene; a region containing the yeast *CRC1* (iso-1-cytochrome C) promoter; the *G418*^{r} gene; an *ori* (origin of replication) region; the pBR322-derived *amp*^{r} (ampicillin resistance) gene; and a second *HO* region. Transformants of the strains grew weakly on minimal lactose medium, and untransformed strains were unable to grow on lactose.

Experiments in which hexokinase activities have been manipulated by genetic engineering of the yeast *S. cerevisiae* may sometimes provide an example of unexpected results. Wild-type yeasts possess three different enzymes that catalyze phosphorylation of glucose: glucokinase (GLK), hexokinase PI (HXK1), and hexokinase PII (HXK2). In order to study the effect of genetic alterations in these activities, three different genetically engineered *S. cerevisiae* strains were investigated. The host cell involved in all three cases is a mutant that lacks all three glucose phosphorylation activities. Each strain investigated carried one of the kinases expressed from the corresponding cloned gene on a multicopy plasmid (Bailey et al., 1990). Analysis suggests that intracellular magnesium-free ATP levels may be sufficiently high to inhibit hexokinase PII, thereby contributing to the disparity between *in vitro* glucose phosphorylation and *in vivo* glucose uptake activities in the engineered strain carrying HXK2.

6.5 TRANSPOSONS

Addition mutations often take place through insertion elements (IS). These elements are about 700 to 1400 base pairs in length; in *E. coli* about five different IS sequences are known and are present on the chromosome. These elements can move on the chromosome from essentially any one site to another. Often they will insert in the middle of a gene, totally destroying its function. A closely related phenomenon is a transposon, which refers to a gene or genes that have the ability to jump from one piece of DNA to another piece of DNA or to another position on the original piece of DNA. The transposon integrates itself into the new position independently of any homology with the recipient piece of DNA. Transposons differ from insertion sequences in that they code for proteins. Many of the transposons encode antibiotic resistance; they appear to arise when a gene becomes bounded on both sides by insertion sequences (Shuler and Kargi, 1992). An excellent discussion of these mobile genetic elements in eucaryotes appears in "Biotechnology Focus 3" (Mohr and Esser, 1992).

A method for introducing genes into the chromosomal or episomal DNA of bacteria comprises insertion of the gene into a nonessential part of a transposon and allowing the transposon to integrate into the bacterial DNA. The *thr* operon was inserted into a plasmid containing a Mu phage from which the transposase A and B genes were deleted. *Escherichia coli* was transformed with this plasmid and a second plasmid supplying the transposase genes. Clones containing 0, 1, 4 and 10 copies of the recombinant Mu were isolated. The clones produced 3, 3.3, 8.9 and 12 grams threonine/L culture, respectively (Richaud et al., 1989). Transposon Tn*10* insertion in the *galS* (ultrainduction factor) gene of *E. coli* allows the *gal* operon to be constitutively expressed at a very high level, equal to that seen in a Δ*galR* strain in the presence of an inducer (Golding et al., 1991). In *B. subtilis*, insertion of an *E. coli*-derived *tet* operator sequence from Tn*10* placed between the -35 and -10 boxes of the very strong *xyl* promoter produced a useful construct (Geissendoerfer and Hillen, 1990). In the presence of the *tetR* gene, this construct is strongly inducible and has high promoter strength. Using the system with a single *tet* operator, inducible expression of glucose dehydrogenase from *B. megaterium* was obtained at a very high level, and inducible expression of human single-chain urokinase-like plasminogen activator was achieved at the same level as in *E. coli*. Unlike the *E. coli* case, the product was not degraded up to four hours after induction in *B. subtilis*.

srfA is an operon required for surfactin production, competence development and efficient sporulation in *Bacillus subtilis*. The *srfA* locus of *B. subtilis* is defined by a transposon Tn*917* insertion and is required for production of the peptide secondary metabolite, surfactin. The *srfA* locus was isolated by cloning the DNA flanking *srfA*::TN*917* insertions followed by chromosome walking. These experiments show that *srfA* gene products function in *B. subtilis* cell specialization and differentiation.

6.6 TRP OPERON

A concise schematic description is provided in the book, "The Operon," by Miller and Reznikoff (1978). In addition to the *trp* operon (*trpEDCBA*), other genes involved include the *trp* repressor (*trpR*), the 3-deoxy-D-arabinoheptulosonic acid-7-phosphate (DAHP) synthetase (*aroH*), the tryptophanyl tRNA synthetase (*trpS*), the tryptophanyl tRNA (*trpT*), tryptophanase (*tna*), and transcription termination factor rho (*rho*).

Expression vectors for *Escherichia coli* use the *trp* operon with a catabolite activator protein-responsive element (CAP sequence) up-stream to express heterologous genes at high level under glucose limitation. Fermentation conditions for most effective use of these plasmids and an expression vector for manufacture of interleukin α_1 are described by Schroeckh et al. (1991).

A method of cultivating *Escherichia coli* k-12 transformed with a recombinant DNA expressing an heterologous gene from the *trp* operon promoter that achieves higher levels of gene expression is described. The method comprises incubating the mid or late log-phase culture at 28-34°, after addition of the gratuitous inducer, indole acrylic acid, to the medium. Improved expression of the tryptophan synthase gene using this method was demonstrated (Terasawa et al., 1991). A recent report by Martin et al. (1990) describes the cloning and physiology and functional characterization of the genes involved in amino acid biosynthesis in corynebacteria. Topics covered include the improvement of promoter-probe vectors; cloning and characterization of corynebacteria promoters in *E. coli*; the *hom* and *thr* gene cluster encoding the threonine biosynthetic pathway; analysis of the promoter, attenuator and terminator of the tryptophan operon; control of expression of the tryptophan operon; expression of cloned genes of the threonine pathway and of the tryptophan operon in *E. coli* and in corynebacteria; and the development of a system for purification of intact RNA from corynebacteria.

Tryptophanase (*tna*) operon expression in *Escherichia coli* is induced by tryptophan. This response is mediated by features of a 319-base-pair leader region preceding the major structural genes of the operon. Site-directed mutagenesis was used to investigate the role of the single Trp codon, at position 12 in *tnaC*, in regulation of the operon. These studies suggest possible models for tryptophan induction of *tna* operon expression involving $tRNA^{Trp}$-mediated frame shifting or readthrough at the *tnaC* stop codon (Gollnick and Yanofsky, 1990). The effects of changing the length and codon content of the *trp* leader peptide coding region on expression of the *trp* operon of *E. coli* were also examined. It had previously been shown that coupling of transcription and translation in the *trp* leader region is essential for both basal level control and tryptophan starvation control of transcription attenuation in this operon. Increasing the length of the leader peptide coding region by 55 codons allowed normal basal level control and normal tryptophan starvation control. As expected, the presence of a nonsense codon early in the leader peptide coding region decreased basal expression and eliminated starvation control. Introducing tandem rare codons had no effect on basal level expression, but eliminated the tryptophan starvation response. Frameshifting at tandem rare codons was tested as the most likely explanation for loss of the tryptophan starvation response, but the results are as yet inconclusive (Roesser and Yanofsky, 1991.

In more applied studies, a new method to firmly stabilize recombinant plasmids was exploited, using *E. coli* Tna (*trpAE1*, *trpR*, *tnaA*) and pSC101trpI15-14 (tetracycline resistance, whole *trp* operon) as a model system (Sakoda and Imanaka, 1990). The Tna strain carrying pSC101trpI15-14 was mutagenized and mutant 6F484, which stably maintained the recombinant plasmid for 100 generations, was isolated. From 6F484, plasmid-free cells (tetracycline sensitive) were screened-for on selective agar plates containing fusaric acid. The host strain FA14 lost the ability for active transport of tryptophan, in addition to the phenotype of Trp^-. Therefore, strain FA14 could not grow normally even in a complete medium. However, when the strain was transformed with the *trp* operon recombinant plasmid, its growth rate was almost restored to the original level. These results suggest that the recombinant plasmid is indispensable for the normal growth of host cells like FA14. Even if plasmid-free segregants appear during the cultivation, they cannot grow so rapidly and are diluted as a minority in the total population. Consequently, owing to the deficiency of both the biosynthesis and uptake of tryptophan in a host strain, the *trp* operon recombinant plasmid can be stably maintained.

A mathematical model of the tryptophan operon is analyzed to investigate the regulatory effects of feedback repression and the demand for tryptophan in the cell. In this model, feedback repression is considered to be a two step process. First, the endproduct tryptophan combines with the inactive repressor produced by the regulatory genes to yield an active complex. This complex subsequently binds to the operator and prevents transcription of the structural genes into mRNA. The demand for tryptophan in the cell is modeled by a hyperbolic saturation function of the Michaelis-Menten type. Results are obtained for the expression of the tryptophan operon in *E. coli*, and their applicability to tryptophan production by microbial fermentation is discussed. It is shown that, depending on the strain level of the operon and the rate of utilization of tryptophan in the cell, an overproduction of tryptophan can be achieved under stable operating conditions; in other circumstances, the operon may become stable or unstable, and may lead to a periodic synthesis (Sen and Liu, 1990).

Similarly, mathematical models for the procaryotic control systems of tryptophan biosynthesis (both normal and with cloned blocks) and arabinose catabolism have been built, using the method of generalized threshold models (Prokudina et al., 1991). Kinetic curves for molecular components (mRNAs, proteins, metabolites) of the systems considered are obtained. Analysis of the functioning of the mechanisms of control of tryptophan biosynthesis demonstrates that feedback inhibition is the most operative mechanism, while repression allows the bacterium to economize intracellular resources. In the control system for arabinose catabolism, modeling showed that induction by arabinose within a wide range of parameter values caused two subsystems (*araBAD* and transport operons) of the arabinose regulon with a low rate of arabinose utilization to pass into a stationary state, and one subsystem (*araC* operon) to pass into a stable periodic state. During effective utilization of arabinose it was shown that under induction by arabinose, stable oscillations with small increases in the concentration of regulatory protein, and oscillations with large increases in the concentrations of arabinose-isomerase and transport protein may occur. The period of the oscillation depends on the mean lifetime of the activator-DNA complex and on the rate constant of arabinose isomerase degradation.

Optimal operating conditions were determined for recombinant *E. coli* cells in fed batch and two stage continuous fermentors. The model expression system used in this work was the *E. coli trp* promoter cloned on plasmids. Model equations for cell growth and cloned-gene expression were formulated and used to evaluate process performances under different operating modes. The operating variables manipulated for maximum performance include the timing of 3β-indoleacrylic acid (IAA) addition to derepress transcription from the *trp* promoter, the total operating period, and the nutrient concentration profile during fermentations. For a fed batch mode, the performance was significantly improved by adjusting the IAA addition (environmental switch) time relative to the total operation period. An optimal switching time exists for a given total operation period. For a two stage continuous fermentation system, the productivity is more sensitive to the combination of the dilution rates than to the volume ratio of the two reactors. In general, as long as the down time is less than the total operation time in the fed batch mode, the latter gives higher productivity than the two stage continuous system (Park et al., 1989).

6.7 APPLICATIONS OF GENETIC ENGINEERING: PHENYLALANINE CASE STUDY

Backman et al. (1990) have completed a project which can serve as a paradigm of the genetic engineering approach. What was attempted was the optimization of the biosynthetic capacity of the shikimate pathway, already introduced in Chapter 1 (Fig. 1.11) and repeated here as Fig 6.6 for ready reference. Note that this same pathway was manipulated in the candicidin study described in detail in Chapter 5.

One mole of phenylalanine is synthesized from two moles of glucose and one mole of ammonia in common bacterial systems. This establishes minimum raw material costs. *E. coli* was chosen as the production organism because of its rapid growth properties and its well established capability toward phenylalanine biosynthesis. The steps in phenylalanine biosynthesis are shown in Table 6.1.

It was confirmed that the *E. coli* strains were capable of rapidly converting high levels of phenylpyruvate to phenylalanine, so this step was left untouched. The next-to-last steps were handled together. Genes for feedback-insensitive derivatives of the chorismate mutase/prephenate dehydratase bifunctional enzyme (CMPD; the *pheA* gene product) of *E. coli* were prepared and attached to strong promoters and the *pheA* attenuator removed. High levels of production of these modified gene products increase the yield of phenylalanine in fermentation (Backman and Balakrishnan, 1989). Sequences for the first 337 N-terminal amino acids of CMPD were fused to coding and regulatory sequences of the *E. coli aroF* gene and the *lac* operon.

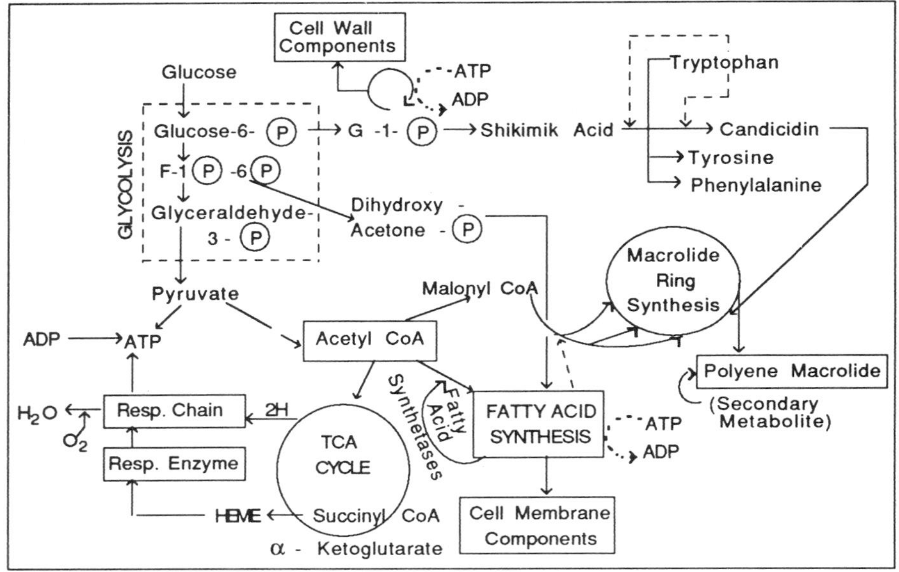

Figure 6.6 Biosynthesis pathways for secondary metabolites (e.g., candicidin phenyalanine).

In *E. coli*, the expression of the phenylalanine biosynthetic enzyme chorismate mutase/prephenate dehydratase, encoded by *pheA*, is likewise elevated in strains carrying *pheR* mutants. The site of action of the *pheR* product on *pheA* expression was determined to be the *pheA* attenuator. The elevated expression of *pheA* and *pheST* in *pheR* mutants apparently is a consequence of a lower frequency of transcription termination in the attenuator caused by lower levels of phenylalanyl-$tRNA^{Phe}$ (Gavine and Davidson, 1991).

Thus, except for the transaminases, all of the genes listed above were cloned. Along the way, it was found that particular genes could prove detrimental; for instance, the optimal plasmid plus *aro"X"* reduced the phenylalanine titer from 45 to 34 g/L.

Because several of the genes relevant to phenylalanine biosynthesis are normally regulated by the repressor protein encoded by the *tyrR* gene (Cornish et al., 1986), *tyrR* mutations were generated and introduced into the production strain. This results in the derepression and consequent constitutive expression of several genes, including *aroF*, *aroG*, *aroL* and genes involved in transport and transamination of phenylalanine. By attacking this regulation issue at the regulatory protein, the authors avoided the more complex task of altering the regulatory sites associated with each of those regulated genes.

Table 6.1 Steps in Phenylalanine Biosynthesis

Intermediate(s)	Enzyme	Gene(s)
Erythrose-4-phosphate + Phosphoenolpyruvate		
	DAHP synthetase	*aroF, G, H*
Deoxyarabinoheptulosonate-7-phosphate		
	Dehydroquinate synthetase	*aroB*
Dehydroquinate		
	Dehydroquinate dehydratase	*aroD*
Dehydroshikimate		
	Dehydroshikimate reductase	*aroE*
Shikimate		
	Shikimate kinase	*aroL*
Shikimate phosphate + Phosphoenolpyruvate		
	EPSP synthetase	*aroA*
Enolpyruvoylshikimate phosphate		
	Chorismate synthetase	*aroC*
Chorismate		
	Chorismate mutase	*pheA*
Prephenate		
	Prephenate dehydratase	*pheA*
Phenylpyruvate		
	Transaminase(s)	*tyrB, ilvE, aspC*
Phenylalanine		

An excision vector carrying a cloned gene (in this case, *tyrA* for tyrosine biosynthesis) was integrated into the bacterial chromosome and made responsive to a temperature signal, allowing tyrosine starvation at an optimal stage of the growth cycle. But, the overall approach was complicated by the fact that the tyrosine-specific DAHP synthase determined by *aroF* is in fact inhibited by phenylalanine. Feedback-inhibition-insensitive mutations in the genes for DAHP synthase were then obtained by means of resistance to toxic amino acid analogs. Low levels of by-products were produced, as shown in Table 6.2.

Table 6.2 By-products Found in Broth

Compound	Concentration (g/L)
Organic acids	
acetate	<2
formate	<1
pyruvate	<0.1
lactate	<0.1
valerate	Not quantified
Aromatic intermediates	
shikimate	<0.2
dihydroxybenzoate	0.03
trihydroxybenzoate	0.11
p-hydroxybenzoate	0.05
prephanate	Below detection limit
phenylpyruvate	Below detection limit
Aromatic amino acids	
tryptophan	Below detection limit
tyrosine	Below detection limit

Table 6.3 Summary of Typical Process Parameters

Final phenylalanine titer	50 g/L
Phenylalanine	0.23 g/g
Fermentation time	36 hours
Final cell density (dry weight)	23 g/L

The genetically engineered strains were productive toward phenylalanine, and plasmid stability was excellent. Typical process parameters are summarized in Table 6.3. Oxygen transfer requirements were determined, and aeration and stirring were controlled to keep phenylalanine production from falling due to anoxia. Automatic control of pH and mechanical foam breakers were also employed in the fermentors. The organism can achieve titers of 50 grams of phenylalanine per liter in about 36 hours of vessel time. In doing this, the strain produces over two grams of product per gram of cell mass formed and about 0.25 grams of product per gram of glucose consumed. Moreover, the fermentation broth is remarkably free from byproducts at the end of the production cycle.

6.8 REFERENCES

Arthur, P. M., B. Duckworth and M. Seidman, *J. Biotechnol.*, **13**, 29-46 (1990).

Arthur, P. M. and B. Duckworth, Eur. Patent EP 355625 A1, Optimized Prokaryotic Promoters and Plasmids Containing Them and Manufacture of Interleukin-1β and Macrophage Colony-Stimulating Factor (1990).

Avots, A., J. Bundulis, V. Ose, N. V. Romanchikova, V. Skrivelis, E. K. Yankevits, A. Cimanis and E. Grens, *Dokl. Adad. Nauk SSSR*, **315**, 994-996 (1990).

Avots, A., N. V. Varivotskaya, N. N. Voitenok, E. Y. Gren, A. E. Martinovich, N. V. Romanchikova, N. V. Sevch, A. Cimanis, U. R. Apsalon, et al., Patent WO 9010069 A1, Recombinant Plasmids for Manufacture of Human Interleukin 2 in *E. coli* (1990).

Backman, K. C. and R. Balakrishnan, Eur. Patent 330773 A1, Phenylalanine-Insensitive Derivatives of the Chorismate Mutase/Prephenate Dehydratase of *Escherichia Coli* for Use in the Manufacture of Phenylalanine (1989).

Backman, K., M. J. O'Connor, A. Maruya, E. Rudd, D. McKay, R. Balakrishnan, M. Radjai, V. DiPasquantonio, D. Shoda, R. Hatch and K. Venkatasubramanian, In "Biochemical Engineering VI," *Ann. N. Y. Acad. Sci.*, **589**, 16-24 (1990).

Bailey, J. E., S. Birnbaum, J. L. Galazzo, C. Khosla and J. V. Shanks, In "Biochemical Engineering VI," *Ann. N. Y. Acad. Sci.*, **589**, 1-15 (1990).

Beppu, T., In "Bioproducts and Bioprocesses," A. Fiechter, H. Okada and R. D. Tanner, Eds., p. 201. Springer-Verlag: Berlin, Heidelberg (1989).

Brabetz, W., W. Liebl and K. H. Schleifer, *Arch. Microbiol.*, **155**, 607-612 (1991).

Burbulys, D., K. A. Trach and J. A. Hoch, *Cell*, **64**, 545-552 (1991).

Chibazakura, T., F. Kawamura and H. Takahashi, *J. Bacteriol.*, **173**, 2625-2632 (1991).

Cornish, E. C., V. P. Argyropoulos, J. Pittard and B. E. Davidson, *J. Biol. Chem.*, **261**, 403 (1986).

Darnell, J., H. Lodish and D. Baltimore, "Molecular Cell Biology," p. 270. Scientific American Books (1986).

Dennis, D. E., Patent WO9118993 A1, Production of Poly-β-Hydroxybutyrate with Transformed *Escherichia coli* (1991).

De Vlieg, J., H. J. Berendsen and W. F. Van Gunsteren, *Proteins: Struct., Funct., Genet.*, **6**, 104-127 (1989).

Feavers, I. M., J. Foulkes, B. Setlow, D. Sun, W. Nicholson, P. Setlow and A. Moir, *Mol. Microbiol.*, **4**, 275-282 (1990).

Fischer, R., R. P. Von Strandmann and W. Hengstenberg, *J. Bacteriol.*, **173**, 3709-3715 (1991).

Freese, E., J. Heinze, T. Mitani and E. B. Freese, In "Spores VII," G. Chambliss and J. C. Vary, Eds., p. 277. American Society for Microbiology: Washington, D. C. (1978).

Furlong, C. E. and J. A. Sundstrom, *Dev. Ind. Microbiol.*, **30**, 141-148 (1989).

Geissendoerfer, M. and W. Hillen, *Appl. Microbiol. Biotechnol.*, **33**, 657-663 (1990).

Georgiou, G. and F. Baneyx, In "Biochemical Engineering VI," *Ann. N. Y. Acad. Sci.*, **589**, 139-147 (1990).
Golding, A., M. J. Weickert, J. P. E. Tokeson, S. Garges and S. Adhya, *J. Bacteriol.*, **173**, 6294-6296 (1991).
Gollnick, P. and C. Yanofsky, *J. Bacteriol.*, **172**, 3100-3107 (1990).
Hartley, B. S., V. Kumar and S. Ramakrishnan, Patent UK 9805674.1, DNA Construct and Modified Yeast (1989).
Hong, E. K., Ph. D. Thesis, Dept. of Chemical and Biochemical Engineering, Rutgers University (1992).
Honjo, M., A. Nakayama, K. Fukazawa, K. Kawamura, K. Ando, M. Hori and Y. Furutani, *J. Bacteriol.*, **172**, 1783-1790 (1990).
Hudson, J. M. and M. G. Fried, *J. Bacteriol.*, **173**, 59-66 (1991).
Jeong, Y. S., W. R. Vieth and T. Matsuura, *I&EC Res.*, **28**, 231 (1989).
Jeong, Y. S., W. R. Vieth and T. Matsuura, *Ann. N. Y. Acad. Sci.*, **589** (Biochem. Eng. 6), 214-228 (1990).
Jeong, Y. S., W. R. Vieth and T. Matsuura, *Biotechnol. Prog.*, **7**, 130-139 (1991).
Jeong, Y. S., W. R. Vieth and T. Matsuura, *Biotechnol. Bioeng.*, **37**, 587-590 (1991).
Johnson, D. F., D. P. Nierlich and A. J. Lusis, *Gene*, **94**, 9-14 (1990).
Katsumata, R., Y. Kikuchi and K. Nakanishi, Jpn. Patent JP01179686 A2, Amino Acids and their Manufacture with Corynebacterium or Brevibacterium Harboring *Escherichia Coli* Lactose Operon (1989).
Khoury, A. M., H. S. Nick and P. Lu, *J. Mol. Biol.*, **219**, 623-634 (1991).
Kisters-Woike, B., N. Lehming, J. Sartorius, B. Von Wilcken-Bergmann and B. Mueller-Hill, *Eur. J. Biochem.*, **198**, 411-419 (1991).
Le Grice, S. F., *Methods Enzymol.*, **185** (Gene Expression Technol.), 201-214 (1990).
Levy, S., G. Q. Zeng and A. Danchin, *Gene*, **86**, 27-33 (1990).
Lunn, C. A., S. K. Narula and R. L. Reim, Patent WO9106655 A1, Production and Secretion of Heterologous Proteins with *Escherichia coli*, Mutants (1991).
Margolis, P., A. Driks and R. Losick, *Science*, **254**, 562-565 (1991).
Martin, J. F., L. M. Mateos, R. F. Cadenas, C. Guerrero, M. Malumbres, A. Colunas and J. A. Gil, In "Microbiol. Appl. Food Biotechnol.," (Proc. Congr. Singapore Soc. Microbiol) 2nd, B. H. Nga and Y. K. Lee, Eds., pp. 20-26. Elsevier: London (1990).
Maruo, B. and H. Yoshikawa, "*Bacillus subtilis*: Molecular Biology and Industrial Application," p. 267. Elsevier Science Publishing Co., Inc.: New York (1989).
Mashko, S. V., V. P. Veiko, A. L. Lapidus, M. I. Lebedeva, A. V. Mochulskii, I. I. Shechter, M. E. Trukhan, K. I. Ratmanova, B. A. Rebentish, et al., *Gene*, **88**, 121-126 (1990).
McCormick, J. R., J. M. Zengel and L. Lindahl, *Nucleic Acids Res.*, **19**, 2767-2776 (1991).
Miller, J. H. and W. S. Reznikoff, "The Operon," p. 265. Cold Spring Harbor Laboratory (1978).

Mohr, S. and K. Esser, In "Biotechnology Focus 3," R. K. Finn and P. Präve, Eds., pp. 5-24. Hanser Publishers: Munich; Dist. in U. S. by Oxford University Press: New York (1992).

Msadek, T., F. Kunst, D. Henner, A. Klier, G. Rapoport and R. Dedonder, *J. Bacteriol.*, **172**, 824-834 (1990).

Msadek, T., F. Kunst, A. Klier, G. Rapoport and R. Dedonder, In "Genet. Biotechnol. Bacilli," Proc. Int. Conf. Bacilli, V, M. M. Zukowski, A. T. Ganesan and J. A. Hoch, Eds., pp. 245-255. Academic: San Diego, CA (1990).

Murakawa, G. J., C. Kwan, J. Yamashita and D. P. Nierlich, *J. Bacteriol.*, **173**, 28-36 (1991).

Nicholson, W. L. and G. H. Chambliss, *J. Bacteriol.*, **169**, 5867-5869 (1987).

Pace, H. C., P. Lu and M. Lewis, *Proc. Natl. Acad. Sci. U. S. A.*, **87**, 1870-1873 (1990).

Park, T. H., J. H. Seo and H. C. Lim, *Biotechnol. Bioeng.*, **34**, 1167-1177 (1989).

Pierce, J. A., C. R. Robertson and T. J. Leighton, *Biotechnol. Prog.*, **8**, 211-218 (1992).

Prokudina, E. I., R. Y. Valeev and R. N., Churaev, *J. Theor. Biol.*, **151**, 89-110 (1991).

Ramasubramanyan, N., Biochem. Eng. Lab. Report, Rutgers Univ., March (1992).

Reddy, P., N. Fredd-Kuldel, E. Liberman and A. Peterkofsky, *Protein Expression Purif.*, **2**, 179-187 (1991).

Resnik, E. and D. C. LaPorte, *Gene*, **107**, 19-25 (1991).

Richaud, F., B. Jarry, K. Takinami, O. Kurahashi and A. Beyou, Fr. Patent 2627508 A1, Use of Transposon to Integrate Genes into Bacterial DNA, (1989).

Roesser, J. R. and C. Yanofsky, *Nucleic Acids Res.*, **19**, 795-800 (1991).

Rudner, D. Z., J. R. LeDeaux, K. Ireton and A. D. Grossman, *J. Bacteriol.*, **173**, 1388-1398 (1991).

Ryan, W. and S. J. Parulekar, Abstract, Biochemical Engineering VII, Engineering Foundation Conf., Santa Barbara, CA, March 3-8 (1991).

Sakoda, H. and T. Imanaka, *J. Ferment. Bioeng.*, **69**, 75-78 (1990).

Schaeffer, P., B. Cami and R. D. Hotchkiss, *Proc. Natl. Acad. Sci. U. S. A.*, 54, 704-711 (1965).

Schroeckh, V., M. Hartmann, E. Birch-Hirschfeld and D. Riesenberg, Ger. Patent DD 292023 A5, Catabolite Derepressible Expression Vectors for *Escherichia coli* (1991).

Sen, A. K. and W. M. Liu, *Biotechnol. Bioeng.*, **35**, 185-194 (1990).

Shuler, M. L. and F. Kargi, "Bioprocess Engineering," p. 222. Prentice Hall: Englewood Cliffs, NJ (1992).

Silhavy, T. J., Paper presented in "Biotechnology: Opportunities, Problems and Solutions " AIChE Mtg., Cen. NJ Sec., Princeton, May (1985).

Terasawa, M., H. Yamagata and H. Yugawa, Jpn. Patent JP 03061479 A2, Culture Method for High-Level Expression of Heterologous Genes in *Escherichia coli* (1991).

Tokeson, J. P., S. Garges and S. Adhya, *J. Bacteriol.*, **173**, 2319-2127 (1991).

Trach, K. and J. A. Hoch, *J. Bacteriol.*, **171**, 1362-1371 (1989).

Weicker, M. J., L. Larson, W. L. Nicholson and G. H. Chambliss, In "Genet. Biotechnol. Bacilli," (Proc. Int. Conf. Bacilli), 5th, Mtg. Date 1989, 237-244, M. M. Zukowski, A. T. Ganesan and J. A. Hoch, Eds. Academic: San Diego (1990).

Wilcox, E. R., Eur. Patent 410665 A1, Improved Control of Expression of Heterologous Genes from Modified *lac* Operon (1991).

Yocum, R. R., S. Hanley, R. West, Jr. and M. Ptashne, *Mol. and Cellular Biol.*, **4**, 1985-1998 (1984).

7

TRANSCRIPTION AND TRANSLATION OF THE GENETIC CODE

7.0 TRANSPORT REGULONS

Transport regulons are widely encountered as important components in the overall mechanism of gene control. For example, Yanofsky's group has done some elegant studies of transport regulation of the *trp* operon. *Escherichia coli* forms three permeases that can transport the amino acid tryptophan: Mtr, AroP and TnaB. The structural genes for these permeases reside in separate operons that are subject to different mechanisms of regulation. The fact that the tryptophanase (*tna*) operon is induced by tryptophan was exploited to infer how tryptophan transport is influenced by the growth medium and by mutations that inactivate each of the permease proteins.

In an acid-hydrolyzed casein medium, high levels of tryptophan are ordinarily required to obtain maximum *tna* operon induction. High levels are necessary because much of the added tryptophan is degraded by tryptophanase. An alternate inducer that is poorly cleaved by tryptophanase, 1-methyltryptophan, induces efficiently at low concentrations in both tna^+ strains and *tna* mutants. In an acid-hydrolyzed casein medium, the TnaB permease is most critical for tryptophan uptake; i.e., only mutations in *tnaB* reduce tryptophanase induction. However, when 1-methyltryptophan replaces tryptophan as the inducer in this medium, mutations in both *mtr* and *tnaB* are required to prevent maximum induction.

In this medium, AroP does not contribute to tryptophan uptake. However, in a medium lacking phenylalanine and tyrosine the AroP permease is active in tryptophan transport; under these conditions it is necessary to inactivate the three permeases to eliminate *tna* operon induction. The Mtr permease is principally responsible for transporting indole, the degradation product of tryptophan produced by tryptophanase action. The TnaB permease is essential for growth on tryptophan as the sole carbon source. When cells with high levels of tryptophanase are transferred to a tryptophan-free growth medium, the expression of the tryptophan (*trp*) operon is elevated. This observation suggests that the tryptophanase present in these cells degrades some of the synthesized tryptophan, thereby creating a mild tryptophan deficiency. In short, these studies assign roles to the three permeases in tryptophan transport under different physiological conditions (Yanofsky et al., 1991).

Aslanidis et al. (1989) describe an active transport of the inducer raffinose as part of the *raf* operon, while Pereira (1989) describes the effect of diffusible factors on the regulation of an operon in *E. coli* K-12. Sweet et al. (1990) point out that the glycerol facilitator is known as the only example of a transport protein that catalyzes facilitated diffusion across the *E. coli* inner membrane. Stirling et al. (1989), Gowrishankar (1989), May et al. (1989) and Lamark et al. (1991) describe the proU loci of *E. coli* and *Salmonella typhimurium* which encode high-affinity glycine betaine transport systems that play an important role in survival under osmotic stress. Tricarboxylates are transported into *S. typhimurium* by a binding protein-dependent transport system known as TctI (Widenhorn et al., 1989). Prince and Villarejo (1990) show that osmotic control of proU transcription in *E. coli* is mediated through direct action of potassium glutamate on the transcription complex. The structural *dctA* and regulatory *dctBD* genes controlling the transport of C_4-dicarboxylic acids from *Rhizobium meliloti* have been cloned and analyzed (Wang et al., 1989) by gene fusions to the enteric *lacZY* reporter gene. It was also recently pointed out by Schnetz and Rak (1990) that the β-glucoside operon in *E. coli* has elements that specifically modulate the transport of β-glucosides.

CHEMICAL MESSENGERS

The evident importance of transport elements as suggested by these studies has led us to develop quantitative models of gene regulation that specifically include a transport system. This chapter demonstrates the approach we have taken, with an emphasis on two systems that have been a long term topic of study in our laboratory. This allows, among other things, for broader interpretation of experimental data and a synthesis of mechanistic features at the molecular level. An ever increasing complexity of interactions governing gene expression is also more manageable within a mathematical framework. Furthermore, as Karube points out (Vieth, 1990), microorganisms are made up of compartments enclosed by various membrane structures. This pattern is so extensive that it is not an exaggeration to say that cells are built of membranes. Therefore, any theory which seeks to describe the widespread role of chemical messengers in carrying out the *central dogma* of biology, i.e., translation and transcription of the genetic code, without explicit consideration of membrane transport steps is certain to fall short of the mark. In contrast, substantial benefits can be obtained through incorporation of such steps in an overall optimization of the fermentation process. For example, Grote (1991) discloses an optimized fermentation process for the production of heterologous proteins with *E. coli* from genes linked to the *lac* promoter or a modified *lac* promoter (e.g., *tac*, *trc*). After the growth phase using glucose as carbon source, the synthesis of product is induced by (i) IPTG with glucose limitation, (ii) by lactose with lactose limitation, or (iii) with IPTG and lactose with lactose limitation. The glucose or lactose limitation is adjusted such that the oxygen partial pressure is set at >10% saturation. The yield of protein product was increased two-fold relative to prior art when using the third procedure. Furthermore, Seaver (1991) asserts that the overall basic question of how the type of bioreactor influences cell productivity has itself still not been answered. We are still learning how to optimize protein production in all types of bioreactors. . . more about this subject in Chapter 8. (Other recent examples show that optimization can sometimes increase productivity over ten-fold.)

7.1 PERMEASE ACTIVITY IN E. COLI

In what follows, expression of the lactose (*lac*) operon in the *Escherichia coli* chromosome is analyzed for both wild-type and recombinant cell cultures under chemostatic conditions. A unified model which involves regulation of active inducer (lactose) transport, promoter-operator regulated expression of the *lac* operon, glucose-mediated inducer exclusion, and catabolite repression is formulated and verified.

The synthesis of α-amylase with a recombinant form of *Bacillus subtilis* is likewise investigated, the *lac* operon model providing a point of departure. While there are similarities in the influence of transport on both regulating models, there are also important differences. In a chemostat system, the synthesis of α-amylase is nongrowth associated, while β-galactosidase is a growth-associated enzyme. Nevertheless, transport regulation is an important feature in both instances.

In probing some of the subtleties of positive and negative modes of gene regulations, one would like a well controlled comparison in which the two systems were identical in every respect except that one has a positive and the other a negative mechanism of regulation. As noted by Savageau (1989), while one might conceive of constructing the alternatives by appropriate genetic engineering, this is not easily accomplished. He recommends that at the present time it is much more practical to simulate comparisons by appropriate mathematical analysis of gene regulation.

Let us briefly summarize the molecular mechanistics which are entailed, as shown in Fig. 7.1. In the *lac* operon model, a specific repressor protein binds at the operator site; this prevents RNA polymerase from binding to the promoter and initiating transcription. Lactose, the inducer, is transported into the cell via a permease and can bind to the repressor, preventing its binding to the operator, permitting transcription of mRNA to commence. In the presence of high levels of cyclic adenosine monophosphate (cAMP), a complex forms between cAMP and the catabolite receptor protein (CRP) which then binds to the P_I region of the promotor gene, facilitating RNA polymerase binding at the P_{II} site (De Crombrugghe et al., 1984). Glucose repression in the form of the catabolite modulator factor (CMF) interferes with this process. Recently, Soegaard-Andersen et al. (1990) described an interaction of the two mechanisms of repression, showing that the cytoplasmic

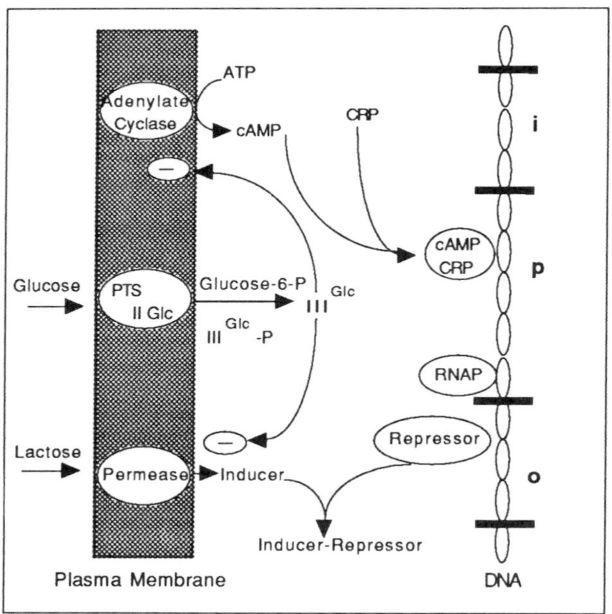

Figure 7.1. *lac* Operon.

repressor protein selectively regulated cAMP-CRP-dependent intiations, although transcription started from the same site in the absence or presence of cAMP-CRP.

Kaback (1990) regards the *lac* permease of *E. coli* as a prototypic energy-transducing membrane protein. Dean describes the kinetics of the permeases and β-galactosidases of six lactose operons from natural isolates of *E. coli* (1989). Oehler et al. (1990) describe the cooperation of the three operators of the *lac* operon in repression. Gardner and Matthews (1990) report on characterization of two mutant lactose repressor proteins containing single tryptophans, while Deuschle et al. (1990) have investigated RNA polymerase II transcription blocked by *E. coli lac* repressor. Kleina and Miller (1990) studied the effects of systematic amino acid replacements in the *lac* repressor protein, while Khoury et al. (1990) report the DNA length dependence of the *lac* repressor-operator interaction. Macromolecular binding equilibria in the *lac* repressor system are the subject of studies using high-pressure fluorescence spectroscopy by Royer et al. (1990).

PTS MEDIATED REGULATION OF INTRACELLULAR CAMP AND LACTOSE TRANSPORT

PTS sugars are transported into bacteria by a group translocation mechanism which requires phosphorylation of the sugars. The sugars which are PTS substrates negatively regulate the induction of the necessary enzyme systems for the metabolism of non-PTS sugars. In *E. coli*, glucose specific enzyme III^{Glc} of the PTS in its unphosphorylated form is believed to be involved in the inhibition of adenylate cyclase activity, and also, in the regulation of non-PTS sugar (e.g., lactose in *E. coli*) permease activity. Therefore, the extent of PTS regulation is dependent on the intracellular concentration of III^{Glc} as shown schematically in Fig. 7.2. Many bacteria, including *Bacillus*, *Streptococcus* and *Salmonella*, share strong sequence homology in some of the genes coding for the sugar phosphotransferase system (Gonzy-Treboul et al., 1989). The model is therefore likely to be applicable to other operons besides the model *lac* operon investigated here. For example, the levanase operon of *Bacillus subtilis* includes a fructose-specific phosphotransferase system regulating the expression of the operon (Martin-Verstraete et al., 1990; Martin et al., 1990). Likewise, mutants of *S. typhimurium* defective in the proteins of the fructose operon [*fruB(MH)KA*], the fructose repressor (*fruR*), the energy-coupling enzymes of the phosphoenolpyruvate:sugar phosphotransferase system (PTS) (*ptsH* and *ptsI*), and the proteins of cAMP action (*cya* and *crp*) were analyzed for their effects on cellular physiological processes and expression of the fructose operon

(Feldheim et al., 1990). Experiments indicated that sugar uptake via the PTS can utilize either the phosphorylated tridomain fusion protein, enzyme III^{Fru}-modulator-FPr, or an alternative labeled HPr-P, but that FPr is preferred for fructose uptake, while HPr is preferred for uptake of all the other sugars.

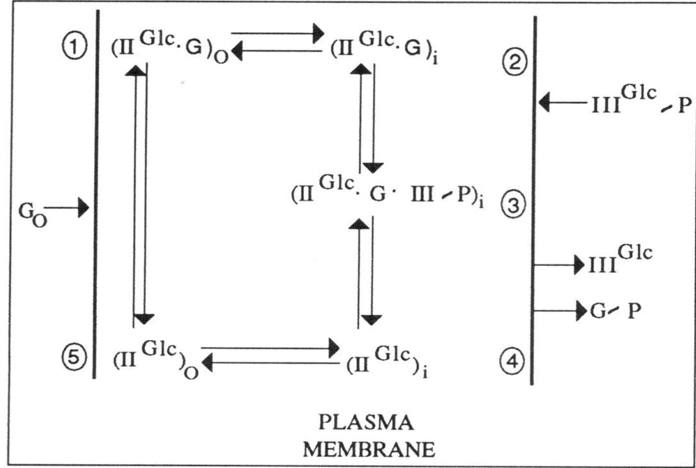

Figure 7.2. Schematic diagram of the group translocation via PTS. G_O represents glucose concentration in the external medium.

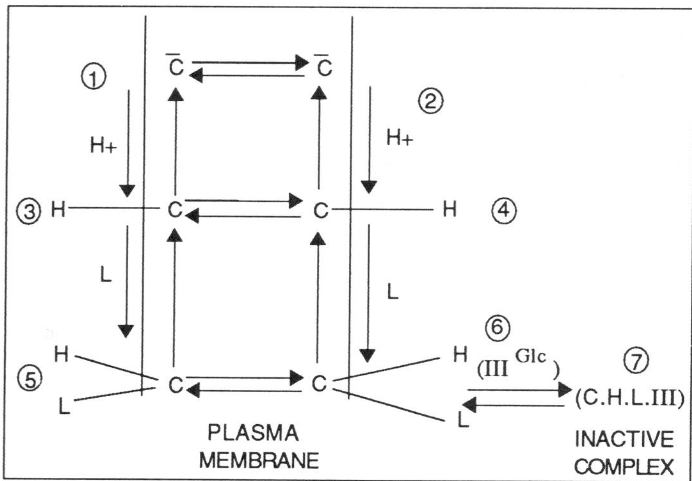

Figure 7.3. Schematic representation of inducer translocation and exclusion. H^+, L and C represent proton, lactose and carrier, respectively.

In developing the mathematical description of inducer exclusion, inhibition of *lac* permease activity by III^{Glc} during active transport of lactose across the plasma membrane is considered as the model system. The general sequence of kinetic steps involved in the lactose-proton symport is shown in Fig. 7.3. It has been found that lactose exclusion by PTS-III^{Glc} does not involve interruption of energy coupling to solute accumulation (Mitchell et al., 1982). Since III^{Glc}, a cytoplasmic protein, interacts with the *lac* carrier only when lactose is bound to the carrier, it is likely that the CHL complex at the cytoplasmic site of the membrane would be the target of III^{Glc}.

INDUCER EXCLUSION

Another of the glucose mediated regulations to prevent induction of the non-PTS sugar enzymes is to inhibit the entry of the inducer into bacteria. Available evidence suggests that glucose-specific enzyme III^{Glc} of the PTS in its non-phosphorylated form directly binds to the various non-PTS transport systems and thus inhibits their activities (Mitchell et al., 1982). Saier et al. (1983) observed that the binding of III^{Glc} to the *lac* permease showed cooperativity in substrate binding. Therefore, binding of lactose to its permease is required for effective regulation by the III^{Glc}. Schnetz and Rak (1990) show that the β-glucoside permease also interacts with glucose-specific enzyme III.

The intracellular lactose concentration is dependent on the bulk lactose concentration. Ray et al. (1987) developed the following equation for active lactose transport.

$$\frac{dL_{in}}{dt} = \frac{1}{1 + B'(t)[III^{Glc}]} \times \frac{A'(t)L_B}{\dfrac{B(t)}{(1 + B'(t)[III^{Glc}])} + L_B} - \mu L_{in} \qquad [7.1]$$

A simplified rate expression is found for $(B'\ [III^{Glc}]) \ll 1$:

$$\frac{dL_{in}}{dt} = \frac{A'(t)L_B}{B(t) + L_B} - \mu L_{in} \qquad [7.2]$$

The fraction of total specific binding sites of the promoter bound by the RNAP is F or f / f_{max}. f is dependent on the levels of cAMP and catabolite modulator factor, or CMF. f_{max} is the point at which catabolite repression is at a minimum, cAMP=$cAMP_{max}$ and CMF=0. Expressions for the time dependent variations of cAMP, CMF as well as III^{Glc}

were developed by Ray (1985). Under chemostatic conditions, the time derivatives will, of course, vanish and the constants A', B, B' replace A'(t), B(t), B'(t), respectively, in eqns. [7.1] and [7.2].

$$\frac{dcAMP}{dt} = \frac{k_a' \mu}{1 + k_b' + k_c' [III^{Glc}]} - k_d [cAMP] - \mu[cAMP] \qquad [7.3]$$

$$\frac{dCMF}{dt} = k_m \frac{\mu}{Y_x} - k_{-m} [CMF] - \mu[CMF] \qquad [7.4]$$

$$\frac{dIII^{Glc}}{dt} = \frac{K_R[G_0]}{K_{RS} + [G_0]}\mu - K_{RD}[III^{Glc}] - \mu[III^{Glc}] \qquad [7.5]$$

INDUCER TRANSPORT

Figure 7.4 shows the steady state experimental data and model prediction for the effect of dilution rates on intracellular lactose concentrations of *E. coli* growing on lactose in a chemostat culture. The saturation trend is expected. As the dilution rate is increased, the residual substrate concentration increases in the culture and the transport model incorporates this through a balance for saturable sites. Notice that the intracellular concentration rises to a saturation value nearly five-fold that of the bulk ("active transport").

Figure 7.5 shows the simulation curves and experimental results for steady state specific β-galactosidase activity as a function of dilution rate. It is interesting to note that even with lactose alone the specific enzyme activity (SEA) goes through a maximum and the optimum dilution rate shifts as the feed lactose concentration is changed. However, above a certain dilution rate, depending on the feed concentration, SEA becomes independent of D and, also, essentially independent of feed lactose. When lactose is present in the culture, it is transported actively in the cells by the lac permease. Part of the intracellular lactose is transformed to allo-lactose (glucose-1, 6-galactose), which acts as inducer, and the larger part is hydrolyzed to glucose and galactose. Both the reactions are catalyzed by β-galactosidase. Glucose and galactose (which is ultimately converted to glucose by the *gal* operon enzymes) are actively metabolized by the organisms. As a result of glucose metabolism, the repressive intermediary catabolite, catabolite modulator factor (CMF), accumulates in the cells.

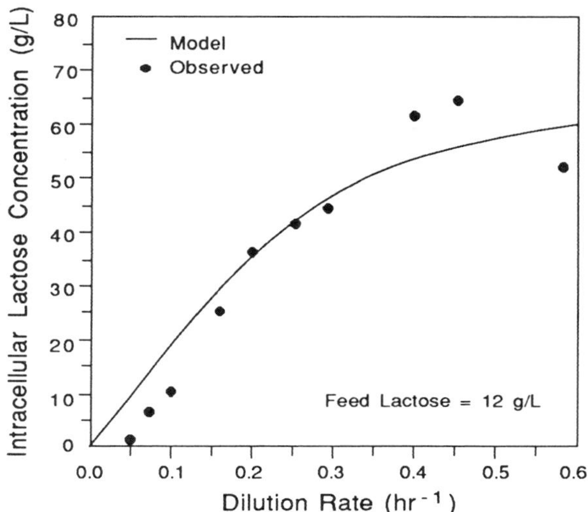

Figure 7.4. Effect of dilution rate on intracellular lactose concentration.

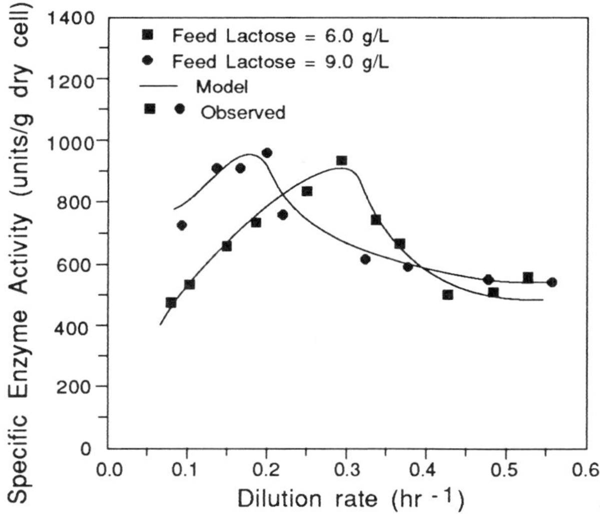

Figure 7.5. Effects of dilution rate and feed lactose concentration on steady state enzyme specific activity in chemostat cultures.

According to the model description, the accumulation of CMF is insignificant below a critical substrate utilization rate. Above the critical value, CMF exerts negative control at the level of transcription. At low dilution rates, below the critical value, only *lac* repressor-inducer controlled regulation at the operator site is dominating and, as the dilution rate is increased, induction predominates over cytoplasmic repression as a result of higher intracellular inducer concentration. The concentration of repressor molecules was assumed constant at a low level, because in a wild-type strain of *E. coli* they are synthesized constitutively and the *lac i* gene, which codes for the repressor protein, has a weak promoter. When the dilution rate exceeds the critical value, corresponding to the critical substrate utilization rate, the regulation of expression of the *lac* operon by CMF becomes the dominating factor and the enzyme activity curves drop sharply. Attainment of constant SEA level at high dilution rates is due, therefore, to the achievement of maximum intracellular CMF concentration, as predicted by the saturation kinetics of CMF production. It has been experimentally observed that the biomass yield coefficient, Y_X, is inversely proportional to the feed sugar concentration. Therefore, the dilution rate corresponding to the critical substrate utilization rate varies as the feed sugar concentration is changed.

BALANCE OF INDUCTION AND REPRESSION

In Fig. 7.6, one sees the comparison between steady state experimental data and model predictions when both glucose and lactose are present. In these experiments, lactose to glucose ratios in the feed medium were changed, but the total sugar was kept at 6.0 g/L. As expected, a higher level of SEA was observed as the ratio was raised. As predicted by the model, the rising parts of the curves at low dilution rates are due to increased levels of induction because of higher intracellular concentration of inducer as the dilution rate is increased, and the declining parts of the curves are due to domination of catabolite repression and inducer exclusion over induction.

Apart from repressive action by CMF, also observed with growth on lactose alone, as the dilution rate is increased, the presence of glucose in the external medium causes a reduced level of intracellular cAMP. These additional regulations in mixed substrate fermentation, exerted by unphosphorylated enzyme III^{Glc} of the PTS during glucose translocation across the plasma membrane, cause enzyme activity to approach zero as D is increased. This is quite different than what is observed with lactose alone. Fig. 7.7 clearly indicates the glucose mediated regulations of inducer entry into the cells.

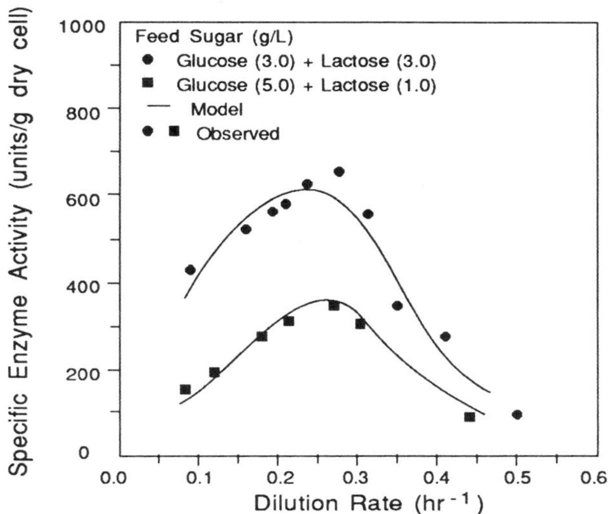

Figure 7.6. Effect of dilution rate and feed compositions on steady state enzyme specific activity in a chemostat culture (mixed sugar medium).

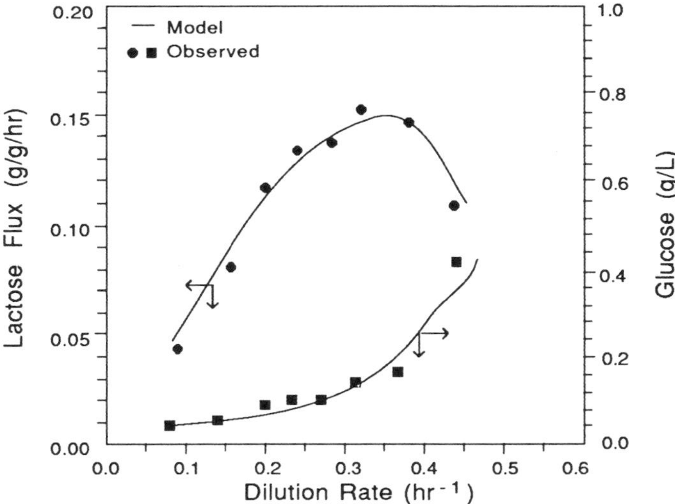

Figure 7.7. Effect of dilution rate on steady state lactose flux in a chemostat culture (mixed sugar medium).

A mechanistic model which considers total mRNA and total protein rather than considering a specific operon was developed by Peretti and Bailey (1986). Similar to the model discussed above, Peretti and Bailey's model is based on the kinetics of transcription and translation. The amount of mRNA transcribed is dependent on RNA polymerase content, the fraction of RNA polymerase transcribing and its distribution. Also Peretti and Bailey included an efficiency term to take into account premature termination of mRNA transcription. The translation of proteins is based on the assumption that chain elongation is constant, as is the amount of mRNA and the availability of nontranslating ribosomes. Finally, the level of total protein content is controlled by the rate of degradation, which is dependent on the state of the culture. For example, starvation for amino acids results in less proteolysis than does energy deprivation.

7.2 RECOMBINANT CELL CULTURE

Product release/separation strategies for recombinant-DNA organisms figure importantly in overall process economics. Among the possibilities are intracellular containment, secretion into the extracellular medium and facilitated/coupled transport from the cell. The first two strategies just mentioned are considered in the succeeding sections of this chapter.

In the work described here (Bailey, 1987), the host was *E. coli* MBM7061 (z^-, y^+), harboring the plasmid pMLB1108 which is *lac* i^q z^+. Thus, while the host strain does contain the gene which codes for the permease (y^+), it is deficient in the z gene for expression of β-galactosidase, in the normal position on the chromosome. This deficiency is made up by the plasmid pMLB1108, which contains the proximal half of the *lac* operon; i.e., the repressor gene (i^q) and the gene coding for β-galactosidase (z^+). The q mutation causes the repressor protein to be overproduced, allowing better control for the switching on and off of protein synthesis in chemostats and in immobilized cells. Thus, one has a very flexible genetic system to study strategies for maintaining plasmids in genetically engineered systems. Lately, an improved vector system has been developed for the *in vitro* construction of transcriptional fusions to *lacZ*, by Linn and St. Pierre (1990).

E. COLI STRAIN DEVELOPMENT

Promoter induction enhances recombinant product yields by increasing the efficiency of transcriptional initiation. Copy number amplification improves recombinant yields by increasing the number of plasmid copies that are available for transcription. Betenbaugh and Dhurjati (1990) found that promoter induction using the *lac* promoter was a more effective tool than plasmid amplification for increasing recombinant product yields for CSH22 with plasmid pVH106/172. Both recombinant gene expression of the *lac* operon genes and plasmid amplification caused detrimental effects on cell growth rate. Simultaneous induction of the *lac* promoter and plasmid amplification produced a combination of the individual induction effects on recombinant product yields and cell growth rate.

The strain employed in the investigation next reported is a *lac* i^q strain, meaning that it overproduces the *lac* repressor protein, which in turn blocks transcription of the *lac* structural gene. A *lac* i^q host can produce as much as ten times the amount of repressor present in a normal cell (Bailey et al., 1983). Overproduction of the *lac* repressor by the resulting recombinant cells is necessary to provide a basis for inducible enzyme biosynthesis.

In the development of a recombinant strain to be used in subsequent bioreactor studies, it was necessary to have a host possessing the *lac* permease gene so that lactose could function as the inducer. The host therefore had to be *lac* z^-, y^+ so that the *lac* z^+ plasmid and host could be easily distinguished when grown on indicator plates. A plasmid/host system was chosen which meets these requirements. The plasmid, pMLB1108, was developed by Michael L. Berman at the National Cancer Institute. It contains *lac* i^q *o*, *p* and *z* genes. The plasmid repressor gene overproduces the *lac* repressor protein, thereby providing the ability to regulate protein biosynthesis by controlling the inducer (i.e., lactose or IPTG) concentration.

The host is a *lac* z^-, y^+ strain, MBM7061 (Berman and Jackson, 1984). The permease activity was obtained through lysogenic incorporation of the transducing phage l p1048 in the *E. coli* host strain MC4100 (Casadaban, 1976). The resulting recombinant strain was tested in small-scale culture and found to possess the desired characteristics. It was used for all subsequent bioreactor studies. Therefore, all experimental results were obtained using this recombinant strain.

GENE EXPRESSION KINETICS

For *plasmid-bearing cells,*

$$\frac{dmRNA}{dt} = K_{+M}FN_pQ\mu - K_{-M}[mRNA] - \mu[mRNA] \tag{7.6}$$

$$\frac{dE}{dt} = K_{+E}[mRNA] - K_{-E}[E] - \mu[E] \tag{7.7}$$

where N_p is gene copy number.

Other equations are used to evaluate the fraction of unbound operator gene, Q, and the binding efficiency, F, of RNA polymerase at the promoter site (Ray et al., 1987). The operator index is:

$$Q = \frac{1 + b_1S_{in}}{1 + b_1S_{in} + b_2[R]_t} = \frac{1 + b_1S_{in}}{1 + b_1S_{in} + b_2} \tag{7.8}$$

The intracellular lactose concentration, S_{in}, is typically a multiple of the lactose level in the extracellular environment. It may be evaluated with the aid of the following equation:

$$S_{in} = \frac{A\, S_F}{B + S_F} \tag{7.9}$$

The catabolic repression index is:

$$F = \frac{K_1}{K_2 + [M]} \tag{7.10}$$

The catabolite modulating factor concentration ([M]) has a negative effect on the RNAP binding efficiency. Although its exact chemical structure and mechanism of action are not yet known, it it responsible for the existence of an optimal dilution rate at which the specific enzyme concentration reaches a maximum. This has been modeled by assuming that a critical growth rate exists such that the concentration of catabolite modulating factor is negligible at growth rates below the critical value, and increases hyperbolically at growth rates above the critical value (Ray, 1985). The catabolite modulating factor is:

$$[M] = \frac{k_m[\mu - \mu_{crit}]}{Y_X[k_{-m} + (\mu - \mu_{crit})]} \qquad [7.11]$$

Thus, the cytoplasmic repressor of *E. coli* cannot independently regulate gene expression. Very recently, Soegaard-Andersen et al. (1991) suggest that CytR forms a bridge between tandem cAMP-CRP complexes, and that cAMP-CRP functions as an adaptor for CytR.

Lastly, the growth rate of cells in the bulk phase may be evaluated using the following growth model (Ray et al., 1987).

$$\mu = \frac{\mu_m S_F}{S_F + K_s[1 + K_i X]} = \frac{\mu_m S_F}{S_F + K_s[1 + K_p(S_0 - S_F)]} \qquad [7.12]$$

7.3 BIOREACTOR STUDIES

The CSTR bioreactor model is a limiting case of a more general immobilized recombinant cell (IMRC) bioreactor model (see Appendix A), and is obtained by setting the liquid fraction (ε) equal to one, and simplifying the component balance equations (Vieth, 1988). The steady state bioreactor simulations are another limiting case, and are obtained by setting the time derivatives equal to zero and solving the component balances to obtain cell, substrate and product concentrations (see Chapter 8).

The parameter values estimated for the following bioreactor simulations were obtained by adjusting them to minimize the sum of squared deviations between the mathematical model and experimental results.

Mathematical modeling results describing CSTR bioreactor dynamics are shown in Figs. 7.8 to 7.10, along with the experimental CSTR bioreactor data. Parameter estimates for the biomass yield factor, (Y_X), and the gene expression parameter, b_2 , were obtained using NONLIN to give the best possible fit between the model and experimental data. The parameter value for μ_{crit} was the dilution rate at which the β-galactosidase specific activity displays its maximum value. The remaining parameter values were taken from previous mathematical modeling studies of wild-type *E. coli*, carried out in this laboratory (Ray, 1985). The parameter values used in the simulation are shown in Table 7.1.

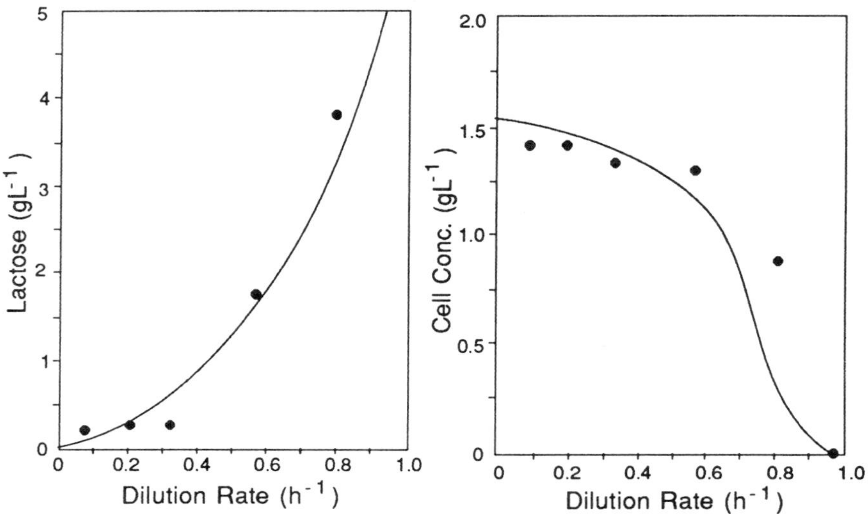

Figure 7.8. Steady state CSTR modeling results: cell and lactose concentration profiles.

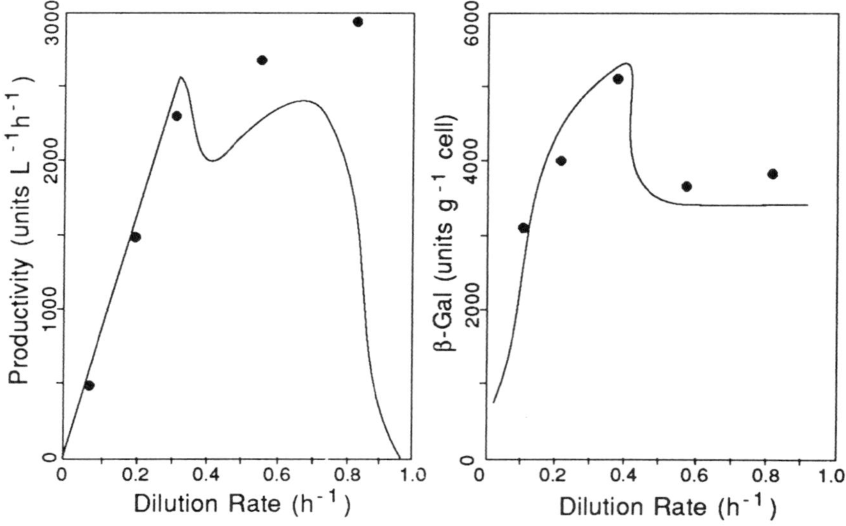

Figure 7.9. Steady state CSTR modeling results: enzyme concentration and productivity profiles.

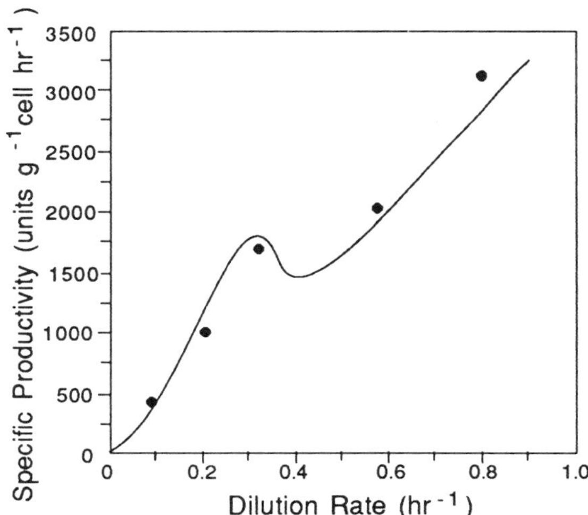

Figure 7.10. Steady state CSTR modeling results: specific productivity profile.

Table 7.1 Parameter Values Used in Steady State Bioreactor Model

Parameters	Values	Parameters	Values
μ_m	0.92 h^{-1}	N_p	7.2 plasmids per cell
μ_{crit}	0.35 h^{-1}	K_{+E}	1000 units mg^{-1} h^{-1}
Y_x	0.32	K_{-E}	0.0 h^{-1}
K_s	0.04 g L^{-1}	b_1	5.0 L g^{-1}
K_p	10.0 L g^{-1}	b_2	245 dimensionless
K_1	4.82 mg L^{-1}	A	60.0 g L^{-1}
K_2	4.89 mg L^{-1}	B	0.40 g L^{-1}
K_{+M}	50.0 mg g^{-1}	k_{+m}	1.48 mg L^{-1}
K_{-M}	27.6 h^{-1}	k_{-m}	0.05 h^{-1}

The fit between the mathematical model and experimental data is good and can be attributed in part to the accuracy of the cell growth and *lac* operon gene expression parameters obtained from previous studies (Ray et al., 1987), and also to the the accuracy with which the plasmid copy number could be estimated from the CSTR stability experiment, shown in Fig. 7.11.

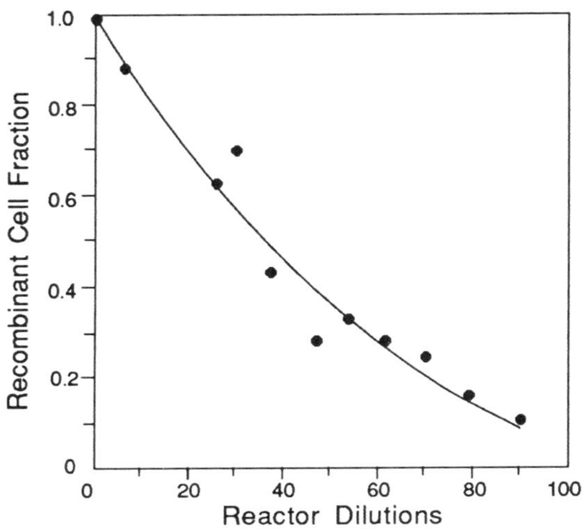

Figure 7.11. CSTR bioreactor stability modeling results.

Regarding stability studies, Porter et al. (1990) likewise call attention to reactor takeover by plasmidless cells in genetically engineered bacteria. They approach the problem by deleting the essential *ssb* gene from the *E. coli* chromosome and placing it on a plasmid. Plasmidless cells do not accumulate even after growing such strains under non-selective continuous culture conditions for extended periods of time.

Returning to this study, estimation of one *lac* operon gene expression parameter value, b_2, was necessary because the plasmid used in this study has a *lac* i^q repressor gene which overproduces the *lac* repressor protein. The additional level of repression is manifested by increasing this dimensionless parameter, which has an estimated value of 245, as compared to a normal *lac i* gene, which has a value of 100.

7.4 α - AMYLASE FROM BACILLUS SUBTILIS

While the *lac* operon is chiefly concerned with the production of intracellular enzymes, α-amylase and its regulatory system, in contrast, constitute a model system for the study of the elaboration of exocellular enzymes. Hybrids are, of course, possible; i.e., the signal sequence that transports and translocates the protein across the cell membrane can be hybridized to foreign proteins. The use of strains of bacilli as a host for production and secretion of foreign proteins can lead to a decrease in purification steps and thus purification costs.

Henner (1990) describes importation of an inducible promoter system from *E. coli* with influence on the *lac* repressor gene. This constitutes a very simple system for placing *B. subtilis* genes under control of an inducible promoter. With respect to native strains of *B. subtilis*, Fujita et al. (1990) describe the gluconate operon, which functions through gluconate kinase and permease. Regarding catabolite repression effects with glucose, as with *E. coli*, the phosphoenolpyruvate:sugar phosphotransferase system (PTS) functions under the control of the *ptsHI* operon in *B. subtilis* (Gonzy-Treboul et al., 1989).

The development of a model to describe the production of α-amylase focuses on transcriptional control of mRNA synthesis, as with the *lac* operon. However, transcriptional control for the extracellular enzymes is not as well understood as in the *lac* operon system. Unlike the production of many enzymes in *E. coli* during the growth phase, extracellular synthesis primarily occurs in the late exponential phase and into the early stationary phase. Coincidentally, Tormo et al. (1990) recently discussed mutations in genes not required for exponential growth but essential for survival in stationary phase, isolating them in an effort to understand the ability of wild-type *E. coli* cells to remain viable during prolonged periods of nutritional deprivation.

TRANSCRIPTIONAL CONTROL

It appears that exoenzyme synthesis is controlled by both catabolite repression and temporal activation, both of which are set at the transcriptional level. Catabolite repression refers to the cessation of inducible or constitutive enzyme synthesis in the presence of glucose or some other rapidly metabolizable carbon source. Catabolite repression is a well understood phenomenon in *E. coli*. To reiterate, it was determined that when catabolite repression is relieved, increased levels of cAMP follow. cAMP binds to a cyclic AMP receptor protein (CRP); the cAMP-CRP complex facilitates the binding of RNA polymerase to the promoter regions of catabolite-sensitive promoters. Bacilli, however, do not contain cAMP or adenylate cyclase (the enzyme involved in cAMP synthesis), although catabolite repression is an important factor in growth and differentiation of bacilli.

As already stated, the mechanics of catabolite repression in *Bacillus* are not wholly understood. However, through mutations of the α-amylase structural gene and its promoter, researchers have developed a plausible scheme. Catabolite repression in *Bacillus* involves the binding of a regulatory protein to a site that is at, or adjacent to, the transcriptional start site. The repression by a regulatory protein is dependent on the presence of a repressing sugar.

The growth rate of *B. subtilis* 168 increases as the carbon source is changed from lactate to glutamate to maltose to glycerol to glucose. The rate of amylase production decreases in the same order (Sekiguchi and Okada, 1972). A similar trend was observed in *B. stearothermophilus* (Welker and Campbell, 1963). As the other carbon sources become easier to metabolize, the rate of carbohydrate conversion to intracellular glucose increases. The higher concentrations of intracellular glucose lead to increased catabolite repression. If growth rate is limited by nitrogen, no effects due to catabolite repression are observed.

Temporal activation refers to the expression of genes as the growth rate of the culture begins to slow down and some initiating signal is produced. Freese et al. (1979) suggested that reductions in GTP or a related metabolite could be the initiating signal for sporulation, as well as exoenzyme synthesis. This was indicated further by the discovery that sporulation of *B. subtilis* can be initiated in the presence of excess glucose, ammonium ions and phosphate by the partial deprivation of guanine nucleotides.

Jeong et al. (1990) measured the changes in concentration of GTP as a function of growth rate in *B. subtilis.* As growth rate decreased the level of GTP decreased. However, at very low growth rates the authors observed an increase in GTP levels.

One difficulty in interpreting the results of continuous culture of *Bacillus* species is the heterogeneity of the culture. Bacilli cultures are comprised of three subpopulations: vegetative cells, sporulating cells and refractile spores. The experimental measurements reflect the average intracellular concentrations of the culture. Based on Dawes and Mandelstam's analysis (1970) the vegetative portions of the culture constitute 100, 90, 70 and 20 percent at dilution rates of .56, .4, .2 and .05 hr^{-1}, respectively.

A direct relationship between levels of GTP, GTP binding protein, sporulation and exoenzyme synthesis does not exist. Further, the evidence suggests that temporal activation can be decoupled from catabolite repression.

Laoide et al. (1989) found that the α-amylase gene, *amy L*, of *B. licheniformis* is subject to catabolite repression and temporal activation. *B. subtilis* which had been transformed with a plasmid containing the *B. licheniformis* amylase gene was grown in a minimal medium with and without glucose. When the culture was grown without glucose, the production of amylase occurred in the late exponential phase of growth. This clearly illustrates the temporal activation of the amylase gene. When the cells were grown in minimal media with glucose, the final amylase activity was ten-fold lower than when the culture was grown under non-repressive conditions. Catabolite repression was apparent.

It was also found, through amino acid deletions, that glucose-mediated repression of a promoterless α-amylase gene did, in fact, occur. The authors suggested that the putative regulatory protein involved in mediating catabolite repression binds to a cis-acting site present on the promoterless amylase fragment, resulting in the attenuation of the *amyL* transcript in the absence of a repressing sugar. (Cis-acting refers to a binding site which is specific to the binding protein; i.e., the protein does not bind randomly to the DNA.)

Nicholson et al. (1987) studied the synthesis of extracellular amylase in *B. subtilis* 168, and its regulation by activation in the early exponential phase, in the absence of induction and catabolite repression, both of which are controlled at the transcriptional level. They concluded that amylase synthesis is activated *de novo* from the P1 promoter of *amy R1* and that glucose mediated repression of amylase synthesis is a consequence of the inability of RNA polymerase-σ^A to initiate transcription from that promoter. They found that the promoter is a consensus form of the vegetative promoter. A mutation 5bp downstream of the *amyE* initiation start site allows for the synthesis of the amylase, despite excess glucose levels. This is further evidence that temporal activation can be separated from glucose-mediated repression.

Stephens et al. (1984) studied the promoter sequences of both *B. licheniformis* and *B. amyloliquefaciens* amylase genes and found their sequences were similar to the consensus sequence for the vegetative holoenzyme, as discussed above. Finally, Gray et al. (1986) studied the genes encoding the thermophilic α-amylase of *B. stearothermophilus* and *B. licheniformis* enzyme. They observed a great deal of homology within the structural genes. Using the evidence presented above, a schematic of the α-amylase genome and its possible regulatory system is presented in Fig. 7.12.

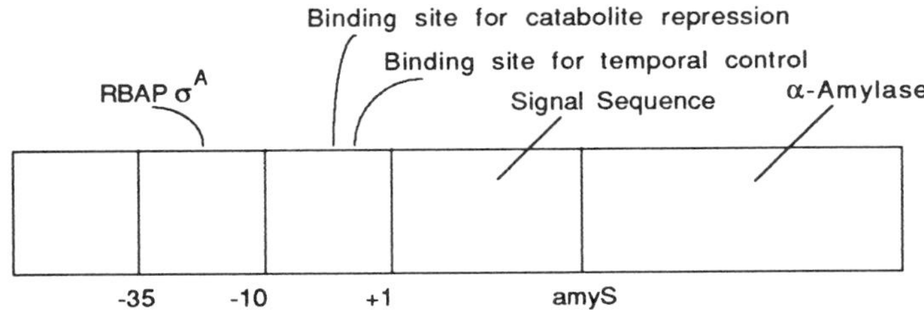

Figure 7.12. Schematic of α-amylase genome and its possible regulatory system.

Taking due account of these facts, through experimentation and analysis (Boyer, 1990; Boyer et al., 1992), a more complete understanding of the metabolic activities during exoenzyme synthesis has resulted. In batch fermentations it was confirmed that α-amylase is produced in the late exponential and stationary phases of growth. Since exoenzyme synthesis is controlled at the transcriptional level by temporal activation (sporulation) and catabolite repression, one is led to the conclusion that enzyme production can be controlled in chemostat cultures by manipulation of dilution rates. Indeed, it was observed that higher rates of α-amylase production occurred at the lower dilution rates, while cell washout took place at a dilution rate near 0.5 h^{-1}. Temporal control, in part, is responsible for the lower production of amylase with respect to increasing dilution rate. As mentioned earlier, as dilution rate decreases the number of cells undergoing sporulation increases; at a dilution rate less than 0.1 hr^{-1}, the percentage of cells sporulating is greater than 80%. Bacteria which are going through the initial stages of sporulation are actively producing enzymes. The change in metabolic activities can thus be attributed to temporal activation, since the growth rate is manipulated solely through variations in dilution rate and not by changes in medium conditions. Yet at the same time, it was demonstrated that the plasmid copy number varies with the growth rate of the culture, leveling out at an upper limit at high dilution rates. Clearly, an opportunity exists for process optimization, involving a trade-off of temporal activation and plasmid replication. Further details are provided in the Appendix A, Section A.1.

Further, it was found that there is an optimum concentration of maltose with respect to α-amylase expression; at very low concentrations, the growth is limited by the lack of readily metabolizable carbon sources, while at high concentrations the effects become more pronounced. This is predicted by a model for the kinetics of α-amylase production incorporating catabolite repression effects (Yoo et al., 1988). It is relevant to point out that maltose is utilized by the bacterium after conversion to glucose by the enzyme α-glucosidase and, hence, the repressive effects of maltose can be attributed to increased intracellular levels of glucose.

7.5 NOMENCLATURE

A' Lactose transport coefficient, nmol (mg protein)$^{-1}$ min^{-1}.

b_1, b_2 Operator index constants, g/L.

B, B' Lactose transport coefficients, nmol (mg protein)$^{-1}$ min^{-1}.

D	Dilution rate, hr^{-1}.
E	Enzyme concentration, g/L.
f	Fraction of total specific binding sites of the lac promoter bound with RNA polymerase to form "open" complex.
F	f/f_{max} , RNAP binding efficiency.
G_i	Intracellular glucose concentration, g/L.
G_o	Bulk glucose concentration, g/L.
k'_a	Equilibrium constant, mg/L.
k'_b	Equilibrium constant, mg/L.
k'_c	Equilibrium constant, mg/L.
k_d	Rate constant, hr^{-1}.
k_m	Rate constant, mg/L.
k_{-m}	Decay constant for catabolite modulator factor, hr^{-1}.
K_{+E}	β-galactosidase rate constant, units/mg/hr.
K_{-E}	Decay constant for enzyme (β-galactosidase), hr^{-1}.
K_i	Cell inhibition constant, L/g.
K_{+M}	mRNA rate constant, mg/g.
K_{-M}	Decay constant for mRNA, hr^{-1}.
K_p	Cell inhibition constant, L/g.
K_S	Monod saturation constant, g/L.
L_{in}	Intracellular lactose concentration, g/L^{-1}.
L_B	Bulk lactose concentration, g/L^{-1}.
M	Intracellular concentration of catabolite modulator factor.
N_p	Plasmid copy number.
Q	Fraction of repressor-free operator.
S	Substrate concentration, g/L.
S_F	Liquid phase substrate concentration in the microenvironment of a cell, g/L.
S_{in}	Intracellular lactose concentration, g/L.
S_o	Inlet substrate concentration, g/L.
t	Time, hr.
X	Cell concentration, g/L.
Y_x	Biomass yield coefficient.
μ	Specific growth rate, hr^{-1}.
μ_m	Maximum specific growth rate, hr^{-1}.

7.6 REFERENCES

Bailey, J. E., M. Hjortso and F. Srienc, *Ann. N. Y. Acad. Sci.*, Biochem. Eng. III, **413**, 71 (1983).

Bailey, K., Ph. D. Thesis, Dept. of Chemical and Biochemical Engineering, Rutgers University (1987).

Berman, M. L. and D. E. Jackson, *J. Bacteriol.*, **159**, 750 (1984).

Betenbaugh, J. and P. Dhurjati, In "Biochemical Engineering VI," *Ann. N. Y. Acad. Sci.*, **589**, 111-120 (1990).

Boyer, J., Ph. D. Thesis, Dept. of Chemical and Biochemical Engineering, Rutgers University (1990).

Boyer, J., W. R. Vieth, K. Bailey and H. Pedersen, *Biotechnol. Adv.*, 25pp. In press (1992).

Casadaban, M. J., *J. Mol. Biol.*, **104**, 541 (1976).

Dawes, I. W. and J. Mandelstam, *J. Bacteriol.*, **103**, 529 (1970).

Dean, A. M., *Genetics*, **123**, 441-54 (1989).

De Crombrugghe, B., S. Busby and H. Bus, *Science*, **224**, 831 (1984).

Deuschle, U., R. A. Hipskind and H. Bujard, *Science*, **248**, 480-483 (1990).

Feldheim, D. A., A. M. Chin, C. T. Nierva, B. U. Feucht, Y. W. Cao, Y. F. Xu, S. L. Sutrina and M. H. Saier, Jr., *J. Bacteriol.*, **172**, 5459-5469 (1990).

Freese, E., J. E. Heinze and E. M. Galliers, *J. Gen. Microbiol.*, **115**, 193-205 (1979).

Fujita, Y., T. Fujita and Y. Miwa, *FEBS Lett.*, **267**, 71-74 (1990).

Gardner, J. A. and K. S. Matthews, *J. Biol. Chem.*, **265**, 21061-21067 (1990).

Gonzy-Treboul, G., M. Zagorec, M. C. Rain-Guion and M. Steinmetz, *Mol. Microbiol.*, **3**, 103-12 (1989).

Gowrishankar, J, *J. Bacteriol.*, **171**, 1923-1931 (1989).

Gray, G. L., S. Mainzer, M. W. Rey, M. H. Lamsa, K. L. Kindle, C. Carmona and C. Requadt, *J. Bacteriol.*, **166**, 635-643 (1986).

Grote, M., Eur. Patent Appl. EP 411501 A1, Optimized Fermentation Processes for the Production of Foreign Proteins in Escherichia coli (1991).

Henner, D. J., *Methods Enzymol.*, **185**, 223-228 (1990).

Kaback, H. R., *Biochim. Biophys. Acta*, **1018**, 160-162 (1990).

Khoury, A. M., H. J. Lee, M. Lillis and P. Lu, *Biochim. Biophys Acta*, **1087**, 55-60 (1990).

Kleina, L. G. and J. H. Miller, *J. Mol. Biol.*, **212**, 295-318 (1990).

Lamark, T., I. Kaasen, M. W. Eshoo, P. Falkenberg, J. McDougall and A. R. Strom, *Mol. Microbiol.*, **5**, 1049-1064 (1991).

Linn, T. and R. St. Pierre, *J. Bacteriol.*, **172**, 1077-1084 (1990).

Martin, I., M. Debarbouille, A. Klier and G. Rapoport, In "Genet. Biotechnol. Bacilli," Proc. Int. Conf. Bacilli, V. M. M. Zukowski, A. T. Ganesan and J. A. Hoch, Eds., 69-79. Academic Press: San Diego, CA (1990).

Martin-Verstraete, I., M. Debarbouille, A. Klier and G. Rapoport, *J. Mol. Biol.*, **214**, 657-671 (1990).

May, G., E. Faatz, J. M. Lucht, M. Haardt, M. Bolliger and E. Bremer, *Mol. Microbiol.*, **3**, 1521-1531 (1989).

Mitchell, W. J., T. P. Misco and S. Roseman, *J. Biol. Chem.*, **257**, 14553 (1982).

Nicholson, W. L., Y. K. Park, T. M. Henkin, M. Won, M. J. Weickert, J. Gaskell and G. H. Chambliss, *J. Mol. Biol.*, **198**, 609-618 (1987).

Oehler, S., E. R. Eismann, H. Kraemer and B. Mueller-Hill, *EMBO J.*, **9**, 973-979 (1990).
Pereira, R. F., Ph. D. Thesis, University of South Carolina (1989).
Peretti, S. W. and J. E. Bailey, *Biotechnol. Bioeng.*, **28**, 1672-1689 (1986).
Porter, R. D., S. Black, S. Pannuri and A. Carlson, *Bio/Technology*, **8**, 47-51 (1990).
Prince, W. S. and M. R. Villarejo, *J. Biol. Chem.*, **265**, 17673-17679 (1990).
Ray, N. G., Ph. D. Thesis, Dept. of Chemical and Biochemical Engineering, Rutgers University (1985).
Ray, N. G., W. R. Vieth and K. Venkatasubramanian, *Biotechnol. Bioeng.*, **29**, 1003 (1987).
Royer, C. A., A. E. Chakerian and K. S. Matthews, *Biochemistry*, **29**, 4959-4966 (1990).
Saier, M. H., J. J. Novotny, D. Comeau-Fuhrman, T. Osumi and J. D. Desai, *J. Bacteriol.*, **155**, 1351 (1983).
Savageau, M. A., In "Theoretical Biology," B. Goodwin and P. Saunders, Eds., pp. 42-66. Edinburgh University Press (1989).
Schnetz, K. and B. Rak, *Proc. Natl. Acad. Sci. U. S. A.*, **87**, 5074-5078 (1990).
Seaver, S. S., *Can. J. Chem. Eng.*, **69**, 403-408 (1991).
Sekiguchi, J. and H. Okada, *J. Ferment. Technol.*, **50**, 801-809 (1972).
Soegaard-Andersen, L., J. Martinussen, N. E. Moellegaard, S. R. Douthwaite and P. Valentin-Hansen, *J. Bacteriol.*, **172**, 5706-5713 (1990).
Soegaard-Andersen, L., H. Pedersen, B. Holst and P. Valentin-Hansen, *Mol. Microbiol.,*, **5**, 969-975 (1991).
Stephens, M., S. A. Ortlepp, J. F. Ollington and D. McConnell, *J. Bacteriol.*, **158**, 369-372 (1984).
Stirling, D. A., C. S. Hulton, L. Waddell, S. F. Park, G. S. Stewart, I. R. Booth and C. F. Higgins, *Mol. Microbiol.*, **3**, 1025-1038 (1989).
Sweet, G., C. Gandor, R. Voegele, N. Wittekindt, J. Beuerle, V. Truniger, E. C. C. Lin and W. Boos, *J. Bacteriol.*, **172**, 424-430 (1990).
Tormo, A., M. Almiron and R. Kolter, *J. Bacteriol.*, **172**, 4339-4347 (1990).
Vieth, W. R., "Membrane Systems: Analysis and Design," pp. 265-297. Hanser Publishers: Munich; Dist. in U. S. by Oxford University Press: New York (1988).
Vieth, W. R., "Biotechnology and Membrane Science," Transl. by Drs. S. Hirose and I. Karube, 290 pp. Asakura Shoten Publishers: Tokyo (1990).
Wang, Y. P., K. Birkenhead, B. Boesten, S. Manian and F. O'Gara, *Gene*, **85**, 135-144 (1989).
Welker, N. F. and L. L. Campbell, *J. Bacteriol.*, **86**, 687-691 (1963).
Widenhorn, K. A., J. M. Somers and W. W. Kay, *J. Bacteriol.*, **171**, 4436-4441 (1989).
Yanofsky, C., V. Horn and P. Gollnick, *J. Bacteriol.*, **173**, 6009-6017 (1991).
Yoo, Y. J., T. W. Cadman, J. Hong and R. T. Hatch, *Biotechnol. Bioeng.*, **31**, 426-432 (1988).

8

IMMOBILIZED RECOMBINANT MICROBIAL CELL CULTURES AND BIOREACTORS

8.0 INTRODUCTION

It is obvious that genetic engineering is having a tremendous impact on the direction of scientific research in both universities and industry; public and private funding of research in molecular biology has increased substantially. This trend is in recognition of the popular view that potentially large gains in protein overproduction reside in the development of superior plasmid-containing cell systems. However, it is imperative that research not be prematurely narrowed at this early stage: superior organisms must of necessity be employed together with optimal and innovative process strategies if genetically engineereed products are to be truly cost-effective in the market place. Therefore, it is surprising that substantially less emphasis has so far been placed upon the efficient employment of recombinant cells in well designed bioreactor systems.

It is rapidly becoming evident that conventional batch and continuous culture techniques will scarcely be adequate to control system variables. As knowledge is gained and the challenges appear more clearly in focus, the responsibilities of the biochemical engineer to provide more sophisticated reactor designs to improve protein productivity, cell viability and plasmid stability become more evident. As an example of the latter, a new plasmid construct was used in conjunction with selective recycle to successfully maintain otherwise unstable plasmid-bearing *Escherichia coli* cells in a continuous bioreactor and to produce significant amounts of the plasmid-encoded protein β-lactamase. The plasmid is constructed so that pilin expression, which leads to bacterial flocculation, is under control of the *tac* operon (Ogden and Davis, 1991). Selective recycle allows for the maintenance of the plasmid-bearing cells by separating flocculent, plasmid-bearing cells from nonflocculent, segregant cells in an inclined settler, and recycling only the plasmid-bearing cells to the reactor. As a result, product expression levels are maintained that are greater than ten-fold the level achieved without selective recycle.

A consensus has emerged and been sustained, identifying bioprocess engineering as a bottleneck for the continued development of biotechnology in the United States (Anon., 1984; Humphrey, 1985). In addition to mounting research in the important area of downstream processing, biochemical engineers, in essence, are being challenged to learn, explore and contribute to recombinant cell processes in two major ways: firstly, to analyze the dynamics of live cell processes to arrive at improved reactor designs and operating strategies, and secondly, to provide meaningful cooperation with molecular biologists to produce improved strains which will facilitate control and optimization of overall processes which employ them.

A report from the EEC (Dixon, 1985) states that the new genetics has overshadowed the old physiology to such a degree that innovation is being held back through want of easily acquired data. The report calls for rigorous analysis of the performance of recombinant organisms under simulated process conditions and concludes with the remark that "advances made in our knowledge of controlled expression of genotypes may well be of critical importance for the industrial application of recombinant organisms."

Recognizing these essential features and research needs, this chapter deals with the study of genetically engineered *E. coli* and *B. subtilis* cells in the immobilized state in continuous bioreactors. *E. coli* has been the choice for a number of industrially important fermentation process developments because the genetics of this host are well studied and several vectors with favorable characteristics have already been developed (Carrier et al., 1983; Muth, 1985; Rodriquez and Tait, 1983). In a study by Shuler and coworkers (Georgiou et al., 1985), continuous production of β-lactamase using an IMRC culture of *E. coli* was maintained for a period of 129 days. The cloned gene in this study contained a signal sequence for excretion of the product which constituted about 40% of the total extracellular protein. Another study demonstrated the ability of an IMRC culture to produce human hormones (Karkare et al., 1986). A genetically engineered strain of *S. cerevisiae*, containing a plasmid for α-human chorionic gonadotropin production, was maintained in an IMRC reactor to 46 days without loss of productivity. A 20-fold increase in the rate of protein production was obtained from the IMRC over that of a batch reactor.

8.1 IMMOBILIZED RECOMBINANT CELLS

Immobilized cells, inside a continuously operating bioreactor, form a relatively permanent population that experiences a different strength of selection than the cell population in the bulk fluid. This effect can be demonstrated neatly in wall-growth studies (Dykhuizen and Hartl, 1983). The cells attached to the wall tend to preserve their initial population distribution, even though the proportions of fast-growing plasmid-bearing and plasmid-free cells in the bulk liquid may change drastically. As cell division occurs, the wall population serves as a constant source of cells characterized by the slower growth rate in contact with the liquid medium. Thus, the slow growing cells can never be reduced to a frequency of occurrence smaller than that provided by the wall population. Two such examples are shown by the solid curves in Fig. 8.1. The ratios of *E. coli* DD323 ($Str^r lac^+$) and DD320 ($Str^s lac^-$) in a maltose-limited chemostat undergo perceptible changes in slope with time after 100 and 140 hours, respectively, indicating a pronounced decrease in the rate of selection (Dykhuizen and Hartl, 1983). Growth rates of immobilized cells, analogous to the cells attached to the bio-reactor walls, are typically one order of magnitude less' than free cell

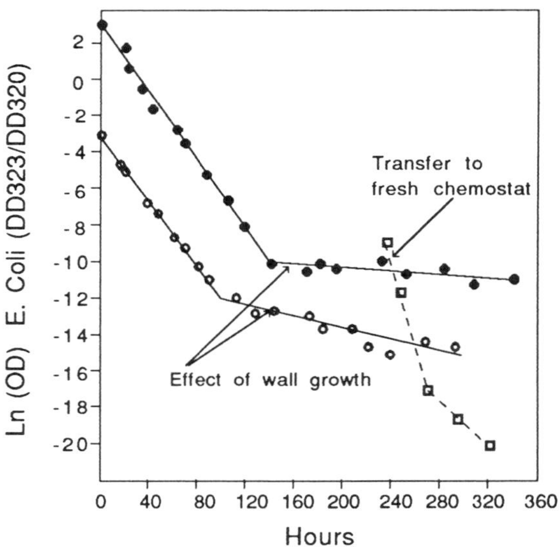

Figure 8.1 *E. coli* self-immobilization.

growth rates. This can improve plasmid stability partly by facilitating plasmid replication, which can go on occurring in the absence of cell division, thus increasing the number of plasmids per cell.

An example of effective plasmid replication is afforded in the work of Chew and Tacon (1990) with *E. coli.* In their work, an expression plasmid in which DNA replication and heterologous gene expression can be simultaneously regulated was constructed to avoid derepression prior to induction. This was achieved by placing a pBR322 origin of replication immediately downstream of an anthranilate synthase-human epidermal growth factor fusion gene (*trpE-hEGF*), both under the control of the promoter from the tryptophan biosynthetic operon. Regulation of plasmid copy number ensured tight repression of the *trp* promoter prior to induction. Upon induction, plasmid copy number increased up to six-fold and the fusion protein accumulated to approximately 12% of total cell protein.

Returning to a consideration of entrapped live cell systems, the most common observation thus far is the accumulation of cells within a small thickness at the outer surface. SEM views show that cells are packed to the maximum possible density in this region, and elsewhere the cell density is reduced dramatically. In the core of such particles, the cell population is very low and close to the distribution at the instant of immobilization. Mean cell concentration in the immobilized state is a function of a number of variables: the major ones being particle size, gelling and filler material concentrations and growth conditions; while minor ones are starting growth state and concentration of cells, immobilizing and incubation conditions. Obviously, an optimal design aims at maximizing the activity expression of entrapped cells, and thereby maximizing volumetric productivity.

Based on the observations made on intraparticle cell distribution, it is considered that live cells form a shell around the cell-free core. Immobilized cells grow to maintain constant cell density in the shell and the net growth results in emergence of cells into the bulk fluid. Bulk fluid values of substrate diffusion coefficients may be expected to be higher than for densely packed cells in calcium alginate gels.

Controlled porosity of the gel particle is of utmost importance when a complex fermentation medium is being used, because the skin layer of the gel surface acts like a semipermeable membrane which allows entry of only low molecular weight compounds (Klein et al., 1983). In this connection, the gel particles of approximately one millimeter in diameter facilitate oxygen transfer to the entrapped cells. For zero order reaction taking place in a spherical particle, the Thiele modulus should be near unity so that nearly 100% effectiveness of the catalyst volume can be achieved. Since oxygen diffusivity in the aquagel is about 3 x 10^{-5} cm^2/sec, while the maximum bulk oxygen concentration in equilibrium with air at normal temperature and pressure is about 0.25 mmol/L and the maximum uptake rate for ten-fold concentrated bacteria is about 100 mmol/L (Enfors and Mattiasson, 1983), the calculated particle radius works out to about 0.5 mm.

8.2 PRODUCTION OF β-GALACTOSIDASE

As described so far, an IMRC reactor housing a genetically engineered microorganism must meet two fundamental requirements:

- increased activity of the target protein;
- lowered susceptibility to plasmid loss or degradation.

For the first requirement, a strong promoter can give up to 25% of the cellular proteins as the product. The second condition can be met by immobilizing the cells and using a plasmid with an inducible promoter, coupled to a hybrid gene fused with the target gene (Carrier et al., 1983; Muth, 1985).

In the work described here, the host was *E. coli* MBM7061 (z^-, y^+), harboring the plasmid pMLB1108 which is *lac* i^q z^+. Thus, while the host strain does contain the gene which codes for the permease (y^+), it is deficient in the *z* gene for expression of β-galactosidase, in the normal position on the chromosome. This deficiency is made up by the plasmid pMLB1108 (see Fig. 8.2), which contains the proximal half of the *lac* operon; i.e., the repressor gene (i^q) and the gene coding for β-galactosidase (z^+). The *q* mutation causes the repressor protein to be overproduced, allowing better control for the switching on and off of protein synthesis in chemostats and in immobilized cells. Thus, one has a very flexible genetic system to study strategies for maintaining plasmids in genetically engineered systems.

The second rather important aspect of this work is the comparative study of plasmid-bearing cells in free suspension cultures and in immobilized states. So-called "plasmid-shedding" occurs as a consequence of natural selection; i.e., competitive growth of wild-type and recombinant strains on the available substrate. If cell division is blocked (by immobilization), differential reproduction cannot take place. Thus, immobilized recombinant cell cultures are expected to be more stable (Dykhuizen and Hartl, 1983; De Taxis du Poët et al., 1986). However, weighing against this advantage is the possibility that immobilized cells may not overproduce the desired metabolite as actively. But that, too, is mitigated by the flexible metabolic switching policies afforded by the inducer/repressor combination which, at least in theory, could be applied at an optimal time in the growing cycle.

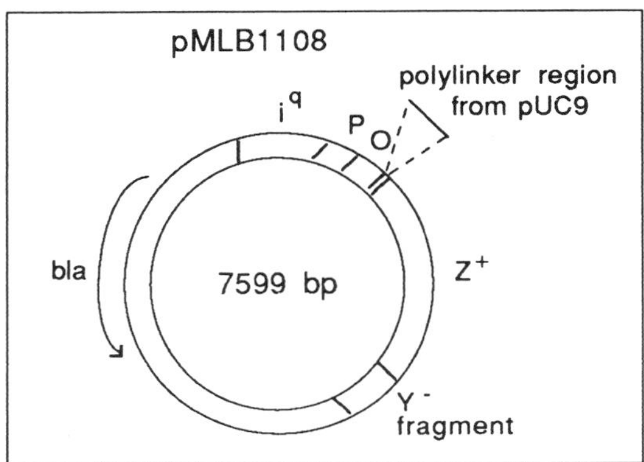

Figure 8.2 Genetic Map of pMLB1108.

8.3 STRAIN DEVELOPMENT

In the development of a recombinant strain to be used in bioreactor studies, it was necessary to have a host possessing the *lac* permease gene so that lactose could be metabolized. The host therefore had to be *lac* z^-, y^+ so that the *lac* z^+ plasmid and host could be easily distinguished when grown on indicator plates. A plasmid/host system was chosen which meets these requirements, and provides for inducible biosynthesis of β-galactosidase. The plasmid, pMLB1108 (shown in Fig. 8.2), was developed by Michael L. Berman at the National Cancer Institute. It contains *lac* i^q *o, p and z* genes. The plasmid repressor gene overproduces the *lac* repressor protein, thereby providing the ability to regulate protein biosynthesis by controlling the inducer (i.e., lactose or IPTG) concentration.

The host is a *lac* z^-, y^+ strain, MBM7061 (Berman and Jackson, 1984). The permease activity was obtained through lysogenic incorporation of the transducing phage λp1048 in the *E. coli* host strain MC4100 (Casadaban, 1976). The resulting recombinant strain was tested in small-scale culture and found to possess the desired characteristics. It was used for all subsequent bioreactor studies. Therefore, all experimental results were obtained using this recombinant strain.

8.4 IMRC BIOREACTORS

The ability of a recombinant culture to maintain high productivity levels for extended time periods is contingent upon several genetic and environmental factors. Although it is relatively easy to control environmental factors in a continuous culture system, it is extremely difficult to obtain complete control over the genetic factors governing gene expression. Plasmid segregational instability problems are common in the operation of continuous fermentation processes. Design and operational strategies for minimizing this problem were initially based on early modeling efforts (Dwivedi et al., 1982; Imanaka, 1983). The use of multi-stage bioreactor systems (Siegel and Ryu, 1985) has been explored as a means of maximizing cell growth and protein biosynthesis, while minimizing plasmid loss.

A continuous IMRC *hybrid* bioreactor is used to study the behavior of the immobilized cells and the cells issuing from the immobilized phase. The ratio of the two kinds of populations can be changed because the immobilized cell contribution becomes negligible as dilution rate is raised to a level which is a multiple of the free cell growth rate. In the latter case, immobilized cell kinetics alone can be isolated if fluid-phase and gel-phase mass transfer resistances are properly controlled and minimized.

Immobilized cells are well-suited for continuous flow operations by offering a ready means of uncoupling the residence time from the cell doubling time. In the same sense, plasmid-harboring cells can be proliferated, despite their growth rate disadvantage, without reaching washout. Some other advantages include: supply rate(s) of any limiting nutrient(s) can be uncoupled from cell growth kinetics; fluid viscosity and foaming can be controlled; toxins can be excluded; and, most importantly, bioreactor design can more readily meet such requirements as: a reliable basis for scale-up; lowered mechanical energy consumption; and, when warranted, sophisticated process control.

An IMRC bioreactor is, usually, a three-phase packed or fluidized bed reaction system, possessing solid, liquid and gas phases. An IMRC bioreactor may be operated in either mode. The disadvantage in using a packed bed reactor is that channeling of air or medium through the bed can occur, leading to stagnant zones in the reactor with very poor mass transfer rates. Most often gas (air) bubbles are used to provide mixing of the solid and liquid phases. However, this type of three-phase system is difficult to scale up which makes it difficult to obtain consistent results among reactor runs, due to differences in fluidization characteristics. Aeration of the fermentation broth in a separate vessel and rapid recycle of the oxygenated liquid to the IMC reactor column provides more gentle and predictable fluidization while obtaining excellent mass transfer (Karkare et al., 1985). As a result, scale-up of the IMC reactor becomes much more predictable.

In a similar vein, *Escherichia coli* K12 with multicopy plasmid (lambda P_R-promoter and temperature-sensitive lambda cI 857 repressor) was cultivated in 60-L bubble column and airlift tower loop reactors. The results evaluated in the 60-L bubble column and airlift tower loop reactors are compared with those evaluated in a 1-L stirred-tank reactor, with regard to scale-up (Kracke-Helm et al., 1991). The medium composition, cell concentration and intracellular enzyme activity were monitored online during batch, fed-batch and continuous cultivations. The specific growth rates, cell mass yield coefficients, plasmid stabilities, productivities of the amount of active fusion protein (β-galactosidase activity), concentrations and yields of acetic acid, and volumetric oxygen transfer coefficients were evaluated for different medium compositions and cultivation conditions. The enzyme activity was also monitored during the temperature induction.

Returning to the case at hand, the bioreactor design (K. Bailey, 1987) actually used for immobilized cell reactor studies is shown schematically in Fig. 8.3. The reactor is a three column system, composed of the IMRC reactor column (shown right), the liquid phase oxygenation column (shown left), and the gas-liquid disengagement column (located in the center). The main feature of this reactor system is that the liquid phase is oxygenated in a separate vessel from the column containing the gel particles. The advantages of this design are:

- The rates of fluidization and oxygenation can be controlled independently.
- The mixing pattern inside the IMRC reactor column is more thorough, yet more gentle due to lower shear forces, thereby reducing the rate of bead attrition.

The total system volume (liquid + gel particles) varied between 650 and 750 mls, depending on the liquid level in the IMRC reactor. The temperature was maintained at 37°C. Approximately 200 mls of gel beads were used in each reactor run. The cell broth was recycled between the IMRC reactor column and the oxygenation unit; a liquid recycle rate of approximately 700 ml/min was maintained.

When using a phosphate buffered medium, it was unnecessary to control the pH in the reactor system. The phosphate salts had an ample buffering capacity. The *E. coli* culture has a pH of approximately 6.8 when grown to stationary phase in a phosphate buffered medium. In contrast, when grown in tris buffered medium, a stationary phase *E. coli* culture has a pH of 4.2. Tris buffered medium had to be used with calcium alginate gel, because high phosphate concentrations dissolve the alginate material, by chelating the calcium ions. It was therefore necessary to control the system pH when using tris buffered media.

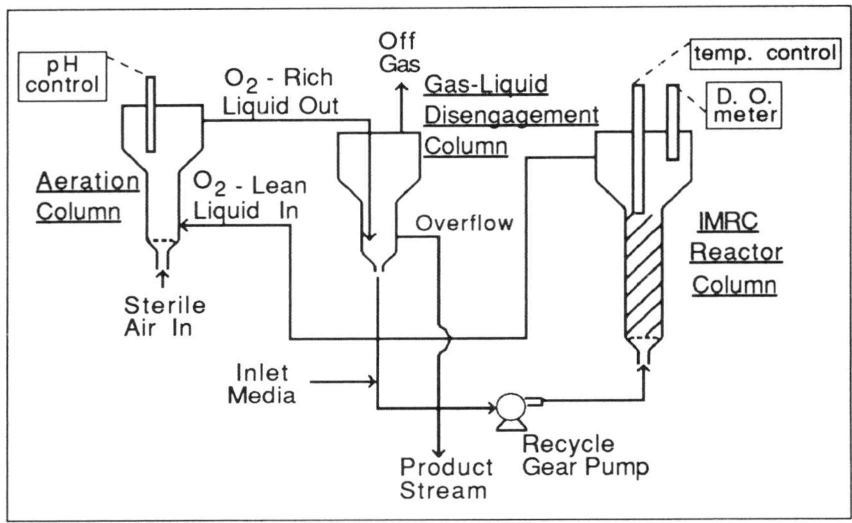

Figure 8.3 IMRC reactor design.

8.5 IMRC BIOREACTOR STUDIES

The IMRC bioreactor system was operated using tris buffered minimal lactose medium as the inlet feed, and calcium alginate as the gel matrix material. Steady state effluent concentrations of cell, lactose and β-galactosidase were obtained as functions of dilution rate, which is based on the total (liquid + gel phase) volume of the three column bioreactor system. A protracted experiment was performed twice, at one week intervals, using the same system. Experimental results are shown in Figs. 8.4 and 8.5, respectively.

These results show that the IMRC bioreactor system can be operated at a dilution rate as high as 1.8 hr^{-1} without incurring the cell washout problem observed in the free cell system at a dilution rate of 0.97 hr^{-1}. The maximum β-galactosidase productivity for a continuous free cell reactor was 2950 units/L/hr with lactose as the carbon source and a dilution rate of 0.8 hr^{-1}. Cell washout occurred at 0.97 hr^{-1}. Fig. 8.5 provides an indication of the rate of β-galactosidase biosynthesis.

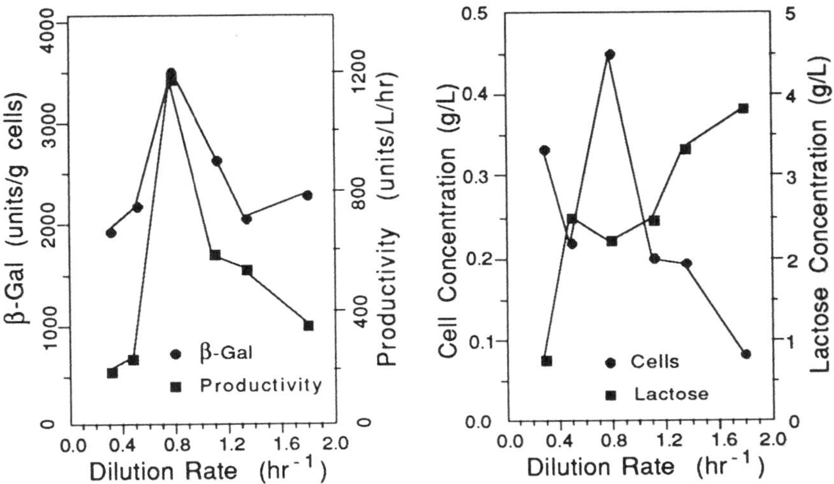

Figure 8.4 Steady state IMRC bioreactor performance: lactose feed.

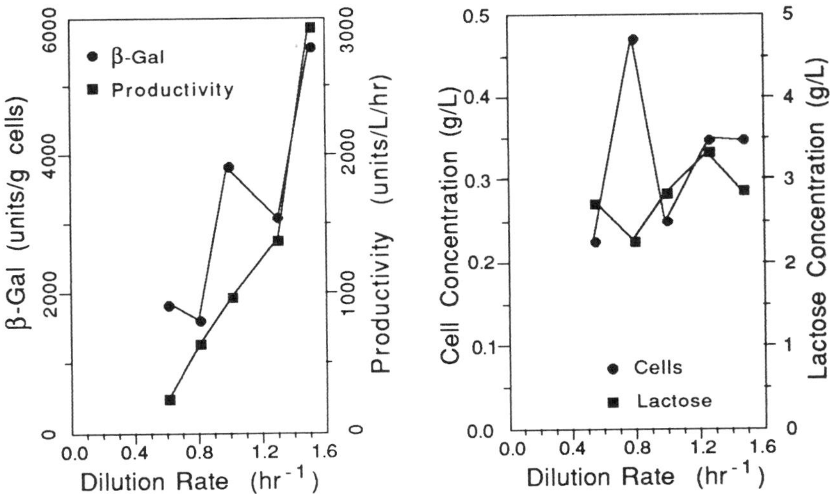

Figure 8.5 Steady state IMRC bioreactor performance: lactose feed.

This rate increases almost linearly with cell growth rate. This type of relationship is expected for production of any protein which is necessary for cell growth. The IMRC bioreactor system did demonstrate that product concentration and volumetric productivities similar to the free cell chemostat system were obtained at the lower dilution rates. Fig. 8.6 indicates that the rate of enzyme biosynthesis in the IMRC is approximately equal to the corresponding rate of enzyme production in the chemostat, under these conditions. Above the maximum specific growth rate there is a two- to three-fold improvement in productivity over the chemostat.

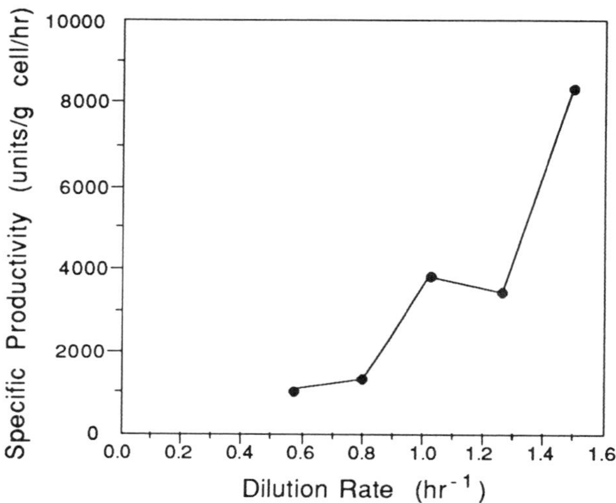

Figure 8.6 Steady state IMRC bioreactor performance: lactose feed, specific productivity data.

8.6 STEADY STATE IMRC BIOREACTOR DYNAMICS WITH SELECTION PRESSURE

When a sufficient quantity of antibiotic is present in the reactor feed, the growth rate and concentration of plasmid-free cells in the bioreactor is negligible. For steady state reactor operation, the IMRC reactor balance equations simplify, allowing for an analytical solution of substrate, recombinant cell and enzyme concentrations as functions of dilution rate. The bulk substrate concentration can be obtained by solving the following quadratic equation.

$$[\nu\delta - \beta] \cdot S_F^2 + [\beta S_0 - \kappa - \nu\delta S_0 - \nu\lambda] \cdot S_F + \kappa \cdot S_0 = 0 \qquad [8.1]$$

where: $$\nu = \frac{\mu_m}{D}$$

$$\beta = 1 - K_s K_p$$

$$\kappa = K_s [1 + K_p S_0]$$

$$\delta = \varepsilon [1 - \theta_F]$$

$$\lambda = \frac{\eta_I [1 - \varepsilon]}{Y_X} \cdot \frac{[1 - \theta_I]}{[1 - \theta_F]} \cdot X_I$$

The substrate concentration is obtained using the quadratic formula; the negative root is the meaningful one.

The free cell growth rate is evaluated from a Monod type model, having also a product inhibition term. (This equation, eqn. [A.16], is also part of the full transient model, described in Appendix A). The free cell concentration is evaluated using the following equation:

$$X_F^+ = Y_X [1 - \theta_F] [S_0 - S_F] \qquad [8.2]$$

The specific activity of β-galactosidase (units/g cells) can be solved directly using the following equation:

$$[E] = \frac{K_{+E} K_{+M} \cdot F \cdot N_p \cdot Q \cdot \mu_F^+}{[K_{-E} + D/\varepsilon] \cdot [K_{-M} + D/\varepsilon]} \qquad [8.3]$$

8.7 ANALYSIS

The IMRC bioreactor model was used to evaluate the experimental results obtained from CSTR and IMRC bioreactor experiments. The model was applied to shorter-term steady state bioreactor kinetics using selection pressure (ampicillin), and longer-term transient bioreactor kinetics, for which stability experiments were conducted using a glucose minimal medium. The CSTR bioreactor model is a limiting case of the more general IMRC bioreactor model, and is obtained by setting the liquid fraction (ε) equal to one, and simplifying the component balance equations (see Appendix A). The steady state bioreactor simulations are another limiting case, and were obtained by setting the time derivatives equal to zero and solving the component balances to obtain cell, substrate and product concentrations.

The parameter values estimated in the following simulations were obtained by adjusting them to minimize the sum of squared deviations between the mathematical model and experimental results. This was accomplished through the use of NONLIN, the nonlinear statistical regression package referred to previously.

Mathematical modeling results describing CSTR bioreactor dynamics have already been shown in Chapter 7, along with the experimental data. Parameter estimates for the biomass yield factor, (Y_x), and the gene expression parameter, b_2, were obtained using NONLIN to give the best possible fit between the model and experimental data. The estimate for plasmid copy number, N_p, was obtained from fitting the transient model to experimental results, shown in Fig. 8.7. The parameter value for μ_{crit} was the dilution rate at which the β-galactosidase specific activity displays its maximum value. The remaining parameter values were taken from previous modeling studies of wild-type *E. coli*, carried out in this laboratory (Ray et al., 1987). The parameter values used in the simulation are shown in Tables 8.1 and 8.2.

Table 8.1 Parameter Values Used in Steady State Bioreactor Model

Parameters	Values	Parameters	Values
μ_m	0.92 hr^{-1}	N_p	7.2 plasmids/cell
μ_{crit}	0.35 hr^{-1}	K_{+E}	1000 units $\cdot mg^{-1} \cdot hr^{-1}$
Y_x	0.32	K_{-E}	0.0 hr^{-1}
K_s	0.04 g/L	b_1	5.0 L/g
K_p	10.0 L/g	b_2	245 dimensionless
K_1	4.82 mg/L	A''	60.0 g/L
K_2	4.89 mg/L	B	0.40 g/L
K_{+M}	50.0 mg/g	k_{+m}	1.48 mg/L
K_{-M}	27.6 hr^{-1}	k_{-m}	0.05 hr^{-1}

The fit between the mathematical model and experimental data is good and can be attributed in part to the accuracy of the cell growth and *lac* operon gene expression parameters obtained from previous studies, and also to the the accuracy with which the plasmid copy number could be estimated from the CSTR stability experiment, shown in Fig. 8.7.

Table 8.2 List of Operational and Estimated Parameter Values Used in Bioreactor Stability Simulations

CSTR		IMRC	
$\bar{\mu}_F = D$	0.67 hr^{-1}	D	1.08 hr^{-1}
ε	1.0	ε	0.71
X_F^+ (t=0)	1.57 g/L	X_I	3.25 g/L
ϕ_F (t=0)	1.0	N_p	7.2 plasmids/cell
α	1.02	X_F^+ (t=0)	0.35 g/L
N_p	7.2 plasmids/cell	$\bar{\mu}_F$	0.98 hr^{-1}
		α	1.10
		η_I	0.22

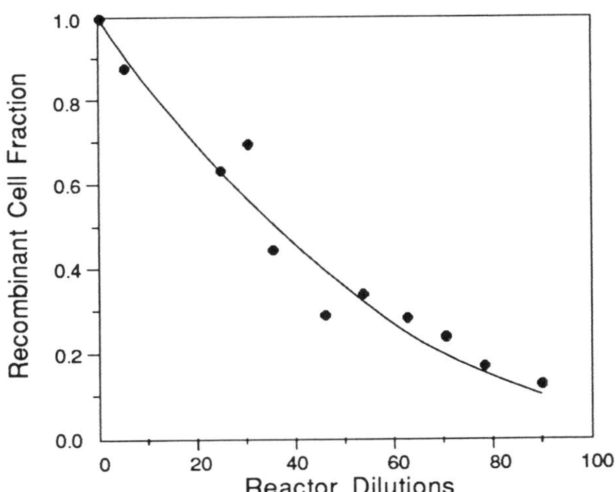

Figure 8.7 CSTR bioreactor stability modeling results.

PLASMID STABILITY

The mathematical modeling of plasmid segregational instability kinetics requires consideration of the cell growth kinetics only. The gene expression kinetics do not enter the calculations, because very little product was produced when the recombinant cells were grown on glucose. The phosphate limitation previously described was obviated by switching to κ-carageenan as the carrier. The focus of this phase of the modeling and experimental study was to analyze the loss of the recombinant cell population from the bioreactor, and the mechanisms involved. There are two main processes which govern the rise in the fraction of wild-type (plasmid-free) cells in the bioreactor. The first is the rate at which recombinant cells divide to give one recombinant and one plasmid-free cell. This process of plasmid partitioning gives an exponential decline in the recombinant cell fraction. The second process involves the ratio of growth rates between the wild-type and recombinant cells, which causes the fraction of recombinant cells to decrease as a result of having a longer doubling time. This growth-rate difference manifests itself more slowly than the plasmid partitioning process, in that it becomes more noticeable as the fraction of wild-type cells increases.

The parameter values used in the simulation, shown in Fig. 8.7, are listed in Table 8.2. The parameters estimated in this simulation are the plasmid copy number (N_p), and the ratio of growth rates between the wild-type and recombinant cells (α). The remaining parameter values are operational constants used in the bioreactor experiment.

The estimated value of the mean plasmid copy number, N_p, was 7.2 copies/cell, which is less than the value of 11 obtained by isolation and quantification of plasmid from shake-flask experimentation. The statistical estimate is likely to be the more accurate of the two methods, due to the inaccuracies of gel electrophoresis and cell plating quantification procedures. The estimated value of α was 1.02, which indicates that the plasmid-free cell growth rate is 2% greater than the recombinant cell growth rate, which is a very small difference. Our results for N_p and α are in good agreement with those of Seo and Bailey (1985) and Shuler (1986). Had a parameter value of 1.0 been used for α, the estimated plasmid copy number would have been 6.5 copies/cell. Therefore, in this experiment, the main event governing the rate of decline in the recombinant cell fraction is unequal plasmid partitioning.

The parameters used in the IMRC bioreactor stability simulation (Fig. 8.8) are shown in Table 8.2. The plasmid copy number estimate from the free cell bioreactor stability experiment was used. The value of η_I was estimated by simulation, using the IMRC bioreactor equations, to obtain the observed free cell concentration at the initial conditions. The specific growth rate of wild-type cells ($\mu_{\bar{F}}$) was obtained by solving the IMRC bioreactor model equations at the dilution rate used in the experiment. The value obtained is the same as the maximum specific growth rate, as a result of the high dilution rate providing ample substrate to the free cells. The ratio of growth rates between wild-type and recombinant cells is very sensitive to the actual growth rate. It has been shown that this ratio increases with decreasing growth rate (Seo and Bailey, 1985). The statistical estimate of α obtained in this simulation was 1.10, which is a significant increase over the value of 1.02 obtained in the free cell bioreactor simulation.

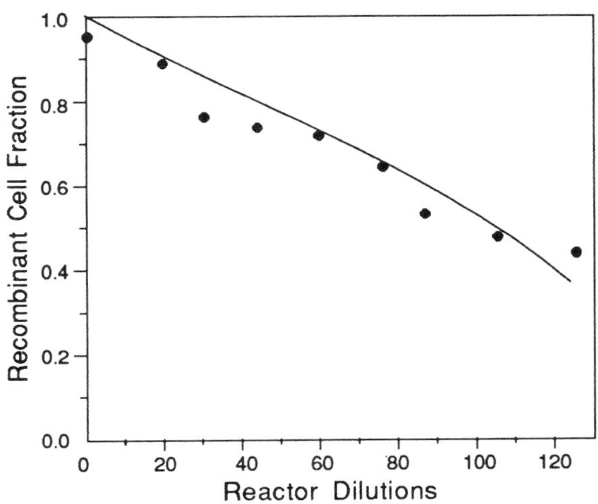

Figure 8.8 IMRC bioreactor stability modeling results.

The difference in growth rates between recombinant and wild-type cells becomes very important in an immobilized cell bioreactor system. Despite the fact that this ratio increases the rate of loss of plasmid-containing cells, there is a marked reduction in the overall rate due to the slower growth rate of immobilized cells. It is this net improvement in the time profile for the recombinant cell fraction that stands out.

In a related study, *Escherichia coli* B/pTG201 was immobilized in carrageenan gel slabs. Spatial distributions of viable cell concentration and plasmid stability were measured by preparing cross-sections of the gel. Results showed a gradient in cell concentration, with a ten-fold greater cell density close to the surface, compared to more distant parts of the gel. Measured cell densities within several millimeters of the surface agreed qualitatively with the profile predicted from a one-dimensional reaction-diffusion model of growth within the gel, although the cell density measured deep within the gel was greater than predicted by the model. Plasmid stability in the immobilized cells declined slightly during the first fifty hours of growth throughout the gel. Following this time, a further decrease in plasmid stability in the top-most layer of the gel was recorded. This latter decrease is apparently due to the growth of cells released from the gel, which eventually form a surface-adsorbed layer of bacteria. The results support the hypothesis that increased stability of gel-immobilized bacteria results from limitation of the number of cell divisions, imposed by the physical constraints of the cavities in the gel (Briasco, et al., 1990).

Table 8.3 Parameter Estimates and Regression Statistics

Parameter	Estimate	σ	95% C.I.
Y_x	0.32	0.04	0.24 - 0.40
N_p	7.2	0.38	6.35 - 8.07
b_2	244.7	22.7	181 - 308
α_{CSTR}	1.02	0.01	0.99 - 1.05
α_{IMRC}	1.10	0.007	1.09 - 1.12

In our work, a total of five parameters were estimated using non-linear regression. The parameter estimates, standard deviations and 95% confidence intervals are shown in Table 8.3. It is apparent from the values of α obtained in each bioreactor system and the confidence intervals associated with each that a clear difference exists in the ratio of wild-type to recombinant cell growth rates between the two bioreactor systems, and this ratio becomes more important in the bioprocess kinetics in the slower-growing immobilized cell system.

8.8 α-AMYLASE FROM B. SUBTILIS

INTRODUCTION

Current research in this phase of our work is similarly focused on the study of recombinant cell kinetics and protein production in both free and immobilized cell bioreactor systems. Building on our previous work on *lac* gene expression, we have chosen a *Bacillus subtilis* strain with a plasmid for constitutive expression of α-amylase. The plasmid, p8D214, contains both the structural gene and the promoter and signal sequences for enzyme secretion. Experimental design and data gathering procedures are integrated with mathematical modeling aspects to uncover a unified picture of gene expression in recombinant cells in prototypical bioreactors, to be of significant utility and applicability.

CATABOLITE REPRESSION, SPORULATION AND α-AMYLASE SYNTHESIS

Quite recently, by mutational analysis, the exact bases on the gene which are crucial for repression have been identified (Weickert et al., 1990). The repressive effects were found to act at the transcriptional level, as measured by the amount of specific mRNA. In another study, the repressive effect of glucose on *amyRi-lacZ* (amylase promoter+*lacZ* +*amyL* from *Bacillus licheniformis*) was investigated in the background of various sporulation mutants and it was observed that the repressive effects were pronounced only with *spo0A* mutants (Laoide and McConnel, 1989). The authors also observed that in the absence of glucose, there was no change in the expression of amylase from the wild-type.

Likewise, the *Bacillus amyloliquefaciens amyF* gene, when expressed in *B. subtilis* in the presence of glucose, was repressed drastically. Interestingly, it was observed that the repressive effect did not occur at the transcriptional level. mRNA analysis suggested that the repression occurred in a post-transcriptional step, most likely secretion. However, the sporulation gene *spo0A* was repressed at the transcriptional level by glucose.

In another recent study (Henkin et al., 1991) aimed at identification of genes whose products are important in glucose repression of α-amylase, a gene designated *ccpA* (catabolite control protein) has been identified. The mutation in this gene was found to result in a loss of glucose repression. *ccpA* was found to share considerable homology with regulatory proteins that are members of the *lac* repressor family, indicating that it is likely to be a DNA-binding protein.

In the same study, the repressive effects of various carbon sources on amylase synthesis has been determined. While glucose repressed 83%, maltose repressed 68% of the α-amylase activity found in wild-type *B. subtilis* grown without sugar. This gives a clearer picture of the effect of maltose as a catabolite repressing sugar in amylase synthesis.

Sporulation initiation has been determined to have an identity of its own, independent of temporal activation, although both have been triggered by similar conditions. However, it has been observed that the mutants of the *spo0A* gene which were unable to phosphorylate were unable to produce α-amylase at the activated levels. One explanation could be that the envirosensory mechanism of the *spo0A* protein through its phosphorylation may be necessary for the temporal activation mechanism. As mentioned earlier, the reduction in GTP levels triggers temporal activation as well as initiation of sporulation, although reduced

GTP levels are unable to overcome the catabolite repression by glucose (Nicholson and Chambliss, 1987).

Temporally coincident with the initiation of sporulation in *spo0A* mutants, at the onset of the stationary phase, *B. subtilis* activates a battery of genes normally silent during vegetative growth. One such gene is the α-amylase gene. Temporal activation causes a manifold increase in the transcription of α-amylase gene and hence an increase in specific mRNA level. Yet it can be shown that the temporal activation mechanism is delinked from the catabolite repression mechanism, which regulates the synthesis of α-amylase (Ramasubramanyan, 1992). Mutations which allow normal temporal activation of amylase synthesis in the presence of a concentration of glucose that causes catabolite repression in the wild-type have been isolated and characterized (Nicholson and Chambliss, 1985). Experiments with these mutants have clearly shown that they are specific for α-amylase - the catabolite repression of sporulation or of the synthesis of extracellular proteases or RNase was not affected by these mutations. Further, three cis-acting alleles of the *amyR* promoter locus, each causing catabolite repression-resistance, were cloned and the functional domains of the *amyR* region responsible for temporal activation and glucose-mediated repression of *amyE* were further defined (Nicholson and Chambliss, 1986). The results of the above studies clearly indicate that the mechanism responsible for activation of *amyE* expression is insensitive to carbon source catabolite repression and that the former is independent of the latter.

Recently, it was shown that the *in vivo* half life of specific mRNA in vegetative cells is about 24 minutes while that of the stationary phase specific mRNA is 10 minutes (Nicholson and Chambliss, 1990). This implies that, contrary to earlier belief that the longer half life of specific mRNA is responsible for increased synthesis of amylase (Tonkova et al., 1989), the increase in the specific mRNA level seems to be important. However, it may be possible that there are differences in mechanisms in different species for the same gene in bacilli.

Finally, in addition to the catabolite repression and temporal activation mechanisms for control of α-amylase synthesis, there is another negative control which is called "temporal turn-off," which occurs at stage II of the sporulation. This essentially causes the halting of the transcription of amylase gene. Inactivation of *spoIIAA* or *spoIIAC* was found to prevent this turn off (Weicker et al., 1990).

MALTOSE UPTAKE

Compared to glucose, maltose is one of the less easily metabolizable carbon sources. Since the carbon source supplied has a major influence on the growth and metabolism of the bacteria, it is relevant to understand the mechanism of maltose uptake and metabolism, and its regulation. In addition, maltose also influences gene expression by catabolite repression.

In nature, one of the sources of maltose is starch. *Bacillus subtilis* produces α-amylase, which degrades starch into maltose, maltotriose, maltotetrose, maltopentose and maltodextrins. Hence, all bacteria which possess α-amylase enzyme have a system for utilization of maltose, etc.

The first step in the utilization of maltose is the transport across the cytoplasmic membrane. Only one specific study has been reported on the maltose regulon of *B. subtilis* (Gonzalez, 1989). It is clear from this study that maltose is a non-PTS sugar in *B. subtilis*. The same has been observed in *Staphylococcus aureus* (Button et al., 1973), another gram positive bacterium. However, there seems to be an indirect involvement of PTS in the uptake and utilization of maltose. When fed maltose, *B. subtilis* with a temperature sensitive *pstI* mutant was neither able to grow nor take up maltose (Gonzalez, 1989). The author suggests that the restrictive temperature might have caused the inactivation of the III^{Glc} enzyme. It has been implicated as being involved in the regulation of glycerol, a non-PTS sugar in *B. subtilis*, when the same phenomenon was observed in that study (Riezer et al., 1984). It has been suggested that the failure of tight *ptsI* mutants to utilize non-PTS sugars can reflect the operation of the PTS mediated control mechanism (Riezer et al., 1988). Poor utilization of maltose cannot be explained exclusively on the basis of deficient glucose utilization, since exogenous glucose is utilized more rapidly than maltose. Hence, the authors conclude that PTS mediated regulation of uptake occurs for some non-PTS sugars in *B. subtilis*.

Gonzalez (1989) found that uptake of maltose was induced by maltose and the induced cells accumulated maltose more effectively than uninduced cells. However, the latter did accumulate detectable amounts of maltose. This can be explained by the fact that the α-glucosidase activity is induced by maltose, which in turn induces the uptake of maltose into the cell. In the same study, it was observed that when cells were treated with CCCP (an inhibitor of active transport by PTS) and an inhibitor of transport by pH gradients, the latter reduced uptake, while the former had no effect. Hence it was concluded that maltose uptake did not involve substrate level phosphorylation and hence PTS. It is suggested that pH gradients play a role in transport. In another study with gram positive *Staphylococcus aureus* cells, the mechanism of transport was determined to be facilitated diffusion (Button et al., 1973).

MALTOSE METABOLISM

The enzymes that degrade polysaccharides made of units of glucose linked by an α-(1,4) linkage (α-glucosides) are known as α-glucosidases. The α-glucosidases with the highest activity toward maltose are known as maltases and those with the highest activity toward aryl-α-glucoside (pNPG) are called by the generic name α-glucosidases. The α-glucosidases can be extracellular or intracellular, varying with the species and strain; e.g., in *B. subtilis* P-11 the enzyme is extracellular, while in *B. subtilis* 168 the enzyme is intracellular.

Those enzymes specifically involved in the metabolism of maltose and maltodextrin are called amylomaltases and maltodextrin phosphorylases, respectively. In *Escherichia coli* and *Staphylococcus pneumoniae* amylodextrin phosphorylase is the main enzyme involved in maltose metabolism. However, in *B. subtilis*, amylodextrin phosphorylase activity is absent, suggesting that maltose is metabolized through a different pathway. In addition it was observed that none of the α-glucosidase activities found in *B. subtilis* play a central role in the metabolism of maltose. This suggests that maltose is converted into glucose rather than glucose phosphate in *B. subtilis*.

Maltose induces α-glucosidase activity in *B. subtilis* and the affinity for pNPG has been found to be greater than for maltose. A low inducer concentration was sufficient to fully induce MTase activity, and full induction was observed within ten minutes of maltose addition.

Glucose not only represses the induction of the maltose uptake system, but it also inhibits the uptake of maltose by already induced cells by 90%. Experiments were conducted to delineate the effects of catabolite repression and inducer exclusion by glucose. When glucose resistant α-glucosidase mutants (*cdh3*) were grown on maltose and glucose, the cells were found to synthesize α-glucosidase, but the maltose uptake was inhibited by over 90% in these cells. It was also observed that glucose inhibited α-glucosidase activity to below uninduced levels in normal cells grown in maltose+glucose. From these results it is clear that both catabolite repression as well as inducer exclusion takes place, and that they are independent of each other.

Finally, the α-glucosidase activity is repressed to a lesser extent than glucose by other sugars. When the cells are placed in a minimal medium with maltose as the sole carbon source, malate and glycerol supress the synthesis of maltose inducible enzymes, indicating that maltose is not a favored substrate for growth in *Bacillus subtilis*.

IMMOBILIZED CELLS

In contrast to β-galactosidase production in *E. coli*, amylase production is a typical nongrowth-associated fermentation, i.e., α-amylase is actively produced when cell growth is repressed by a shortage of nutrient. Shinmyo et al. (1982) studied *B. amyloliquefaciens* immobilized in κ-carageenan. The entrapped cells displayed suppressed respiratory and growth activities but normal to higher rates of α-amylase synthesis. After the exponential phase, cell growth and nucleic acid synthesis are repressed but protein synthesis is still potentially active. Proteins relating to stable or accumulated mRNA, such as α-amylase, are preferentially expressed. Shinmyo et al. found that the immobilized cells displayed both decreased respiratory rate and growth rate to levels of 1/2 and 1/6, respectively. The authors attribute changes in metabolism and composition to shifts in DNA replication and cell wall synthesis.

When shifting the focus to an organism which naturally secretes protein products (e.g., exoenzymes), *B. subtilis* is an excellent choice. Virtually all of the 48 species secrete a variety of soluble extracellular enzymes. Secretion of the desired products obviates purification steps: cells are retained within the IMC matrix while the product is eluted with the effluent stream. Also many of the enzymes are secreted while in the late exponential and entering into the stationary phase of growth. This also enhances the prospect of using an IMRC for *B. subtilis*. Simple nutrient media might be applied as metabolic switches for the reactor after a rich growth medium has allowed a dense population to develop. The nutrient concentration and dilution rate of influent can be adjusted so that maximum production of enzyme occurs, while physical retention by immobilization prevents cell washout. Thus, immobilization of *B. subtilis* has prospects to optimally utilize the advantages of the IMC bioreactor.

It was to be expected that experimentation would probably establish an optimum with respect to the maltose concentration and uptake rates. This optimum would be representative of the condition in which the temporal activation can just outdo the repression. Therefore, the use of maltose as a carbon source is attractive because of its lesser repressive effect on amylase, while still being adequately metabolizable. Hence, the use of maltose may obviate the need for a two-stage process; growth and production phases might be combined in a single stage.

IMMOBILIZED RECOMBINANT CELLS

Simultaneous with the free cell work in our laboratory, immobilized recombinant cell (IMRC) bioreactor systems have been developed.

An improved technique to immobilize the bacterial cells was devised, using barium alginate or calcium alginate as the immobilization medium. The methods provided a gel bead that can be maintained for long periods of time without appreciable dissolution or attrition (Kawakami, 1990). In one study, *B. subtilis* cells were immobilized and cultured in a bioreactor which is an improved version of one described above for immobilized recombinant *E. coli*. IMRC operation displayed increased plasmid stability retention, as determined by plating methods, compared to the continuous free cell experiments discussed above. A doubling of the effective continuous production operational time profile was achieved for the bioreactor. Likewise, with increased plasmid stabilities it is possible to delete the antibiotic while maintaining the plasmid in the *B. subtilis*.

Genetically engineered yeast cells which are able to survive at high ethanol concentrations have been introduced (Yocum et al., 1984). Two kinds of transformant strains were developed to convert lactose into ethanol by inserting plasmid pSH096 and pSH206 into a yeast integrating vector and *S. cerevisiae* DBY745, respectively. In a study in our laboratory parallel to the *B. subtilis* work (Jeong et al., 1991), these recombinant yeast strains were employed. A novel concept of membrane bioreactor in which a thin layer or coating of living cells is sandwiched between ultrafiltration (UF) and reverse osmosis (RO) membranes was applied.

Plasmid pSH206 differs from pSH096 primarily in that pSH206 is capable of autonomously replicating in constitutively high copy number in yeast cells. However, replicating plasmids (multicopy plasmids) sometimes have the disadvantage of being unstable, so they may be more readily lost from a population of cells in the absence of specific selection for the plasmid. Two recombinant yeast strains, called Lactophile 13B (13B) and Lactophile 13D (13D) were developed. 13B contains an integrated plasmid, and 13D contains an autonomously replicating plasmid.

The productivity of the Lactophile 13B strains was higher than that of the Lactophile 13D strains. In both cases performance data similar to those for glucose fermentation to ethanol by *Saccharomyces cerevisiae* were obtained. However, the operational stability of recombinant yeast cells was improved in the new bioreactor, in comparison to the stability of these cells in a shake flask.

Throughout this research, we have kept our focus on the goals stated in section 8.2:

- increased activity of the target protein;
- lowered susceptibility to plasmid loss or degradation.

On the basis of the preponderance of the evidence to date, we next decided to develop a coating process for recombinant cells immobilized in calcium alginate to be applied to polysulfone hollow fiber modules. This work is described in the section which follows. Coincidentally, Swope and Flickinger (1992) briefly describe a model system for investigation of enzyme activity in films of *E. coli* in order to increase the specific activity of immobilized whole cell biocatalysts. Thin films of viable cells are immobilized onto nonporous solid supports using permeable coatings, resulting in a catalyst carrier with minimal mass transfer limitations for substrate, product and oxygen.

8.9 COATED HOLLOW FIBERS

The coating method used in our laboratory involves a modification of a hollow fiber system (e.g., Inloes et al., 1983). There, actively growing cells are immobilized within the macroporous matrix of assymetric hollow fiber wall structures; high cell density is achieved within the free pore volume. In contrast, in this study (Suazo, 1992) cells are immobilized in a thin layer of calcium alginate of 80-100 μm thickness. The thin layer approach was devised in an attempt to take into consideration existing research results; e.g., those of de Taxis du Poët et al. (1986) and Shinmyo et al. (1982), who found that recombinant bacteria cultured in gel beads concentrate their growth in the outer 50 μm of the particle. This selective growth is attributed to the lack of oxygen in the interior of the beads.

After sterilizing the hollow fiber cartridge with a solution of 4% w/w sodium hypochlorite and 0.1N NaOH over a four hour period, the shell space was partially filled with an alginate/cells mixture, while a 0.05 M calcium chloride solution was passed through the lumen. As the precipitant, calcium ion, diffused across the microporous membrane, a thin layer of alginate, containing cells, was formed over the outer membrane surface. Afterwards, the excess of alginate was drained from the shell side. The thin film of alginate was hardened by contact with residual calcium ion. All these manipulations were conducted under aseptic conditions.

BIOREACTOR EQUIPMENT

The experimental set up shown in Fig. 8.9 was operated in two modes, depending on the type of immobilized cell system to be tested; i.e., beads (fluidized bed recombinant immobilized cells, or FBRIC) or coated hollow fibers (HFRIC). The reaction conditions were the same in both cases: pH was set at 7.0, temperature at 37°C and dissolved oxygen concentration was held above 30% of saturation. The liquid recycle rate was set at 400 ml/min for the FBRIC and 350 ml/min for the HFRIC; the liquid volume of the system was 900 ml. The main difference between the two bioreactor setups was that in the case of the fluidized bed reactor, 100 ml of beads were used in the reactor column, while in the other system the column was used as a heat exchanger for preheating the liquid to 37°C before entering the hollow fiber reactor. An anisotropic polysulfone hollow fiber membrane furnished by Amicon, Inc. (model MP01-43) was used in the fabrication of the HFRIC.

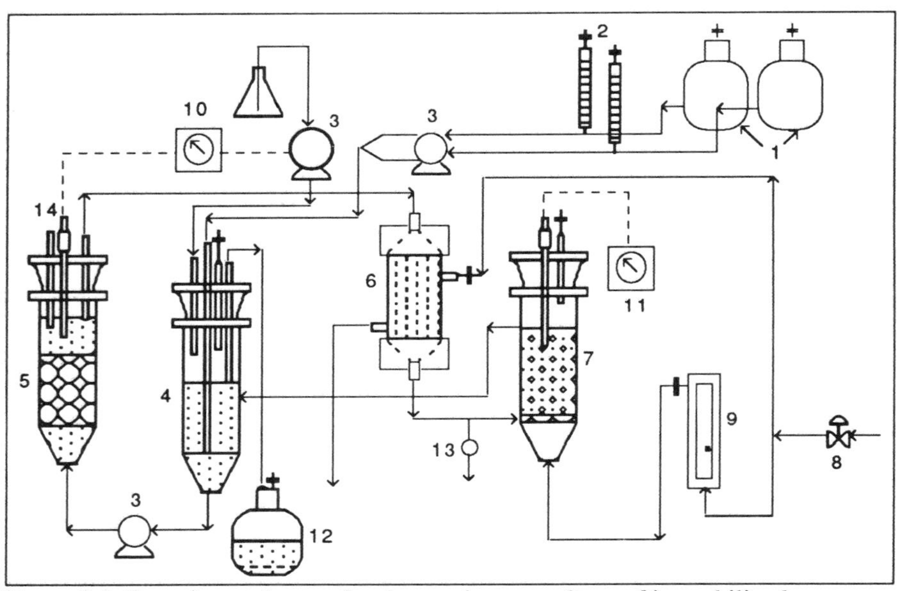

Figure 8.9 Experimental setup for the continuous culture of immobilized recombinant *Bacillus subtilis* cells. 1) Medium reservoir; 2) Liquid flow meter; 3)Peristaltic pump; 4) Disengagement column; 5) FBRIC; 6) HFRIC; 7) Aeration column; 8) Needle valve; 9) Rotameter; 10) pH controller; 11) D.O. probe; 12) Effluent reservoir; 13) Sampling port; 14) pH electrode.

BIOREACTOR PERFORMANCE: FBRIC

Fig. 8.10 shows the results obtained at a dilution rate of 0.50 h^{-1} using a nonselective medium with the FBRIC reactor. Cell mass, maltose and amylase concentrations in the effluent were monitored. In all such experiments, amylase concentration displayed transient behavior due to plasmid loss by the recombinant microorganism, and a steady state was never attained. However, even when the cell mass was declining, amylase was released as a consequence of the lytic process.

Thus, the apparent steady state displayed by the maltose and cell concentrations was not mirrored by the amylase activity. After increasing during approximately ten hours, the amylase activity began to decline to reach a final value lower than the original.

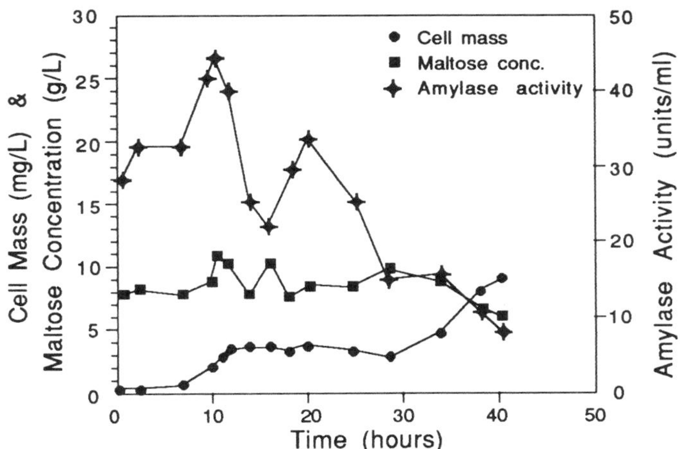

Figure 8.10 Combined plot of the variation of cell mass, maltose concentration and amylase activity with time. Immobilization on beads; without antibiotic pressure; dilution rate 0.50 hr^{-1}.

BIOREACTOR PERFORMANCE: HFRIC

Taking advantage of the characteristics of the HFRIC system, an experiment was performed in order to evaluate a strategy of operation for obtaining higher productivities of the cloned gene protein. Operating under selective pressure with 50 μg/ml of chloramphenicol in the feed medium, the immobilized cells were multiplied for fifteen hours at a dilution rate of 0.43 h^{-1} (close to the washout condition for the free cells as pointed out earlier, in Section 7.4). Afterwards, sterilized $CaCl_2$ solution (0.008 M) was pumped through the equipment for three hours, in order to wash out the antibiotic and to harden the alginate layer. Lastly, feeding was continued with medium without antibiotic so as to maintain the same dilution rate as above. Figure 8.11 shows the results obtained.

It can be seen that during the time interval that the reactor operated with medium containing chloramphenicol, little amylase was detected in the effluent; possibly the carbon source being used by the microorganism for growth exerted repressive effects. Once the antibiotic was removed from the system and feeding resumed with maltose-containing medium, the amylase activity increased rapidly from 0.2 to 28 units/ml, followed by a decline to an apparently stable value of about ten units/ml. (Unfortunately, it was not possible to continue the experiment beyond about two days because of operational problems; it would, of course, be interesting to know the capacity for protein synthesis of the strain for longer periods of time.)

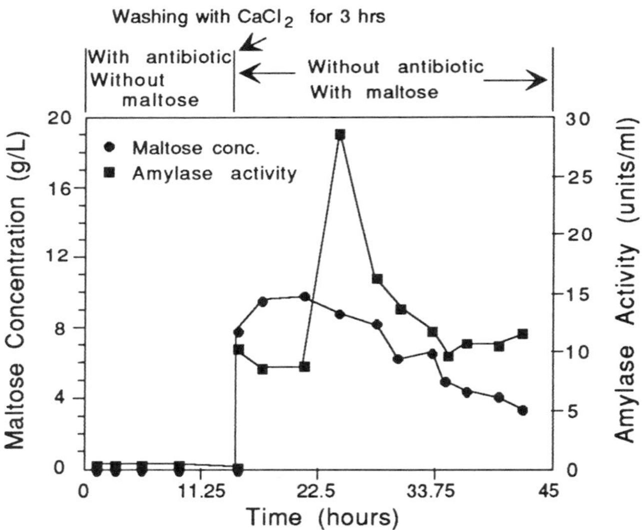

Figure 8.11 Combined plot of the variation of maltose concentration and amylase activity with time. Immobilization on alginate coated hollow fibers; inoculum propagated with antibiotic pressure 950 μg/ml chloramphenicol); dilution rate 0.43 hr^{-1}.

BIOSYNTHETIC PRODUCTIVITY

The ratio of the loading factors (g cells/liter fluid volume) for the beads and coated fibers works out to a value of 15.7 in favor of the beads. Yet, the peak α-amylase activity which was attained for the FBRIC was only about 1.5-fold higher than that for the HFRIC. Therefore, it is evident that the cells in the coating were nearly an order magnitude more productive as catalysts for biosynthesis than those in the beads. Thus, the bioreactor arrangement employed in this work was advantageous for fuller expression of the loading factor.

8.10 NOMENCLATURE (Additional to Chap. 7 List)

SUBSCRIPTS

F Free (liquid) phase.
I Immobilized phase.

SUPERSCRIPTS

+ Recombinant (plasmid-containing).
- Wild-type (plasmid-free).

GREEK SYMBOLS

ε Liquid fraction.
η_I Ratio of IMC to free cell growth rate $[\mu_I/\mu_F]$.
θ Probability of obtaining plasmid-free segregant.

8.11 REFERENCES

Anonymous, Office of Technol. Assess. Rep., "Commercial Biotechnology: An International Analysis," U. S. Govnmt. Printing Office (1984).

Bailey, K., Ph. D. Thesis, Dept. of Chemical and Biochemical Engineering, Rutgers University (1987).

Berman, M. L. and D. E. Jackson, *J. Bacteriol.*, **159**, 750 (1984).

Briasco, C. A., J. -N. Barbotin and D. Thomas, In "Physiology of Immobilized Cells," J. A. M. deBont, J. Visser, B. Mattiasson and J. Tramper, Eds. Elsevier Science Publishers: Amsterdam (1990).

Button, D. K., B. J. Egan, W. Hengstenberg and M. L. Morse, *Biochem. Biophys. Res. Comm.*, **52**, 850-855 (1973).

Carrier, M. J., M. E. Nugent, W. C. A. Tacon and S. B. Primrose, *Trends in Biotechnology* (1983).

Casadaban, M. J., *J. Mol. Biol.*, **104**, 541 (1976).

Chew, L. C. and W. C. Tacon, *J. Biotechnol.* **13**, 47-60 (1990).

De Taxis du Poët, P., P. Dhulster, J. N. Barbotin and D. Thomas, *J. Bacteriol.*, **165**, 871-877 (1986).

Dixon, B., *Biotechnology*, **3**, (1985)

Dwivedi, C. B., T. Imanaka and S. Aiba, *Biotech. Bioeng.*, **24**, 1465 (1982).

Dykhuizen, D. E. and D. L. Hartl, *Microbiol. Rev.*, **47**, 150 (1983).

Enfors, S. O. and B. Mattiasson, In "Immobilized Cells and Organelles," B. Mattiasson, Ed. CRC Press: Boca Raton, FL (1983).

Georgiou, G., J. J. Chalmers, M. L. Shuler and D. B. Wilson, *Biotechnol. Prog.*, **1**, 75-79 (1985).

Gonzalez, D. S., Ph. D. Thesis, University of Wisconsin, Madison (1989).

Henkin, T. M., F. J. Grundy, W. L. Nicholson and G. H. Chambliss, *Mol. Microbiol.*, **5**, 575-584 (1991).

Humphrey, A. E., Personal communication (1985).
Imanaka, T., *Biotechnol. Adv.*, **1**, 279-288 (1983).
Inloes, D. S., J. W. Smith, D. P. Taylor, S. N. Cohen, A. S. Michaels and C. R. Robertson, *Biotechnol. Bioeng.*, **25**, 2653-2681 (1983).
Jeong, Y. S., W. R. Vieth and T. Matsuura, *Biotechnol. Bioeng.*, **37**, 587-590 (1991).
Karkare, S. B., R. C. Dean and K. Venkatasubramanian, *Biotechnology*, **3**, 247 (1985).
Karkare, S. B., D. H. Burke, R. C. Dean, J. Lemontt, P. Souw and K. Venkatasubramanian, *Ann. N. Y. Acad. Sci.*, Biochem. Eng. IV, **469**, 91-96 (1986).
Kawakami, M., Biochem. Eng. Lab. Report, Rutgers Univ., March (1990).
Klein J., J. Stock and K. D. Vorlop, *Appl. Micro. Biotech.*, **18**, (1983).
Kracke-Helm, H. A., U. Rinas, B. Hitzmann and K. Schuegerl, *Enzyme Microb. Technol.*, **13**, 554-564 (1991).
Laoide, B. M. and D. J. McConnel, *J. Bacteriol.*, **171**, 2443-2450 (1989).
Muth, W. L., In "Biotechnology: Applications and Research," P. N. Cheremisinoff, R. P. Ouellette, Eds. Technomic Publishing Co., Inc.: Lancaster (1985).
Nicholson, W. L. and G. H. Chambliss, *J. Bacteriol.*, **161**, 875-881 (1985).
Nicholson, W. L. and G. H. Chambliss, *J. Bacteriol.*, **165**, 663-670 (1986).
Nicholson, W. L. and G. H. Chambliss, *J. Bacteriol.*, **169**, 5867-5869 (1987).
Nicholson, W. L. and G. H. Chambliss, In "Genetics and Biotechnology of Bacilli," M. M. Zukowski, A. T. Ganesan and J. A. Hoch, Eds., 237-257. Academic Press: San Diego, CA (1990).
Ogden, K. L. and R. H. Davis, *Biotechnol. Bioeng.*, **37**, 325-33 (1991).
Ramasubramanyan, N., Biochem. Eng. Lab. Report, Rutgers Univ., March (1992).
Ray, N. G., W. R. Vieth and K. Venkatasubramanian, *Biotechnol. Bioeng.*, **29**, 1003 (1987).
Riezer, J., M. J. Novotny, I. Stuiver and M. H. J. Saier, *J. Bacteriol.*, **159**, 243-250 (1984).
Riezer, J., M. H. J. Saier, J. Deutscher, F. Grenier, J. Thompson and W. Hengstenberg, *CRC Crit. Rev. in Microbiol.*, **15**, 297-338 (1988).
Rodriquez, R. L. and R. C. Tait, "Recombinant DNA Techniques." Addison-Welsey Publishing Co. (1983).
Seo, J. H. and J. E. Bailey, *Biotechnol. Bioeng.*, **27**, 156 (1985).
Shinmyo, A., H. Kimura and H. Okada, *Eur. J. Appl. Micro. Biotech.*, **14**, 7 (1982).
Shuler, M., *Biochem. Eng.V Abstracts*, Engineering Foundation Conferences: New York (1986).
Siegel, R. and D. D. Y. Ryu, *Biotechnol. Bioeng.*, **27**, 28-33 (1985).
Suazo, C. T., Biochem. Eng. Lab. Report, Rutgers Univ., January (1992).
Swope, K. L. and M. C. Flickinger, Abstract, Biochemical Technology Div., American Chemical Society Mtg., San Francisco, CA, April (1992).
Tonkova, A., J. Pazlarova, E. Emanuilova and N. Stoeva, *J. Basic. Microbiol.*, **29**, 55-60 (1989).

Weicker, M. J., L. Larson, W. L. Nicholson and G. H. Chambliss, In "Genetics and Biotechnology of Bacilli," M. M. Zukowski, A. T. Ganesan and J. A. Hoch, Eds., 237-257. Academaic Press, Inc.: San Diego, CA (1990).
Yocum, R. R., S. Hanley, R. West, Jr. and M. Ptashne, *Mol. and Cellular Biol.*, **4**, 1985-1998 (1984).

9

GENE EXPRESSION WITH ANIMAL CELLS

9.0 ANIMAL CELL CULTURES

DECISIVE FACTORS

There are several overriding factors which influence the decision to seek animal cell expression systems as alternatives to bacterial cells. In many cases, heterologous protein production in bacteria can be driven to such high levels that the desired product is found in precipitate form (inclusion bodies) in the cytoplasm of the bacteria. These aggregates can only be solubilized by harsh treatments and this often results in irreversible denaturation. In contrast, many cell types, such as those derived from multicellular organisms as well as some microbial strains, have specialized pathways for protein secretion (Silhavy, 1985).

Furthermore, many medically important products (see Table 9.1) must be modified extensively, principally through glycosylation and acylation, after their synthesis in a eucaryotic cell. *E. coli* does not have the cellular machinery required to accomplish these modifications. In cases where they are essential, this presents a major stumbling block (Silhavy, 1985). Alternatively, animal cells are chosen so that desired post-translational modifications can be achieved through direct manipulation of cell lines, genes and process conditions. This also means, however, a higher degree of complexity, such as special nutrient requirements and delicate handling.

Protein folding itself is an extremely important and intricate subject at the present time. King (1989) describes the sequence whereby a signal peptide guides cellular synthesis of proteins for export. Synthesis and export must be carefully coordinated. A signal peptide, which will not be part of the final protein, forms at the beginning of the polypeptide chain. When the chain itself is long enough to extend beyond the ribosome, protein synthesis is temporarily arrested by the formation of a courier particle/ribosome complex which moves to the cell membrane. There the ribosome binds to a membrane receptor, releasing the signal recognition particle. Peptide synthesis by the ribosome is then resumed, allowing the growing peptide chain to pass through the membrane. Once outside, the signal peptide is cleaved from the chain and the growing polypeptide folds into its correct configuration outside the cell.

Table 9.1 Some Therapeutic Proteins and their Market Size in the 1990s (after Chee, 1987)

Product	Application	Projected Yearly Sales ($Million)
Hepatitis B Vaccine	Prevention of hepatitis & liver cancer	100.0
Interferon, Alpha	Treatment of cancer & virus infections	100.0
Interferon, Gamma	Treatment of cancer & rheumatoid arthritis	100.0
Erythropoietin (EPO)	A major hormone that stimulates red blood cell production & could replace transfusions for dialysis patients	200.0
Factor VII	Treatment of hemophiliacs	200.0
Human Growth Hormone (HGH)	Treatment of growth disorders in children & obesity	200.0
Insulin	Treatment of diabetes	400.0
Interleukin-2 (I l-2	Treatment of human immune deficiencies, AIDS, childhood diseases, & aging	500.0
Tumor Necrosis Factor (TNF)	Treatment of cancer	500.0
Superoxide Dismutase	Treatment of tissue damage from heart attack, stroke, organ transplant, vascular surgery, & kidney disease	1,000.0
Tissue Plasminogen Activator (TPA)	Treatment of heart attacks & strokes	1,000.0
Atrial Natriuretic Factor	Hormone for lowering blood pressure without side effects	5,000.0
Monoclonal Antibodies	Diagnosis & treatment of cancer & serious infectious diseases	5,000.0

GENERAL FEATURES

Animal cells contain a collection of membrane-enclosed compartments: the endoplasmic reticulum, Golgi apparatus, endosome, nucleus, lysosome, mitochondrion and peroxisome. Respiration and ATP production take place in the mitochondrion, while hydrolytic enzymes are enclosed in the lysosomes. A few steps are involved in obtaining a cell line; one such, obtained from a primary culture of normal cells derived from specific organ or tissue culture of limited time span, is referred to as a secondary culture. Such special, *transformed* cell lines, or those produced by cell fusion technology to generate hybrids, are those which do not undergo the senescence of mortal cells but can be indefinitely propagated. Hybridomas are the result of the fusion of an immortal cell line (such as a myeloma) and an antibody-producing lymphocyte to generate a hybridoma cell line secreting monoclonal antibodies. The antibody-making cells are frequently produced in mice as a reaction to immunization with an antigen. With the exception of blood cells, animal cells in their native environment are immobilized by extracellular matrix material to form a well-defined tissue architecture, within which the individual cells retain a characteristic morphology. This anchoring phenomenon then becomes a requirement for growth that must be fulfilled *in vitro* (Feder and Tolbert, 1987).

Mammalian cells grow under physiological conditions, 37°C, pH 7.3; product formation kinetics are described by the Leudeking-Piret model, shown previously in Chapter 5, as eqn. [5.1]. A typical growth medium for mammalian cells contains serum (5% to 20%), carbon and energy sources, nitrogen sources, inorganic salts, buffers, trace elements, growth factors and vitamins. Dean (1989) notes that an important cost reduction factor is the replacement of fetal calf serum by nutrient media which are at least an order magnitude less costly. The latter permit the operation of processes with lower yields than formerly necessary. An indication of the current product spectrum in relation to industrial activity is shown in Table 9.2.

Table 9.2 Segment of Current Product Spectrum (after C&EN, 1992)

Company	Products	1991 Sales ($Million)
Amgen	Erythropoietin & granulocyte colony-stimulating factor	$645.3
Genentech	Tissue plasminogen activator	196.5
	Human growth hormone	185.1
Genzyme	Glucocerebrosidase	81.1
Biogen	α-, β-, λ-interferons; hepatitis B vaccines & diagnostics	56.5
Centocor	Antisepsis monoclonal antibody; cancer diagnosis	44.3
Immunex	Granulocyte macrophase colony-stimulating factor	34.2
Chiron	Diagnostics and ophthalmics	28.2
Genetics Institute	Erythropoietin	18.4
Genentech	λ-interferon	1.5

9.1 CELL PHYSIOLOGY

ENERGY METABOLISM

Hu and Himes (1989) have provided a detailed description of the physiology of mammalian cells in culture. The primary energy sources include glucose and glutamine. This analysis shows that the state of mammalian cell energy metabolism can be estimated from the measurement of six species of substrates/metabolites in the bioreactor. Under normal culture conditions, both glucose and glutamine are essential for cell growth. At a high glucose concentration, most of the glucose consumed is converted to lactate, a toxic waste product. High conversion of glucose to lactate can be reduced by controlling the glucose concentration at a low level via programmed feeding; the rate of glucose conversion to lactate is then decreased and the rate of glutamine consumption is increased. Under steady state conditions in a continuous culture, the carbon from glucose and glutamine ends up along one of three pathways: assimilated into cell mass, converted to lactate or oxidized to carbon dioxide. Thus, the carbon balance on glucose and glutamine results in six fluxes: two associated with assimilation, complementary in nature, as well as two other pairs which are substitutable in principle. Specific consumption or production rates of glucose, glutamine, lactate, ammonium and carbon dioxide, as well as oxygen, can all be estimated from the measurement of the corresponding species. Furthermore, the stoichiometric coefficients may be estimated from the metabolic pathways.

Handa-Corrigan et al. (1992) describe a simple optimization strategy which enables monoclonal antibody (MCA) production in hollow fiber bioreactors to be controlled and predicted. The MCA production rate increases linearly with the uptake rates of glucose and glutamine and with the production rates of lactate and ammonia. The uptake and production rates of these metabolites can, in turn, be predicted from the pumping rates of basal medium to the bioreactor. A recommendation is made for a period of two weeks at the start of the cultivation when intensive assaying and monitoring should be carried out. After this period, the medium flow rate and MCA production rate may be predicted by linear extrapolation.

PRODUCT FORMATION

It is anticipated that optimization of cell culture processing *vis a' vis* cell growth and product expression can be achieved through adjustment and/or manipulation of intracellular metabolic fluxes. Cell proliferation is influenced by cell cycle kinetics, growth factor regulation and cell density effects (Lauffenburger, 1989). Metabolic fluxes are influenced by levels of hormones and dissolved oxygen; pH and redox potential; and nutrient and waste product concentrations. Controlled variations of medium composition can produce substantial effects; i.e., cell growth stimulation, as well as enhancement of cell attachment potential (to collagen, fibronectin, etc.).

As an interesting variation within the above description, it has been observed that Chinese hamster ovary (CHO) cells cultured in collagen microspheres in the presence of serum can readily adapt to serum-free medium. The same is not true for the suspension culture. It seems possible that the intra-matrix microenvironment resulting from the cell-matrix and cell-cell interactions, and the relatively high concentration of cell-derived protein factors has allowed growth in the absence of serum-supplied growth factors. Altered cell physiology in response to serum removal has also been observed from the significantly increased oxygen to glucose uptake ratio.

Hybridoma cells grown in suspension as well as in fluidized collagen microspheres (Fig. 9.1) have also exhibited significantly different growth, metabolic and product formation kinetics. At a constant specific medium feed rate, the specific growth rate of the immobilized population was considerably lower (see Fig. 9.2). However at equal specific growth rates, the metabolic rates (glucose and glutamine consumption, lactate and ammonia production) and the specific antibody secretion rates of the immobilized hybridomas were significantly higher (Fig. 9.3). It appears that the suspension cells have directed more of the substrate resources toward cell growth relative to the immobilized cells, which have allocated more resources toward other cellular functions, including product synthesis and secretion. The suspension culture was also found to have somewhat higher apparent lactate yield on glucose and significantly higher ammonia yield on glutamine. The differences in cell physiology are attributed to the culture biochemical environment (Ray et al., 1990).

In this connection, Wu et al. (1991) have developed a single cell computer model for pH regulation in CHO cells. Its basis, therefore, implies culture synchrony, or at least a smoothed-out population balance. In this way, the response of a large population can be estimated from the response of single cells. The modeling method is based on the application of transient mass balances to key cellular components in an individual cell. The model accounts for Na/H antiport, proton leakage, and the effects of formation and build-up in the culture of weak acids and bases (e.g., lactate and ammonia), and intracellular buffering capacity. The model has successfully predicted both the steady state intracellular pH as a function of external pH, external

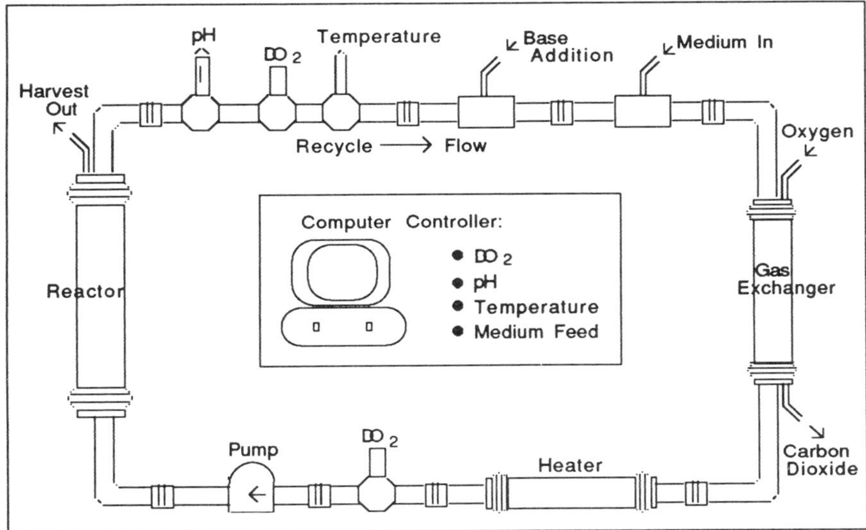

Figure 9.1 Schematic diagram of a fluidized bed bioreactor.

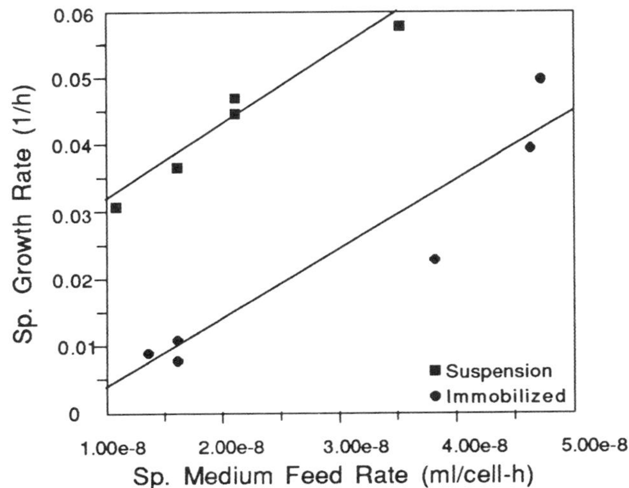

Figure 9.2 Specific growth rate of hybridomas as a function of specific medium feed rate in suspension and immobilized cultures.

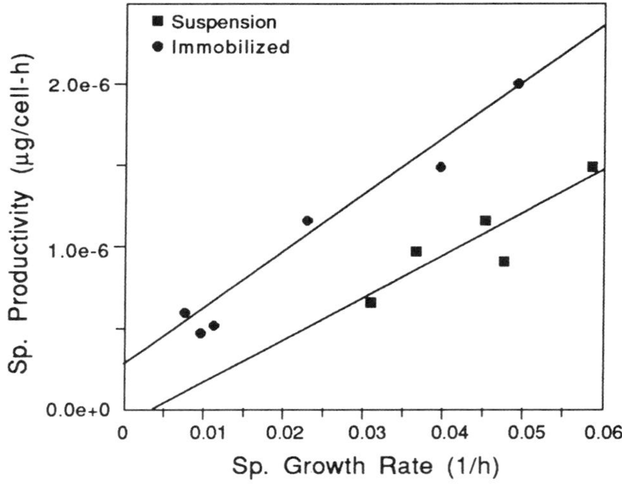

Figure 9.3 Hybridoma cells in suspension and immobilized fluidized bed bioreactors. Specific productivity in these two cultures is compared as a function of specific growth rate.

levels of glucose, glutamine, ammonia, and lactate; and the transient intracellular pH changes in response to "acid load" and "alkaline load," which are induced by shifting cells between ammonia-containing media and ammonia-free media. As already noted, high cell density immobilized systems are aimed at high productivities, but mass transfer limits on substrates/inhibitors come into play. Inhibitory effects are exerted mainly through shifts in internal pH; with Na/H antiport corresponding to competitive inhibition, too much glucose can exceed the respiratory requirement, leading to lactate formation, as also reported by Clark (1991).

9.2 MIXING AND MECHANICAL STABILITY FOR ANCHORAGE DEPENDENT AND INDEPENDENT CELLS

Shuler and Kargi (1992) point out that the oxygen requirement of animal cells is only about 20% that of most plant cells and a small fraction of that for microbial cells. The requirement is in the order of .1 x 10^{-12} gmol oxygen/hr/cell. For a typical cell culture density of 10^6 cells/ml, the oxygen requirement is on the order of .3 mmol oxygen/ L/hr. Denser cultures require forced aeration but air bubbles may cause shear damage, as described below. Microcarrier cultures with high surface to volume ratios permit an order magnitude higher cell concentration to be achieved, with a concomitant larger oxygen transfer requirement. This usually dictates the use of a somewhat more advanced type of contacting scheme, such as a stirred tank or fluidized bed or an airlift concept.

Hofmann (1990) asserts that tank reactors are ideally suited for batch operation. Production from start to finish of a protein batch requires two to three weeks. The first one or two weeks are occupied with proliferating the cells in logarithmic growth phase through a seeding reactor cascade. The reactor volume in this sequence of reactors usually increases by a factor of ten. For hybridomas the final production reactor is generally operated for five to seven days in terminal batch mode, as most hybridomas exhibit highest cell productivity during stationary growth phase. Manufacturing under conditions of batch production demands clockwork-like operation to utilize optimally the expensive reactor equipment. The high number of production unit processes involved makes batch production extremely labor intensive. However, the tank reactor can also be continuously operated under fed-batch or cytostatic conditions. Under fed-batch conditions 80-90% of the cell suspension is harvested when the reactor reaches the end of the log growth phase. The remainder is left in the reactor and utilized as seed culture for the new batch by diluting it to starting titer with fresh nutrient. Alternatively, the reactor can be perfused with fresh nutrient commensurate to the cell generation time. This results in continuous cytostatic production. Obviously, both modes of operation require the cells to stay in logarithmic growth phase, which results in maximum titers of only about 10^6 cells per milliliter.

Recently, recombinant mammalian cells were cultured in a 20 liter cell-recycle perfusion reactor in order to achieve increased volumetric productivity. Viable cell concentration in excess of 10^7 cells/ml was maintained for over a month, with a perfusion rate of 1 vvd. A small purge rate was used to maintain consistently high viability in the reactor (Karkare et al., 1991).

HYDRODYNAMIC SHEAR

By now it is well established that the fluid motions occurring in stirred tank bioreactors under high shear conditions can bring about irreversible damage to animal cells. This occurs at a microstructural level higher than the protein denaturation effects already alluded to in Chapter 2.

Papoutsakis (1986) and Wang (1986) have discussed the necessity of low shear oxygenation in relation to the culturing of relatively fragile mammalian cell systems. Agitation and air sparging are used to transfer oxygen and other nutrients to dense cultures of animal cells. The former author attributed cell injury to air entrainment and bubble breakup. However, Murhammer and Goochee (1990b), in their studies of the role of the protective agent, pluronic F68, have determined that bubble size and gas flow rate are not the only considerations for cell damage in sparged bioreactors. Their results strongly suggest that substantial cell damage occurs in the vicinity of the gas distributor itself. Recent work by Kunas and Papoutsakis (1990a) helps to clarify the issue. When freely suspended hybridoma cells are cultured in an agitated bioreactor, two fluid-mechanical mechanisms can cause cell damage and growth retardation. The first is present only when there is a gas phase and is associated with vortex formation accompanied by bubble entrainment and breakup. In the absence of a vortex and bubble entrainment, cells can be damaged only at very high agitation rates, above about 700 rpm, by stresses in the bulk turbulent liquid. Cell damage then correlates with Kolmogorov eddy sizes similar to or smaller than the cell size. In the absence of a vortex, the entrainment and motion of very fine bubbles cause no growth retardation even at agitation rates ≤600 rpm. Abu-Reesh and Kargi (1991) investigated the effects of long-term hydrodynamic shear on hybridoma cells in a 250-ml continuous stirred-tank reactor (CSTR). Shear damage to cells in a CSTR was approximated by first-order kinetics. The death rate constant increased sharply at impeller tip speeds >40 cm/s.

In studies regarding the role of protective agents *per se*, bioreactor and viscometric studies were used to examine the mechanism by which three additives, fetal bovine serum (FBS), pluronic F68 and polyethylene glycol (PEG), protect freely suspended CRL-8018 animal cells from damage due to interactions with bubbles in agitated bioreactors (Michaels et al., 1991). It was concluded that the protective effect of FBS is both physical and biological in nature, while the protective effect of F68 and PEG is purely physical. In a related study, finding a polyol that protected cells was synonymous with finding one that did not inhibit cell growth (Murhammer and Goochee, 1990a).

The effect of serum on the growth rate and metabolism of CRL-8018 hybridoma cells in an agitated, surface-aerated bioreactor was examined (Kunas and Papoutsakis, 1990b). The results are incorporated into a simple model in which the apparent growth rate is the sum of an invariable growth rate and a changing death rate.

Regarding immobilized cells, researchers have likewise delved into the mechanisms of shear damage and have identified three potentially harmful collision mechanisms: eddy-bead, cell loaded bead-bead and bead-impeller (Cherry and Papoutsakis, 1988; 1989; 1990; Croughan et al., 1987; 1988; Croughan and Wang, 1989). In the case of the mechanism first mentioned, size-matching with turbulent eddies on the Kolmogorov-scale is again important.

A concentric-cylinder airlift reactor, in which the annulus is a packed bed of glass fibers, has been developed in order to facilitate the scale-up and enhance the volumetric productivity of anchorage-dependent animal cell cultures. In this bioreactor, oxygen-containing gas is sparged through the inner draft tube, causing bubble-free medium to flow through the fiber bed in the outer cylinder, and providing both oxygenation and convective nutrient transfer to the cells. A simple hydrodynamic model is found to describe adequately the airlift fiber-bed bioreactor; liquid flow rates and volumetric mass transfer coefficients are predicted and found to be in agreement with experimental measurements. Consequently, the optimal reactor configuration to provide maximum oxygen supply is derived (Chiou et al., 1991; Murakami et al., 1991). Scaled-up versions of airlift bioreactors are also under study (Petrossian and Cortessis, 1990). Forty- and ninety-liter airlift bioreactors have been used successfully to grow hybridoma cell lines in chemically defined serum-free media. In the airlift bioreactor, hybridoma cell growth and monoclonal antibody productivity are comparable to that obtained by conventional cell culture. At sparging rates of 0.60-1.20 vols/h, the airlift bioreactor achieves rapid mixing and adequate mass transfer. Foaming is minimal and inconsequential for serum-free media and media supplemented with 5-10% fetal bovine serum. The use of serum-free media facilitates monoclonal antibody purification and enhances the purity of the final product. As an interesting twist, enhanced oxygen transfer was observed in surface-aeration bioreactors when stabilized foams were generated to increase the gas-liquid interfacial area by slowly introducing coarse bubbles into media containing fetal bovine serum. The enhancing effect of foam on oxygen transfer in surface aeration bioreactors was demonstrated with hybridoma cultures simultaneously grown in three identical bioreactors, with and without stabilized foams (Ju and Armiger, 1990).

9.3 BIOREACTORS FOR RECOMBINANT CELLS AND HYBRIDOMA CELLS

In view of the special requirements and/or limitations outlined above, a variety of bioreactors has been developed to meet the existing demands, as shown in Fig. 9.4.

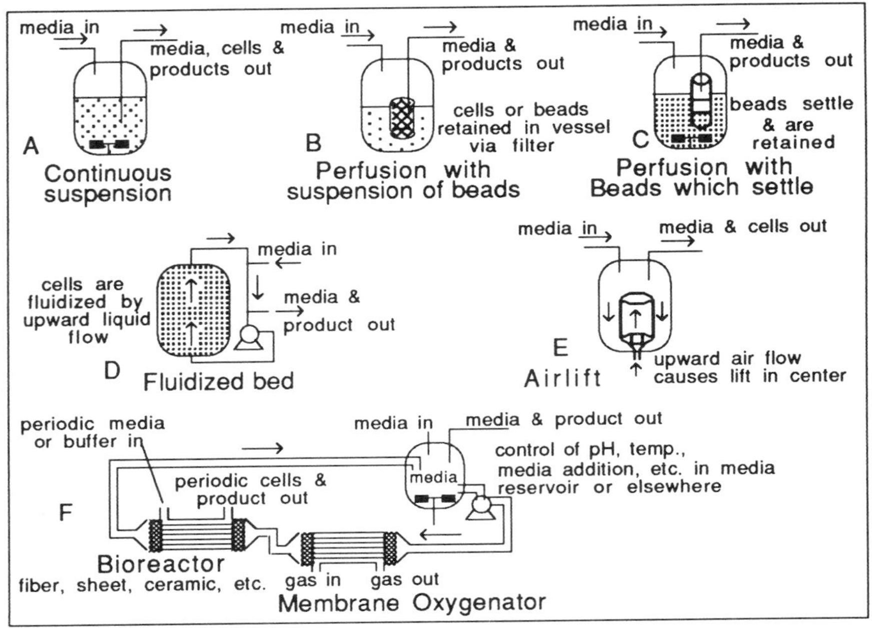

Figure 9.4 Typical bioreactor configurations (after Dean et al., 1987).

McKillip et al. (1991) describe a comparison of bioreactors for large-scale TPA production through the growth of a transformed C-127 mouse mammary tumor cell line. Soft-agar cloning was successfully used to provide selective pressure for the production of fully transformed cells. The transformed C-127 is a surface-dependent cell line, obviating the possibility of suspension culture. The authors chose to evaluate several perfusion-type reactors as follows: the CD Biomedical Cell-Pharm bioreactor (a hollow fiber system), the Bellco (Vineland, NJ) bioreactor (containing either stainless steel matrices or alginate beads), the Charles River (Wilmington, MA) Opticell 5200 (a ceramic core reactor) and the CelliGen bioreactor (with alginate beads) from New Brunswick Scientific (Edison, NJ). Full experimental details of the growth characteristics (too voluminous to be included here) are provided in Table 2 of the subject reference. Table 9.3, below, lists the other performance characteristics.

Table 9.3 Bioreactor Performance Characteristics (After McKillip et al., 1991)

Bioreactor	Growth substrate	Surface Area cm^2	Conc. TPA mg L^{-1}	Productivity of TPA mg day^{-1}	Projected yield mg day^{-1}
Charles R.	AD51 ceramic core	4,250	25	25	224
Bellco	10 stainless steel matrix pads	34,000	32.5	130	145
CD Medical	IV-L hollow fibre core	17,500	2	2	4.34
T-150 flasks	Coated plastic	150	-	0.03-0.12	7.6-30.4
Celligen	125 ml alginate	-	10	10	80
Bellco	1000 ml alginate	-	20	80	80

In general, immobilized cell bioreactor technology has already found a home in the biotechnology industry; e.g., for monoclonal antibody production using hybridoma cells. In addition to the fact that certain anchorage-dependent cell lines cannot be grown in suspension culture, immobilized cell bioreactors are well suited for the following reasons:

1. The fragile hybridoma cells are protected in the immobilized phase from the damaging effects of shear forces present in the liquid environment.
2. The immobilized cell environment closely approximates that of natural tissue, producing high cell density while maintaining adequate levels of nutrient mass transfer.
3. IMC bioreactors can be operated for long time periods (weeks to months) while maintaining excellent cell viability and protein productivity levels.

The ultimate goal in developing any system to produce biological molecules through cell culture is that it should be genetically limited. This implies that cell productivity should be governed by the genetic parameters such as gene copy number, DNA and RNA turnover rate, protein synthesis and processing rate, rather than process parameters such as temperature, pH, and nutrient concentrations, mass transfer rates, cell density, and kinetics. The bioreactor should thus usually be designed to provide conditions similar to physiological conditions and be able to maintain high cell densities that are close to tissue density (about 10^9 cells/ml). Indeed, as already mentioned, for large-scale cultivation of anchorage dependent cells, some form of immobilization is the only method. Immobilization is essentially a wày to decrease the relative mobility of the cells by attaching them on or inside a matrix such as microcarriers, porous beads, etc. Besides drastically cutting down the escape of cells in the harvest liquor during continuous operation, a limitation in normal chemostat cultures, immobilization also provides a high surface area per unit reactor volume that is conducive to cell attachment and growth (Ramasubramanyan, 1990).

Griffiths (1991) identifies scale-up factors at three levels. At the first level, the CSTR and airlift reactors are designed to solve mixing and oxygen transfer problems. At the second level, cell retention and nutrient transfer (diffusional) problems are handled via perfusion systems such as fixed beds of immobilized cells, hollow fibers, etc. At the third level are the high productivity, high cell density processes, with optimized carrier systems. Such systems are optimized with respect to catalyst effectiveness and lower pressure drops via Peclet number/ Reynolds number considerations. Hu and Peshwa (1991) have provided a recent review of scale-up factors; the need for suitable kinetic models to describe animal cell growth is emphasized.

HOLLOW FIBER BIOREACTORS

Recent studies have investigated critical scale-limiting factors in the design of cell culture hollow fiber bioreactors. Oxygen depletion is expected to be the major limiting factor for both the axial and radial scale-up of hollow fiber bioreactors. This was confirmed via experimental measurements of axial cell distributions in a severely oxygen-limited case (Piret and Cooney, 1991; Piret et al., 1991). In related work, heterogeneous protein and cell distributions on the shell side of hollow fiber bioreactors were reduced significantly by periodic alternation of the direction of recycle flow; the reactor productivities were doubled (Piret and Cooney, 1990b). The same authors provide a useful review of immobilized mammalian cell cultivation in hollow fiber bioreactors (Piret and Cooney, 1990a).

FLUIDIZED BED BIOREACTORS

Biogen, Inc. utilizes fluidized bed bioreactors that employ cells immobilized in macroporous weighted collagen microspheres (the Verax system) to produce a variety of animal cell culture products such as antibodies, Factor VIII, TPA, Mullerian Inhibitory hormone, Beta Interferon and soluble CD4 (Galliher, 1991b). One square foot of collagen membrane area supports one liter of reactor volume. Oxygen transfer rate is decoupled from shear via an artificial lung-type oxygenator in the recycle loop, so damage due to aeration, pumping and mixing is obviated. No air-liquid interface comparable to that in a CSTR is present, so foaming is not a problem. Medium development and feed control strategy can be optimized, there is no need for cell recycle, and the fluidized bed can be operated under optimum steady state conditions. The fractionation of the coexisting colonies is as follows: free cells (1-2%), biofilm cells (1-10%) and matrix cells (88-98%). The technology has been successfully demonstrated at production scale, as measured by daily output, product yield, product quality and production cost, by making tissue-type plasminogen activator and monoclonal antibodies (Runstadler et al., 1990).

Regarding the oxygenation loop, (adjunctive to the fluidized bed) *per se*, the oxygen transfer capabilities of a hollow fiber oxygenator have been recently evaluated experimentally (Kalogerakis and Behie, 1991). The oxygenator (480 ml), which is the main component of an artificial heart/lung unit used routinely in cardiopulmonary bypass procedures, was assessed for use in a recycle stream to determine if the oxygen requirements in bubble-free hybridoma cell bioreactors could be supported on a large scale. Oxygen transfer to simulated medium in a fifteen liter bioreactor was found to depend primarily on the liquid recirculation rate (1-5 L/min), and was not seriously affected by the air flow rate in the oxygenator (0.5 to 15 L/min). Based on the experimentally determined mass transfer coefficient across the membrane, it was found that the oxygenator could support a 100 liter fermentor at an average cell density of 2 x 10^6 cells/ml with a specific uptake rate of 4.8 mg oxygen (10^9 cell·h). Furthermore, if oxygen enriched air were to be used in the oxygenator, a single unit could support cell densities beyond 8 x 10^6 cells/ml in a 100 liter working volume fermentor.

A theoretical analysis of oxygen transport across a tubular microporous membrane is described by Su et al. (1992). For fixed inlet pressure, it was found that a tubular segment should not exceed a critical length. Thus, manifolded banks of shorter tube segments are expected to yield a higher oxygen transfer rate per unit tube length; however, this advantage is counterbalanced by the fact that gas distribution into large numbers of tubes in parallel array might not be uniform.

In another study, the production of recombinant human interleukin-2 in a fluidized bed bioreactor containing porous glass carriers is described (Kratje and Wagner, 1992). The production rate showed a maximum of a 1.9-fold decrease compared with a homogeneous stirred bubble-free aerated system. This result was in contrast to that achieved with hybridoma cell lines, where better performance was obtained with the fluidized bed bioreactor. The situation may reflect the problems caused by the dense cell culture with adherent cells, as previously shown in a hollow fiber bioreactor with the same cell line.

9.4 COLLAGEN TECHNOLOGY: SECOND GENERATION

Animal cells are typical eucaryotes, lacking a cell wall but surrounded by a thin and fragile plasma membrane. They occur naturally as dense populations in the immobilized form in tissues, girded by biopolymeric supports such as elastin and collagen. Recognition of the latter point led the author originally to propose a first generation whole cell carrier technology based on collagen, as introduced in Chapter 3. With the rise of mammalian cell culture, it occurred to the author that the formation of "primitive reconstituted tissues" employing the collagen carrier might be an ideal point of departure for a second generation of collagen technology. Such structures would indeed comprise dense cell populations, though they would be lacking in gap junctions for intercommunication. Even so, bioreactors could be stimulated or switched on and off by externally supplied metabolic effectors to secrete desired products.

A little further along, perhaps the genetic machinery for producing gap junctions could be cloned, as well, into animal cell lines, enabling the formation of "artificial glands." These tissues could then be switched on and off with chemical messengers of the usual type(s) (Vieth, 1991) to control the manufacture of desired products.

Employing 8Br(cAMP) as a metabolic switch for secretion (i.e., as an inducer), Stephanopoulos et al. (1991) recently studied the interrelations of signal transduction mechanisms and secretory kinetics for BTC3 cells, a specialized cell line capable of regulated protein secretion in the form of native murine insulin. Using synergistic combinations of agents, the authors were able to cycle BTC3 cells between secretion and recharging media with high efficiency. In experiments with up to 12 complete cycles, greater than 65% of the secreted insulin was recovered in less than 10% of the total process time.

Collagen immobilization of animal cells was discussed at the Engineering Foundation Conference on Biochemical Engineering in Galway, Ireland in 1984. In 1986, on the occasion of the next Biochemical Engineering Conference, in Henniker, New Hampshire, Verax Corporation disclosed its fluidized bed process, employing animal cells immobilized on macroporous collagen beads (See sketch below, Fig. 9.5). An alternative arrangement could involve a flat sheet configuration which could be organized into the spiral wound geometry.

For instance, hybridoma growth and monoclonal antibody formation are reported for a flat sheet membrane bioreactor. As with hollow fiber systems, membrane reactors provide for continuous addition of nutrients and continuous removal of toxic products from the cell culture. This results in higher cell concentrations and specific antibody productivities than obtained with batch culture without membranes. The HDP1 cell line produced antibodies at higher rates during periods of slow growth and adverse culture conditions. A membrane bioreactor with a 100,000 MW cutoff membrane achieved higher cell concentrations and specific antibody productivities than those obtained with a 50,000 MW cutoff membrane reactor or a batch reactor without a membrane. With use of a 100,000 MW cutoff membrane reactor, final monoclonal antibody concentrations were improved by a factor of fourteen as compared to a membrane-free culture (Hagedorn and Kargi, 1990).

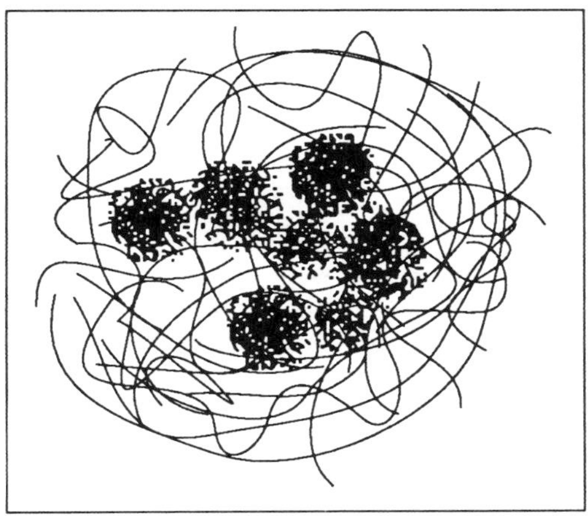

Figure 9.5 Facsimile of wire-like matrix structure.

Collagen in nature appears to have natural affinity to bind a variety of cells. Therefore, it is a logical choice for a matrix material for the *in vitro* culturing of mammalian cells. Runstadler (1991) has identified critical factors in the scale-up of commercial animal cell bioreactors. The production rate is given as the product of the cell specific productivity (CSP) multiplied by the volume and the cell density, insofar as the number of cells is equal to the product of the last two factors. The author calls attention to a possible pivotal role of the extracellular matrix (ECM) based on collagen, which contains receptors, glycoproteins and glycosamineglycans. Binding alters cell geometry for deformable cells. Energy is required for a cell to deform; therefore, it makes sense to consider an ATP-dependent mechanism of cell immobilization to collagen (Galliher, 1991a). As a working hypothesis, the ECM communicates through the cytoskeleton with connections to the nuclear matrix mediated by cell shape changes and may possibly influence gene regulation. In this manner, genetic factors such as transcription, translation and mRNA processing might be involved.

The Verax process uses porous beads made of reconstituted bovine-hide collagen to immobilize whole cells. It has been shown (Saini, 1974) from experiments done on attachment of enzymes and whole microbial cells to collagen, that the physico-chemical linkages between collagen, a protein, and the enzyme or the cell-surface proteins are characterized by salt-linkages, hydrogen bonds and Van der Waals interactions. It has also been shown that the average molecular weight between the network linkages is reduced by 30,000, indicating a highly stable bonding.

Mammalian cells cultured in collagen matrices form a culture physiological environment expected to be quite different from that in a suspension culture. The microenvironment generated inside the collagen microspheres is a result of a complex set of interactions and regulations brought about by the intimate cell-matrix and cell-cell interactions, and the intra-matrix concentration of substrates and cell secreted products, including protein factors (e.g., autocrine growth factors). Therefore, the immobilized cells in the microspheres are believed to experience a significantly different culture environment compared to their counterparts in suspension. The differences in microenvironment in suspension and immobilized cultures are reflected in the observed difference in cell physiology (Ray et al., 1990).

The fundamental element of the Verax continuous culture process is the porous microsphere weighted with a non-cytotoxic material. This endows the bead with a specific gravity around 1.6 so that it will be maintained in a suspended state by a high upward-flowing culture liquor velocity (about 1 cm/s). A typical bead has a diameter of 500 μm with interconnected pores and channels of about 20-40 μm in average individual size. The virtually empty interior of the microsphere allows the cells to enter easily and populate a large fraction of the bead volume. The leaf-like morphology and the high porosity of the microsphere (about 85%) provides a large surface to volume ratio necessary for high density cell growth, and generally facilitates high mass transfer rates.

Post-inoculated microspheres are fluidized as a thick slurry in the bioreactor vessel. A uniform fluid environment is provided for the immobilized cells, due to the free circulation of the microspheres inside the fluidized bed bioreactor. The process parameters include temperature, pH and dissolved oxygen concentration, which are monitored and controlled continuously in a recycle loop.

Mass transfer limitations are commonly encountered at high cell densities. In addition to the probable diffusion limitations with respect to nutrients and metabolites, oxygen transfer is often a key factor in scale-up. Slow diffusion of toxic substances out of the dense cell mass could also be responsible for lowered productivity. Hence a knowledge of the overall mass transfer processes associated with nutrients and products through densely packed cell layers is vital.

This realization stimulated the careful determination of the diffusion coefficients of some common nutrients and metabolites under conditions of high-density immobilization. The apparent diffusivities, measured under static conditions with thin slabs and balanced pressures, were found to be about 12% of their values in water. Though not considered explicitly, the highly porous nature of such beads may also result in intraparticle convection under dynamic conditions (e.g., under hydrodynamic gradients), which could substantially increase mass transfer to the interior of the beads (Jeong et al., 1989; Shuler and Kargi, 1992). The diffusivity values thus measured experimentally were used in conjunction with the kinetic data from reactor operations utilizing immobilized cells, to generate steady state profiles for glucose, oxygen, lactic acid and a protein. It was readily apparent that oxygen is the likely rate limiting nutrient. It was also possible to predict the development of a necrotic core due to oxygen depletion (Ramasubramanyan, 1990).

ARTIFICIAL TISSUE

The liver is the main bioreactor in the body. In order to provide the design for a temporary liver support, artificial tissue reactors are under study (Yarmush, 1991). Hepatocytes sandwiched between two layers of collagen gel have been shown to maintain many liver-specific functions, including the secretion of albumin, for greater than six weeks, when cells are perfused with a serum-like medium containing hormones.

Hepatocytic polysomes are disrupted when such cells are isolated, but the cells can be reaggregated into a proper matrix, *de novo*, with new collagen fibril boundaries when in a collagen double sandwich, so the cells behave properly with respect to protein secretion, etc. Furthermore, the bioreactor works better than static culture, perhaps due to better oxygen transfer.

As alternative carriers for hepatocytes, cellulosics, polyamides and polypropylene have been investigated (Gerlach et al., 1990). If the adhesion of hepatocytes prolongs their metabolic function, a large adhesion surface in bioreactors is necessary. Coating with collagen or fibronectin improves the attachment and spreading on all membranes. Differences between collagen and fibronectin were detected: on collagen, most of the cells spread, while on fibronectin, most of the cells spread and flattened polygonally. Hepatocytes attached to microcarriers (Arnaout et al., 1990) or cultured on collagen-coated culture dishes (Hamada, 1990) have also displayed intact functionalities over varying time intervals.

9.5 REFERENCES

Abu-Reesh, I. and F. Kargi, *Enzyme Microb. Technol.*, **13**, 913-919 (1991).

Anonymous, *C&EN*, 11, April (1992).

Arnaout, W. S., A. D. Moscioni, R. L. Barbour and A. A. Demetriou, *J. Surg. Res.*, **48**, 379-382 (1990).

Chee, D. O., *Genetic Eng. News*, June (1987).

Cherry, R. S. and E. T. Papoutsakis, *Biotechnol. Bioeng.*, **32**, 1001-1014 (1988).

Cherry, R. S. and E. T. Papoutsakis, *Bioproc Eng.*, **4**, 81-89 (1989).

Cherry, R. S. and E. T. Papoutsakis, *Animal Cell Biotechnol.*, **4**, 71-121 (1990).

Chiou, T. W., S. Murakami, D. I. C. Wang and W. T. Wu, *Biotechnol. Bioeng.*, **37**, 755-761 (1991).

Clark, D. S., Abstract, Biochemical Engineering VII, Engineering Foundation Conf., Santa Barbara, CA, March 3-8 (1991)

Croughan, M. S., J. -F. Hamel and D. I. C. Wang, *Biotechnol. Bioeng.*, **29**, 130-141 (1987).

Croughan, M. S., J. -F. Hamel and D. I. C. Wang, *Biotechnol. Bioeng.*, **32**, 975-982 (1988).

Croughan, M. S. and D. I. C. Wang, *Biotechnol. Bioeng.*, **33**, 731-744 (1989).

Dean, R. C., Jr., S. B. Karkare, N. G. Ray, P. W. Runstadler, Jr. and K. Venkatasubramanian, *Ann. N. Y. Acad. Sci.*, **506** (Biochem. Eng. 5) 129-146 (1987).

Dean, R. C., Jr., *Genetic Eng. News*, 4, Sept. (1989).

Feder, H, and W. R. Tolbert , In "Enzyme Engineering 8," A. I. Laskin, K. Mosbach, D. Thomas and L. B. Wingard, Eds. *Ann. N. Y. Acad. Sci.*, **501**, 522-533 (1987).

Galliher, P. M., Personal Communication (1991).

Galliher, P. M., Abstract, Biochemical Engineering VII, Engineering Foundation Conf., Santa Barbara, CA, March 3-8 (1991).

Gerlach, J., P. Stoll, N. Schnoy and E. S. Buecherl, *Int. J. Artif. Organs*, **13**, 436-441 (1990).

Griffiths, B., Abstract, Biochemical Engineering VII, Engineering Foundation Conf., Santa Barbara, CA, March 3-8 (1991).

Hagedorn, J. and F. Kargi, *Biotechnol. Prog.*, **6**, 220-224 (1990).

Hamada, T., *Kanzo*, **31**, 669-677 (1990).

Handa-Corrigan, A., S. Nikolay, D. Jeffrey, B. Heffernan and A. Young, *Enzyme Microb. Technol.*, **14**, 58-63 (1992).

Hofmann, F., In "Trends in Animal Cell Culture Technology," Proc. Second Ann. Mtg. Japanese Assn. for Animal Cell Technol., Japan Nov. 20-22, 1989. H. Murakami, Ed., p. 121 (1990).

Hu, W. S. and V. B. Himes, In "Bioproducts and Bioprocesses," A. Fiechter, H. Okada, R. D. Tanner, Eds., Second Conf. to Promote Japan/U. S. Joint Projects and Cooperation in Biotechnology, Lake Biwa, Japan. pp. 33-46. Springer-Verlag: Berlin (1989).

Hu, W. S. and M. V. Peshwa, *Can. J. Chem. Eng.*, **69**, 409-420 (1991).

Jeong, Y. S., W. R. Vieth and T. Matsuura, *I&EC Res.*, **28**, 231 (1989).

Ju, L. K. and W. B. Armiger, *Biotechnol. Prog.*, **6**, 262-265 (1990).

Kalogerakis, N. and L. A. Behie, *Can. J. Chem. Eng.*, **69**, 444-449 (1991).

Karkare, S. B., L. R. Williams and D. J. Hettwer, Abstract, Biochemical Technology Div., ACS Mtg., New York, Dec. (1991).

King, J., *C&EN*, April, 32-54 (1989).

Kratje, R. B. and R. Wagner *Biotechnol. Bioeng.*, **39**, 233-242 (1992).

Kunas, K. T. and E. T. Papoutsakis, *Biotechnol. Bioeng.*, **36**, 476-483 (1990).

Kunas, K. T. and E. T. Papoutsakis, *J. Biotechnol.*, **15**, 57-69 (1990).

Lauffenburger, D. A., *Chem. Eng. Education*, **23**, 208-213 (1989).

McKillip, E. R., A. S. Giles, M. H. Levner, P. P. Hung and R. N. Hjorth, *Bio/Technology*, **9**, 805-810 (1991).

Michaels, J. D., J. F. Petersen, L. V. McIntire and E. T. Papoutsakis, *Biotechnol. Bioeng.*, **38**, 169-180 (1991).

Murakami, S., T. W. Chiou and D. I. C. Wang, *Biotechnol. Bioeng.*, **37**, 762-769 (1991).

Murhammer, D. W. and C. F. Goochee, *Biotechnol. Prog.*, **6**, 142-148 (1990).

Murhammer, D. W. and C. F. Goochee, *Biotechnol. Prog.*, **6**, 391-397 (1990).

Papoutsakis, E. T., Proceedings, III World Congress on Chemical Engineering, Tokyo (1986).

Petrossian, A. and G. P. Cortessis, *BioTechniques*, **8**, 414-16, 418, 420-422 (1990).

Piret, J. M. and C. L. Cooney, *Biotechnol. Adv.*, **8**, 763-783 (1990).

Piret, J. M. and C. L. Cooney, *Biotechnol. Bioeng.*, **36**, 902-910 (1990).

Piret, J. M. and C. L. Cooney, *Biotechnol. Bioeng.*, **37**, 80-92 (1991).

Piret, J. M., D. A. Devens and C. L. Cooney, *Can. J. Chem. Eng.*, **69**, 421-428 (1991).

Ramasubramanyan, K., Ph. D. Thesis, Dept. of Chemical and Biochemical Engineering, Rutgers University (1990).

Ray, N. G., J. N. Vournakis, P. W. Runstadler, Jr., A. S. Tung and K. Venkatasubramanian, In "Physiology of Immobilized Cells," J. A. M. deBont, J. Visser, B. Mattiasson and J. Tramper, Eds. Elsevier Science Publishers: Amsterdam (1990).

Runstadler, P. W., Jr., A. S. Tung, E. G. Hayman, N. G. Ray, J. von G. Sample and D. E. DeLucia, *Bioprocess Technol.*, **10** (Large-Scale Mamm. Cell Cult. Technol.) 363-391 (1990).

Runstadler, P. W., Jr., Abstract, Biochemical Engineering VII, Engineering Foundation Conf., Santa Barbara, CA, March 3-8 (1991).
Saini, R., Ph. D. Thesis, Dept. of Chemical and Biochemical Engineering, Rutgers University (1974).
Shuler, M. L. and F. Kargi, "Bioprocess Engineering," p. 222. Prentice Hall: Englewood Cliffs, NJ (1992).
Silhavy, T. J., Paper presented in "Biotechnology: Opportunities, Problems and Solutions " AIChE Mtg., Cen. NJ Sec., Princeton, May (1985).
Stephanopoulos, G. N, G. E. Grampp and H. F. Lodish, Abstract, Biochemical Engineering VII, Engineering Foundation Conf., Santa Barbara, CA, March 3-8 (1991).
Su, W. W., H. S. Caram and A. E. Humphrey, *Biotechnol. Prog.*, **8**, 19-24 (1992).
Vieth, W. R., "Diffusion In and Through Polymers," p. 261. Hanser Publishers: Munich; Dist. in U. S. by Oxford University Press: New York (1991).
Wang, D., Proceedings, III World Congress on Chemical Engineering, Tokyo (1986).
Wu, P., N. G. Ray and M. L. Shuler, Abstract, Biochemical Engineering VII, Engineering Foundation Conf., Santa Barbara, CA, March 3-8 (1991).
Yarmush, M. L., Abstract, Biochemical Engineering VII, Engineering Foundation Conf., Santa Barbara, CA, March 3-8 (1991).

10

GENE EXPRESSION WITH PLANT CELLS

10.0 PATHWAYS

CHEMICALS PRODUCTION FROM PLANT CELL AND TISSUE CULTURES

Higher plants are the source of a diverse range of important chemicals including pharmaceuticals, essential oils (flavors and fragrances), insecticides and other fine chemicals. The current critical shortage of taxol, a promising anti-cancer agent, has given new urgency to the development of a commercially viable plant cell culture processing technology. The biosynthetic origin and the names of some compounds are shown in Fig. 10.1. These products are currently obtained by labor-intensive farming techniques in many parts of the world. Often, the

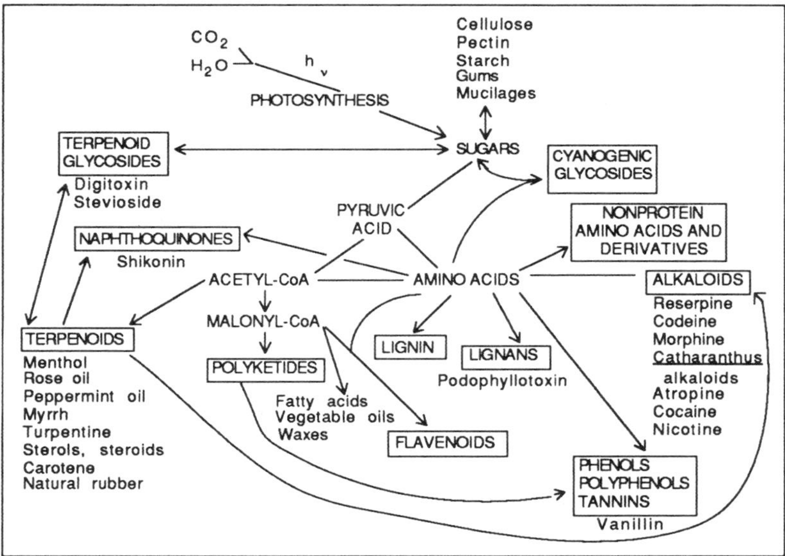

Figure 10.1 Biosynthetic routes in higher plants to some important chemicals. (After Balandrin et al., 1985).

particular plant species that have commercial value are native to very special regions; for instance, many important plants are native to tropical environments. Accessibility to the plants may also be compromised by the usual factors influencing any agricultural crop, such as unpredictable weather patterns and seasonal variation.

For these reasons and others, there is a continuing interest in cell suspension culture as an alternative "source" of plant natural products. In an ideal situation, this would provide for the continuous, controllable and reliable manufacture of industrially important chemicals from higher plants in any location.

It has been estimated that plant-derived pharmaceuticals represent an annual market to $9 billion in the U. S. alone (OTA, 1983). Some examples of these types of compounds and their sources are given in Table 10.1. Flavors and fragrances have a current worldwide market in excess of $1.5 billion. Worldwide markets for insecticides and other fine chemicals such as pigments are more difficult to gauge; however,

Table 10.1 Commercially Important Pharmaceuticals from Plants (Balandrin et al., 1985)

Compound or Class	Botanical Source
Steroids	
Hormones (95% from diosgenin)	*Dioscorea* spp. (Mexican yams)
Digitalis glycosides (digitoxin, digoxin)	*Digitalis purpurea, D. lanata* (foxglove)
Alkaloids	
Belladonna alkaloids (atropine, hyoscyamine, scopolamine)	*Atropa belladonna, Datura stramonium*
Opium alkaloids (codeine, morphine)	*Papaver somniferum* (opium poppy)
Reserpine	*Rauvolfia serpentina*
Catharanthus alkaloids (vincristine, vinblastine)	*Catharanthus roseus* (Madagascar periwinkle)
Physostigmine	*Physostigma venenosum* (Calabar bean)
Pilocarpine	*Pilocarpus* spp.
Quinidine, quinine	*Cinchona* spp.
Colchicine	*Colchicum autumnale* (autumn crocus)
Cocaine	*Erythroxylon coca*
d-Tubocurarine (curare)	*Strychnos* spp., *Chondodendron tomentosum*

note that pyrethrins (a group of natural insecticides) are annually extracted from about 25,000 tons of pyrethrin flowers (Levy, 1981). The first reported commercial processes based on plant cell culture are for the manufacture in Japan of berberine, ginseng and shikonin (Payne et al., 1991), the last of which has both medicinal properties and is also used as a pigment (Curtin, 1983; Fujita et al., 1981a; b).

BIOTECHNOLOGY AND AGRICULTURAL IMPROVEMENT

The potential role of biotechnology in the improvement of crop species is enormous and promises significant economic advantages within the next few decades. Some of the ways that biotechnology will interface with agriculture include (Ammirato et al., 1984):

- the large scale and rapid production of genetically uniform plants that can be introduced to the field in an automated, efficient and controlled fashion;
- the selection of novel and improved varieties of plants using somaclonal and gametoclonal variation techniques;
- the application of protoplast fusion technology for the generation of novel hybrids between cultivars and between species;
- the introduction of new genetic material to plants using recombinant DNA technology.

These technologies are also of interest in the development of improved plant species for chemicals production, particularly in the case of the last three examples mentioned above. The underlying scientific knowledge of the biochemical and genetic mechanisms that allow for the use of these technologies is still rather poorly understood at the present time. Yet, all of the techniques mentioned above have been clearly demonstrated for at least a few agriculturally important species. One of the great attractions of all the technologies is the speed and scale at which they can be carried out, relative to conventional methods of crop improvement and crop breeding. An example of the kind of system required by the first technology mentioned above is shown in Fig. 10.2:

i. A unique genetic variant or hybrid plant is selected.
ii. Cell cultures are established.
iii. Cells may be transferred to liquid medium during first scale-up to increase the number of regenerated plants.
iv. Cells can then be transferred to bioreactors for further scale-up.
v. Embryos produced in bioreactors or cell-suspension cultures can be staged for reliable production of plants. (At this point, it is necessary to develop methods that permit induction of dormancy if artificial seed-delivery systems are to be used.)

vi. Young plants are removed from tissue culture and transferred to the greenhouse, nursery or field. This step is delicate and may require up to one month, even for annual crops. Experimentation continues, to develop efficient delivery systems; these include encapsulation, use of gel suspensions and use of seed tapes.

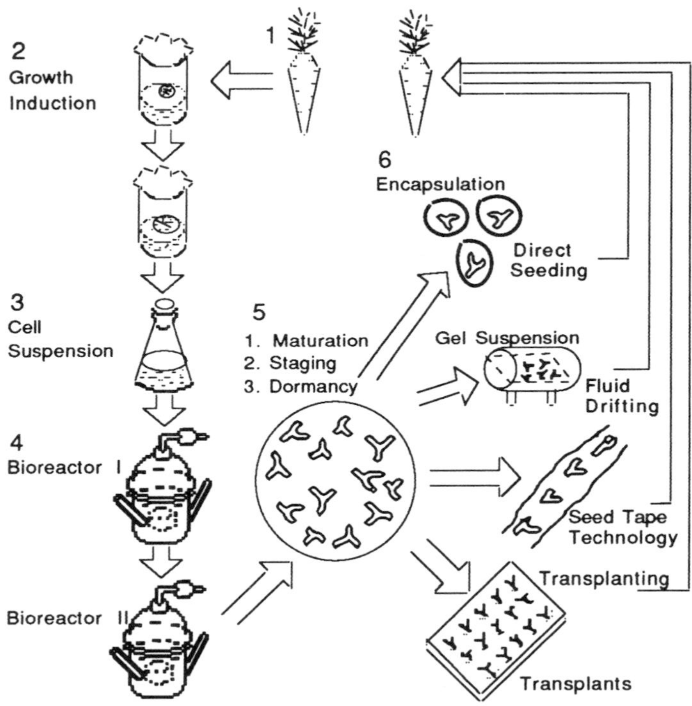

Figure 10.2 Large scale propagation system for higher plants (After Sharp et al., 1984).

The relationship between crop improvement and chemicals production can be viewed on a simple level as whether one is interested in biomass (crop improvement) or products (chemicals production) from cell and tissue culture. This perspective helps to bring the two fields into focus. The main distinction is in the particular plant species utilized, that are obviously quite different for crops and for chemicals manufacture.

BIOPROCESSES RELEVANT TO PLANT BIOTECHNOLOGY

The process engineering requirements of plant cell, tissue or organ systems are quite different than those typically seen with bacteria or non-plant eucaryotic systems. The sizes of plant cells or cell aggregates, for example, are almost four orders of magnitude larger than a characteristic bacterial cell. Organized plant structures in culture are macroscopic and easily seen with the naked eye. The continuum of possible sizes of plant structures in culture follows the embryogenic development of a plant, even if the cell arises from somatic tissues. Some stages are not part of the embryogenic process, such as callus formation, but rather precede it. The ability to handle effectively the variety of possible plant structures is a formidable task, but one which is necessary for successful commercial exploitation both for chemicals production and for crop improvement.

One of the most important enabling technologies for agricultural improvement is large-scale micropropagation with various types of plant cells and tissues. Micropropagation processes that utilize organized plant structures derived from meristem, shoot tip or axillary bud cultures are, in fact, routinely carried out at thousands of research and industrial laboratories around the world. The techniques are extremely labor intensive and have not been successfully automated. The use of visibly organized plant structures for chemicals production has been carried out on a laboratory scale, but few scale-up ideas have been put forth. At a smaller level of plant structure sizes, such as with callus or embryo cultures, scale-up of the processes is more advanced and of great current interest. The development of a large-scale system for micropropagation utilizing somatic embryos, as shown schematically in Fig. 10.2, would allow for the clonal propagation of plants on a huge scale in place of their propagation by seed. Seeds are a problem in some species due to their poor germination rates, and, of course, plants propagated by seed are not truly clonal. Also, the production of hybrid seed is a problem in plants that do not express male sterility. Callus and embryo structures may also exhibit different metabolic pathways than those expressed in cell suspensions and are thus valuable for the production of compounds that require these pathways.

The greatest amount of work on scale-up of bioprocesses relevant to plants has been with cell suspension cultures. The potential for chemicals production and biomass generation from large scale plant suspension cultures has indeed been recognized for some time (Tulecke and Nickell, 1959; Wang and Staba, 1963; Carew and Stabe, 1965); however, many factors remain to be resolved. Some of the factors limiting scale-up are associated with:

- the growth rates and growth characteristics of the cells in suspension;
- mixing and mass transfer in large scale reactor systems;
- the long-term variation, i.e. instability, of cells in culture;
- product formation, release and associated downstream processing steps;
- process modeling, synthesis and optimization.

In some instances, cell suspension culture is used to carry out a single step bioconversion. Two processes have been reported along these lines - the bioconversion of codeinone to codeine (Furuya et al., 1984) and the hydroxylation of digitoxin to digoxin (Alfermann et al., 1980).

Downstream steps are also an important process component and have received only limited attention. In general, the types of downstream processes needed to extract chemicals from cell culture would be similar to the steps involved in their extraction from whole plants. The technology used in the downstream processing of products from recombinant microorganisms or animal cell culture are not appropriate here unless the product happens to be a protein. The extraction, separation and purification steps can generally use harsher conditions than those usually employed in the biotechnology industry. A major emphasis is needed, however, in the integration of these steps into an overall process system.

LARGE-SCALE MICROPROPAGATION VIA SOMATIC EMBRYOGENESIS

Somatic embryos are immature plants derived from nonreproductive (somatic) tissue and were first observed in culture more than 30 years ago (Steward et al., 1958). The embryos possess, ideally, a bipolar structure so that a shoot and root apex appear simultaneously. This is necessary for normal subsequent development into a whole plant. Large quantities of embryos can be grown from cell suspensions and, if the cell suspension is originally derived from a superior plant, then it is possible to generate large quantities of this plant for later field growth. Unfortunately, not all plant species have been induced into generating somatic embryos or have been able to regenerate plants from somatic embryos if they do develop, especially in the important family Graminae (cereals and grasses). Later results suggest, however, that much progress is being made in this area (Vasil, 1985). Embryos have also been used for the production of flavor compounds in the laboratory (Al-Abta et al., 1979) and thus the technology that allows for propagation may also be used for chemicals production on a large scale.

Flores (1991) points out that isolated root cultures can be excised from *in vitro* grown plants, or obtained by genetic transformation with the plant pathogen *Agrobacterium rhizogenes* ("hairy roots"). The latter clones can reproduce the biochemistry of the actual plant root and stable production of secondary metabolites is possible *in vitro*. Unlimited growth is possible at growth rates comparable to those of cell suspensions.

The synthesis of bioactive root metabolites can also be induced by treatment with fungal elicitors, providing a strategy for enhanced production as well as a system for study of metabolic regulation. *In vitro* cultured roots can also express metabolic functions not generally associated with an underground organ, namely, photosynthesis and photoautotropy.

Those embryos that can be regenerated in culture and will successfully give rise to plants are the prime candidates for further development. Successful scale-up requires precise control, so that all embryos develop synchronously. Effects of growth hormones (primarily abscisic acid) and mixing have been studied (Kessel and Carr, 1972; Ammirato, 1983) in species of carrot (*Daucus carota*) and caraway (*Carum carvi*). The extension of such studies to other crops and the development of rigorous quantitative models need to be carried out so that the scale-up criteria can be more firmly established. The correlation of embryo variants, timing of development and induction of dormancy with media hormone levels and oxygen transfer in vessels of different sizes has not yet been fully reduced to quantitative measures that allow for logical process design. None of the scale-up correlations that exist in the fermentation literature are expected to be immediately applicable here because of the unique developmental stages that take place in cultures of plant tissue. However, the engineering principles and methodologies that have previously proved so successful can be applied.

Rodriguez-Mendiola et al. (1991) found that design modifications of 9-L airlift and 9-L column-mesh bioreactors permitted the easy inoculation and evenly distributed growth of hairy root cultures of *T. foenum-graecum* and *Nicotiana rustica*. Growth properties of carrot hairy root cells were investigated by Kondo et al. (1989). A turbine-blade reactor and an immobilized rotating drum reactor system were found to be advantageous because of high oxygen transfer coefficients. Maximum growth rates were nearly identical in the two reactors. Park et al. (1989) report that *Artemisia* shoots can be cultivated in a modified RT (revised tobacco) medium containing one carbon source (glucose) and one nitrogen source (potassium nitrate) under completely submerged conditions with aeration and illumination. In a nonmechanically agitated bioreactor, shoots can be cultivated to increase the biomass up to eight-fold in four weeks.

Buitelaar et al. (1991) studied growth and thiophene production by hairy root cultures of *Tagetes patula* in various two-liquid-phase bioreactors. The cells used were hair root cultures obtained by the transformation of *T. patula* with *Agrobacterium rhizogenes*. Hexadecane and FC40 were the solvents of choice for further experiments; the best results were obtained using hexadecane as the dispersed phase in a bubble column, with excretion of 30 to 70% of the total thiophenes into the hexadecane.

Whitney (1992) points out that growth in culture of plant roots transformed with *A. rhizogenes* provides a method for the production of many commercially valuable products. The yield achieved is critical for the financial viability of the method, and the design of the bioreactor can have a profound effect on the yield and growth characteristics. Most bioreactors are designed for the growth of microorganisms, animal cells or plant cells in suspension, and are not ideally suited for the growth of transformed roots. Novel bioreactors have therefore been designed specifically for the culture of transformed roots. This comparison of novel and conventional culture vessels indicates that the growth of transformed roots can be greater in a mist of culture medium than when submerged or in a trickling film of medium.

An alternative design that is amenable to scale-up involves the development and growth of the embryos on miniature membrane rafts in large numbers of compact multi-well systems. Hamilton et al. (1985) have demonstrated the possibility and ease of this procedure on a limited scale with *Atropa belladonna*. This novel system is scaled-up by simply increasing the number of wells containing plant embryos. Prenosil et al. (1987) obtained excellent cell growth results with membrane rafts in the production of purine alkaloid, using *Coffea arabica* cells. The authors attribute these results to better oxygen supply, enabling the construction of a continuous process.

The handling of the embryos for transfer to the field is also an important research area. Hamilton et al. (1984) have demonstrated the ability to encapsulate plant cells in gels; this can be extended to embryo cultures. Such so-called *artificial seeds* have been field tested; e. g., with celery crops by Plant Genetics, Inc. (Davis, CA).

In a more recent development, Pedersen (1990) has reported that bioreactors for growth of somatic embryos can serve as massive cloning systems in a plant propagation program. The developmental process is conveniently followed using image analysis hardware and software. Embryo development is modeled according to standard classification schemes, as previously developed by plant physiologists. Dependent variables include carbohydrate concentrations (sucrose, fructose and glucose) and biomass apportioned among viable and nonviable embryo types, cell suspension aggregates and morphologically distinct embryo types. A quantitative formulation of embryo development from cell suspension cultures has been developed based on the data obtained by the image analysis system. This allows, for the first time, a rigorous approach to media optimization and reactor scale-up for embryo formation.

GROWTH RATES AND GROWTH CHARACTERISTICS OF PLANT CELLS IN CULTURE

Growth of large amounts of plant cells or tissues in culture is hampered by extremely slow growth rates. Doubling times for plant cells in cultures vary from 20 to 100 hours, and this often necessitates the use of large reactors to obtain reasonable productivities when one is concerned with chemicals production (see Table 10.2). Contaminating organisms are a problem under these conditions and must be dealt with by extreme attention to aseptic conditions. Cell suspensions also tend to grow as aggregates of varying sizes. Cells are also sensitive to fluid shear, and the power input to the reactor is thus limited. This restriction works against the need for uniform conditions inside the reactor and often precludes the direct use of fermentors that have been designed for microbial fermentations. Nevertheless, some large scale systems have been successfully operated, including a two-stage 950 liter fermentor for the production of shikonin, which enables separation of growth and product formation phases, as shown in Fig. 10.3 (Curtin, 1983) and a 750 liter fermentor for the production of *Catharanthus* alkaloids (Berlin, 1984). Fermentors as large as 20,000 liters have been reported, as well, for the production of tobacco cell mass (Noguchi et al., 1977).

The use of air lift fermentors has been advocated as a means of overcoming the sensitivity to shear while still maintaining adequate oxygen and mixing characteristics (Wagner and Vogelmann, 1977). Scale-up of such fermentors has shown mixed results in that the productivity has either decreased (Fuller, 1984; Berlin, 1984) or increased (Wagner and Vogelmann, 1977) in going to larger sizes. The hydrodynamics of such systems are complex to model.

Table 10.2 Volumetric Productivities for Some Products Made in Plant Cell Culture and their Corresponding Value (Payne et al., 1991)

Product	Cell line	Productivity [g/(Lxday)]	Investigators	Bulk price to give 15¢/(Lxday) revenue ¢/g	Bulk price (Fall,1989) ¢/g
Anthocyanin	Grape (Bailey Alicant A, *Vitis* hybrid)	0.06	Yamakawa et al. (1983)	250	-
		0.15	Hirasuna (1990)	100	-
Berberine	*Thalictrum minus*[a]	0.05	Kobayashi et al. (1988)	300	325[b]
	Coptis japonica[c]	0.60	Fujita et al. (1989)	25	325
Diosgenin	*Dioscorea* spp.	0.75	Fowler & Scragg (1988)	20	105[d]
Podoverine	*Podophyllum versipelle*	0.15	Ulbrich et al. (1988)	100	-
Rosmarinic acid	*Coleus blumei*	0.91	Ulbrich et al. (1985)	16	-
Sanguinarine	*Papaver somniferum*	0.034	Park et al. (1990)	440	-
Shikonin	*Lithospermum erythrorhizon*	0.15	Tabata & Fujita (1985)	100	400[e]

[a]Excreted into the medium, where it crystallizes.
[b]Special quote from Sigma Chemical Co. (St. Louis, MO) for 1 kg of berberine.
[c]Continuous flow system with cell retention; intracellular product.
[d]Sigma Chemical Co. (St. Louis, MO) $105/100g; $2,420/kg from ICN Biochemicals.
[e]From Curtin 1983, for 1983 wholesale prices.

Kreis and Reinhard (1990) describe a two-stage cultivation method employed to develop a semicontinuous biotransformation process for the production of deacetyllanoside C, a cardenolide of the important digoxin series. Digitoxin was used as the substrate for biotransformation. The process was optimized in 1-liter shake flasks and then established on the 20-L scale using two airlift bioreactors, one for cell growth (working volume 12 L) and another for deacetyllanatoside C production (working volume 18L). Growth and production phases were synchronized and the process ran successfully in several semicontinuous one week cycles. Matsushita et al. (1991) found that the use of fine bubble sparging in an airlift column permitted the high density culture of carrot cells, these results being superior to those obtained in other types of culture vessels.

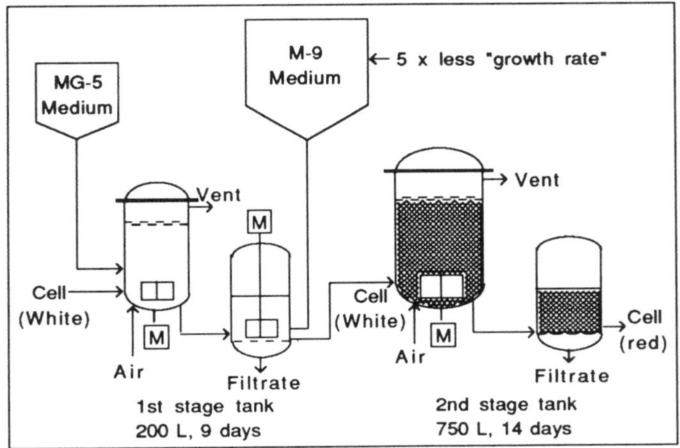

Figure 10.3 Shikonin process.

Several factors enhancing berberine level in an airlift reactor system were found (Cho et al., 1988). These are continuous illumination, utilization of indole 3-acetic acid (IAA) as a growth regulator instead of 2,4-dichlorophenoxy acetic acid (2,4-D), lowering initial phosphate level, addition of ethephon in the early stage, application of 8% sucrose solution as a production medium, and $CuSo_4$ treatment at the late stage of the culture. Process strategies were tested to use the above conditions collectively and in an optimal manner. Combination of positive effectors was discovered to have a synergistic effect and increased product formation significantly. An airlift reactor system was favorable for growth unless the cell density was too high. For berberine production, continuous gas-stripping played a significant negative role. Carbon dioxide and ethylene proved to be very important for the formation of berberine.

In analyzing the growth of airlift cultured plant cells, Buech et al. (1991) present a simple cell growth model to describe the growth behavior as a function of inoculum size, invertase activity, and specific growth rate μ. When C-heterotrophic pear cells were cultivated in a 5-L airlift reactor, the effect of the inoculum size on the growth curve was as predicted. By contrast, a comparative study using peppermint and orange cells did not confirm a positive correlation of higher invertase activity and reduction of the lag phase. Applying standard cultivation protocols, most of the relevant ionic constituents of the nutrient medium and, in the initial cultivation phase, pO_2, are by far overdosed, thus reducing μ and maximum fresh weights.

Kim et al. (1991) describe a hybrid bioreactor, which was developed for the production of secondary metabolites from high density cultivation of plant cell suspensions. Some of the advantages of both airlift and cell-lift by agitation were combined. The addition of a decanting column also made it possible to run a perfusion system for high density culture or to run a two-stage culture efficiently. Cell growth and the production of berberine from *Thalictrum rugosum* in the hybrid bioreactor are reported. A cell density ≤31g/L was obtained by perfusion without problems in mixing or loss of cell viability. The specific berberine productivity was comparable to that in shake flasks. The maximum berberine concentration was 88 mg/L at 3 weeks and declined thereafter.

A novel bioreactor using magnetically stabilized fluidized bed (MSFB) technology was developed (Bramble et al., 1990) that has certain advantages for cultivating cells continuously. In this system, the cells are protected from shear and are constrained to move through the fermentor in lock-step fashion by being immobilized in calcium alginate beads. The authors claim that the MSFB permits good mass transfer, minimizes particle collisions, and allows for the production of cells while maintaining a controlled cell residence time.

The aggregation of cells in suspension culture leads to a heterogeneous population that confounds the analysis and operation of the reactor types mentioned above in many cases. For nongrowth associated products, immobilization of cells provides a means to circumvent various problems, including aggregation. Immobilization can be carried out to lead to a uniform (homogeneous) degree of aggregation or to eliminate aggregation as a confounding factor. Furthermore, in continuous systems, the slow growth rates limit the dilution rates for free cells in suspension, whereas for immobilized systems the dilution rate is not similarly limited. Thus, there are a number of advantages in working with immobilized plant cell systems. Reviews of the techniques and reactor configurations for immobilization of plant cells have been presented (Prenosil and Pedersen, 1983; Brodelius, 1983; Shuler, 1981); these include membrane reactors (Prenosil and Pedersen, 1983) and packed beds (Hamilton et al., 1984). Encapsulation methods such as those mentioned previously for the preparation of artificial seeds also offer distinct advantages in the immobilization of cells for chemicals production.

Another important observation concerning growth characteristics of plant cells deals with the separation of growth and production phases, especially for a large number of secondary metabolites. Certain media components will favor growth while other components will favor secondary metabolite production. This has been extensively documented with pigment formation by *Lithospermum erythrorhizon* (Yamada and Fujita, 1983). Whereas the reported work on *L. erythrorhizon* utilized two-stage reactors to physically separate growth and production phases, an immobilized system might be used to simply switch or cycle the media input streams in an optimal manner.

MIXING AND MASS TRANSFER

The aggregation of cells due to their natural growth characteristics in suspension culture, or due to an imposed condition by various immobilization schemes leads to mass transfer and mixing constraints. An important parameter in non-photoautotrophic cultures is the oxygen transfer coefficient that is typically used as the scale-up criterion for the air lift systems mentioned previously. Other factors can also be expected to be important, such as the local carbon to nitrogen balance. These factors are made even more important in plant cell cultures since the relative balance of nutrients can lead to distinct morphological changes and even to differentiation of the plant cells. It can certainly be expected that the different cell types will possess different biochemical or regenerative abilities. In cell cultures that are selected to be photoautotrophic and can thus utilize photosynthetic pathways in culture, carbon dioxide transport is a major design factor. The possibility of oxygen toxicity exists.

Leckie et al. (1991a; b) examined the effect of long-term high shear and impeller design on the growth and alkaloid accumulation by *C. roseus* cultures grown in stirred-tank bioreactors. As the impeller speed was increased from 100 to 300 rpm, the growth rate increased and the maximum biomass and alkaloid accumulation decreased. An inclined impeller reduced the decrease in aggregate size as the impeller speed was increased to 300 rpm, but the expansion index was not affected. *C. roseus* also proved capable of growth at an impeller speed of 1000 rpm, and inclined impellers increased the growth rate, alkaloid accumulation and expansion index. In addition, experimental results indicated that cultures were adversely affected by a certain level of shear stress which resulted in lower biomass and alkaloid accumulation. The increased levels of shear stress present in a baffled bioreactor, compared to those in an unbaffled one, meant that the cultures grown within the former displayed these effects at a lower speed than the cultures within the latter. Cultures grown in the bioreactor with baffles also appeared to have an increased level of cell breakage with a concomitant release of alkaloid into the medium. The medium alkaloid kept overall alkaloid production levels higher than those achieved by cultures grown in the unbaffled bioreactor.

A perfusion fermentation of *A. officinalis* was carried out in a stirred tank bioreactor integrated with an internal cross-flow filter (Su and Humphrey, 1991). Bubble-free aeration via microporous membrane fibers was used to provide oxygen. A two-stage culture was successfully conducted in this reactor without filter fouling. In a seventeen-day fermentation, a cell density of 26 g dw/L and a rosmarinic acid productivity of 94 mg/L-day were achieved. This productivity is three-fold that obtained in a batch culture.

Alfalfa cells were grown in 500 ml, aerated and stirred batch bioreactors using Schenk and Hildebrant medium by McDonald and Jackman (1989). For cultures in which the pH was allowed to vary, two fairly distinct growth phases were observed. Evidence is presented which indicates that the two-phase growth is most likely a result of the two nitrogen sources in the medium. The ammonium present in the medium is directly utilized during the first growth phase, and ammonium resulting from intracellular nitrate reduction is utilized during the second phase. During the first growth phase, sucrose is completely hydrolyzed to glucose and fructose with some glucose and fructose consumption. In the second growth phase, glucose is consumed preferentially over fructose.

It has been observed for gel immobilized microorganisms (e.g., Chotani, 1984) that mass transfer limits the growth of cells to extremely narrow shells near the surface of the gel beads. Plant cells also show preferential growth near the surface, where nutrient levels can be expected to be highest (Hamilton et al., 1984). This poor utilization of the reactor volume leads to less than optimal biomass productivities. By developing smaller diameter beads or by encapsulating cells in an environment that allows for greater mass transport, productivities are increased.

Sonomoto et al. (1990) constructed a system for the production of blue pigments by gel-entrapped cultured cells of *L. vera*. Blue pigments were synthesized *de novo* in Linsmaier-Skoog (LS) medium in the presence of L-cysteine by the calcium alginate-entrapped cells as well as by the free counterparts. The entrapped cells could be employed for the repeated production of the pigments over seven months, by alternating the growth (activation) and production phases.

CELL STABILITY

It is well known that plant cells in suspension undergo gradual but continuous changes in their genetic and, ultimately, biochemical characteristics. In operation of reactors for chemicals production or in the cloning of elite cell lines, stability is to be encouraged. This is in contrast to the purposeful variability induced in cell cultures in order to select for such elite cell lines. One method of maintaining stable cell lines is to limit their growth, and this has been observed in immobilized plant cell systems. Apparently the high density culture that is achieved by the immobilization process slows cell growth even further than its usual low value. If the cells can be reused and maintained in active nongrowth states, then this is a preferred mechanism of operating plant cell reactors. In other cases where cell growth is a requisite activity, such as in somatic embryogenesis, other factors for selecting against instability, or controlling stability specifically, must be sought. A very significant fraction of the current research activity in plant cell biotechnology is focused on immobilized cell procedures and processes. Some results are summarized in Table 10.3.

To provide some added flavor, we note the studies of Prenosil and Hegglin (1990) and Archambault et al. (1990). In the former, it was found that cell suspension cultures of *Coffea arabica* and *Nicotiana tabacum* could be transformed by a simple procedure into cell lines growing in dense aggregates (self-immobilization). These aggregates are large enough to be used in an expanded-bed reactor with external aeration. The aggregates in such a reactor are practically stationary, with negligible shear stress. The cell mass grows very fast, with a doubling rate of less than 48 hours, and the lag phase could be virtually eliminated by leaving a part of the aggregates and medium in the reactor for the next run. This renders the expanded-bed reactor especially useful for plant cell biomass cultivation, for example, as a first stage in a plant cell culture reactor system. In the latter study, the scale-up of the technique of plant cell surface immobilization was performed successfully in specifically designed laboratory scale bioreactors. The immobilizing matrix was formed into a vertically wound spiral providing for a high immobilizing area-to-volume ratio (0.8-1.2/cm). A modified airlift and a mechanically stirred vessel delivered a bioreactor performance characterized by low biomass frothing and highly efficient plant cell attachment and retention (≥96%). The growth of *Catharanthus roseus* cells investigated in these bioreactors was not mass transfer-limited; it required but mild mixing and aeration levels. The biomass formation pattern of surface immobilized plant cells generally exhibited a linear growth phase, followed by a stationary phase characterized by the

Table 10.3 Immobilized Plant Cells

Species	Immobilization Method	Products	References
Apium graveolens	Entrpment in alginate, carrageenan, chitosan	(test for viability)	Beaumont & Knorr, 1987
Beta vulgaris	Entrp. in recticulate polyurethane foam	(test for growth)	Rhodes et al., 1987
Brassica napus	Adsorption to Cytodex 1 treated with lectins	(test for adsorption)	Bornman & Zachrisson, 1987
	Adsorp. to Cytodex 1,2,3 Biosilon, Serparon, Sorfix and Sepharose	(test for adsorption)	Vankova & Bornman, 1987
Capsicum frutescens	Entrp. in recticulate polyurethane foam	capsaicin	Mavituna et al., 1987
Catharanthus roseus	Adsorp.on glass, SPS, PET, PS and FEP	(test for adsorption)	DiCosmo et al., 1988
	Adsorption to alginate bead	ajmalicine isomers	Kargi, 1988
	Adsorp. to alginate,agar, gelatin, polypropylene bead & glass	(test for adsorption)	Kargi & Freidel, 1988
	Entrp. in membrane	ajmalicine isomers	Payne et al., 1988
	Entrp. in hollow fiber	phenolics	Shuler et al., 1986
	Adsorp. on glass, SPS, PET, PS and FEP	(test for adsorption)	Facchini et al., 1988a
			Facchini et al., 1988b
			Facchini et al., 1989
	Entrp. in alginate	(test for diffusion of sucrose)	Pu & Yang, 1988
Chenopodium rubrum	Entrp. in alginate & chitosan	amarathin	Knorr & Berlin, 1987
Cinchona ledgeriana	Entrp. in recticulate polyurethane foam	(test for growth)	Rhodes et al., 1987
Coffea arabica	Entrp. in alginate	caffeine	Haldimann & Brodelius, 1987
	Entrp. in alginate	caffeine	Kim, 1987
Cynara Cardunculus	Entrp. in alginate	coagulation of milk	Esquivel et al., 1988
Digitalis lunata	Entrp. in alginate	digoxin, β-methyl-digoxin	Petersen et al., 1988

Table 10.3 Cont.

Species	Immobilization Method	Products	References
Dioscorea deltoidea	Entrp. in polyurethane foam	diosgenin	Ishida, 1988
	Entrp. in alginate	cis- and trans-glucoside	Vanek et al., 1989
Humulus lupulus	Entrp. in recticulate polyurethane foam	(test for growth)	Rhodes et al., 1987
Lithospermum erythrorhizon	Entrp. in hollow fiber	phenolics	Kim et al., 1989
Mald. domestica	Entrp. in alginate	(test for diffusion of sucrose and yohimbine)	Pu & Yang, 1988
Nicotiana tabacum	Entrp. in alginate	2,3,4-Decanol	Hamada et al., 1988
	Entrp. in alginate	phenolics	Schmidt et al., 1989
Papaver somniferum	Entrp. in alginate	codeine	Furusaki et al., 1988
Rosa sp.	Adsorp. to Cytodex 1,2,3 Biosilon, Serparon, Sorfix & Sepharose	(test for adsorption)	Vankova & Bornman, 1987
Silybum marianum	Entrp. in alginate, agar gelatin & polyacrilamide	clotting of milk	Cabral & Kennedy, 1987
Solanum surrattense	Entrp. in alginate	solasodine	Barnabas & David, 1988
Tagetes minuta	Entrp. in alginate, agar, agarose, carrageenan	non-polar thiophenes	Ketel et al., 1987
	Entrp. in agar, gellan and carrageenan	(test for viability)	Buitelaar et al., 1988
Thalictrum minus	Entrp. in alginate	berberine	Kobayashi et al., 1987
	Entrp. in alginate	berberine	Kobayashi et al., 1988

presence of residual carbohydrates in the medium, contrary to suspension cultures. This behavior depended on the plant cell type and/or line cultured, as well as on the inoculum age. The space restriction and uni-directional growth of the surface immobilized plant cell biofilm, combined with the limited availability of essential intra-cellular nutrients rapidly accumulated from the medium by the stationary phase inoculated plant cells, all probably contributed to departure of the culture from usual behavior.

Facchini and DiCosmo (1991) describe a plant cell bioreactor for the production of protoberberine alkaloids from immobilized *Thalictrum rugosum* cultures. Maximum biomass and protoberberine alkaloid levels were maintained for more than 14 days in immobilized cultures. In contrast, fresh weight, dry weight and total alkaloid content decreased in suspension cultures following the linear growth phase. The scale-up potential of an immobilization strategy based on the spontaneous adhesion of cultured plant cells to glass fibers was demonstrated. *Erwinia chrysanthemi* cells were used to study the possibility of producing bacterial enzymes in a bioreactor coupled with a membrane filtration unit (Denis and Boyaval, 1991). Continuous fermentations with total cell recycle failed to give good production of pectate lyase. Enzymic mechanical and physicochemical damage was involved. With a sequential recycle mode, productivity of 1.5 units/L-h was obtained with a high enzyme concentration. Protease accumulation occurred when the bioreactor was coupled to a filtration unit. There was no loss of activity due to high shear stress caused by pumping.

PRODUCT FORMATION, RELEASE AND DOWNSTREAM PROCESSING

Products formed in plant cells are usually intracellular storage products and must therefore be released from the cells if the product is to be continuously collected in the medium. This is especially important in immobilized cell systems. There are several environmental factors that will allow for release of products, such as addition of solvents that permeabilize the cells (Felix et al., 1981) and modifications in the media pH (Renaudin, 1981).

Regarding the pH effect, the chemiosmotic theory of energy conservation based on the Mitchell hypothesis (already discussed for bacteria in Chapter 7) provides the basis for the movement of solutes across plant cell membranes. For instance, in the case of the chloroplast thylakoid membrane, the oxidation of water and the resulting photosynthetic electron transport activity result in the generation of NADPH, the reducing power required for the synthesis of reduced carbon compounds. As pointed out by Uribe and Lüttge (1984), electron transport is necessarily linked to the vectorial transport of hydrogen ions across the thylakoid membrane, which has a reduced permeability for protons. A direct consequence of electron transport is thus a differential in proton concentration across the membrane; in this case, the interior of the granum enclosed by the membrane is enriched in protons. This establishes the electrochemical proton gradient $\Delta\bar{u}_{H+}$, the energy of which may be used to bring about the synthesis of high-energy phosphate bonds. ATP photophosphorylation by spinach thylakoid was examined to evaluate its use as an ATP regeneration reaction in biosynthetic reactors that consume ATP. When phosphoryl transfer reactions were coupled to cyclic photophosphorylation, ATP was continuously regenerated by thylakoids over 14-24 cycles in batch reactors (Yu and Hosono, 1991).

In the area of cell permeabilization, it has also been demonstrated that operation in two-phase reactors where the products are extracted into the nonaqueous phase can be effective. Both liquid-liquid (Bisson et al., 1983) and liquid-solid (Knoop and Beiderbeck, 1983) systems have been used. The continuous release of products in this manner would also favor their continuous production since the storage capacity of the cells would not become saturated. Lipophilic products, such as fragrances (essential oils) and alkaloids, can be produced efficiently by such methods. Downstream processing is also simplified by the preliminary extraction into a second phase. Subsequent purification of the product and general downstream processing for the secondary metabolites can use relatively harsh techniques that are familiar to the chemical process industries.

PROCESS MODELING, SYNTHESIS AND CONTROL

Successful commercialization of processes from plant cells requires the ability to control and optimize the overall process scheme. All of the studies mentioned above have to be judged finally on the basis of process economics and this, of course, requires that all stages of the process be examined together. With respect to chemicals production, the two basic parts of the process, as in classical fermentations, are the reactor section and the downstream processing section. Unlike the current processes envisioned for recombinant microorganisms, the reactor section is likely to be the major cost variable. This is due to the inherent low productivities currently seen in plant tissue culture systems that translate into higher capital costs for large reactors. With respect to propagating plants in reactor-type systems, the preliminary background work is not sufficiently advanced even to speculate on the cost effectiveness. What is clear, however, is that the potential market is tremendous.

Some studies have been reported with hypothetical plant production processes in order to estimate production and capital costs (Kudo, 1985; Sahai and Knuth, 1985). The processes are arranged in conventional flowsheets, although some attention has been given to the possibility of working with immobilized cell systems. These exercises highlight the areas where further research can be expected to have the greatest impact on process economics. The ability to increase product yields and to excrete the product into the media are of paramount importance.

A hypothetical process for manufacturing ajmalicine and serpentine alkaloids from an immobilized plant cell culture (Kudo, 1985) is shown schematically in Fig. 10.4. A gel (e.g., calcium alginate) is used to entrap the plant cells, and downstream processing follows the organic extraction steps that are used in the laboratory. The corresponding production costs are shown in Fig. 10.5 and demonstrate the advantage of immobilization even when a continuous process is not used. Profitability can only be realized at high production levels and with yields greater than 5% (Kudo, 1985), versus a highest reported yield of only 1.3% (Zenk et al., 1977). Other plant cell systems have been looked at (Sahai and Knuth, 1985) and the conclusions regarding yield and product excretion, along with the advantages of cell immobilization, are similar.

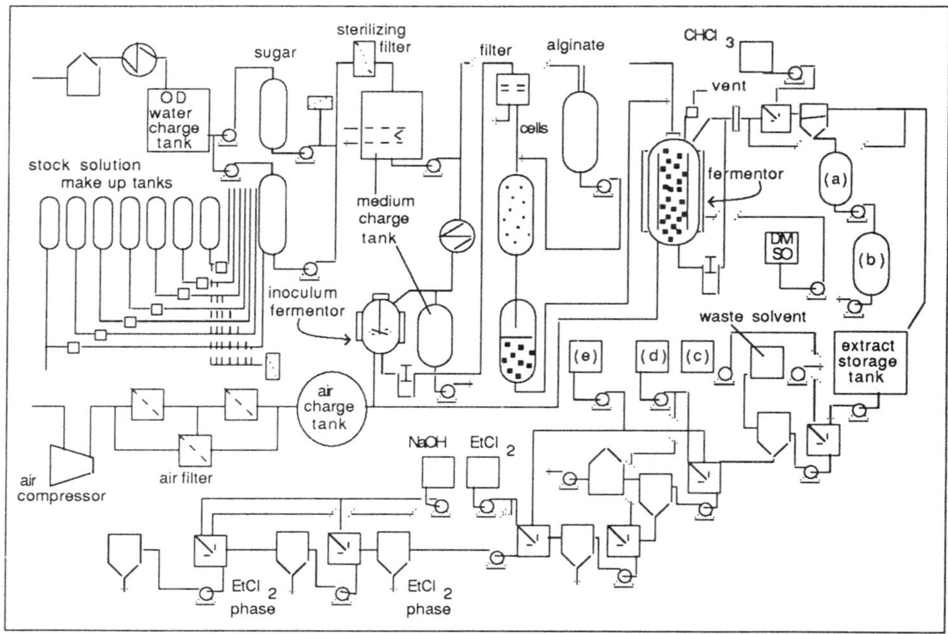

Figure 10.4 Hypothetical immobilized cell process: (a) recovered medium; (b) medium make-up tank; (c) nHexane; (d) benzene; (e) tartaric acid.

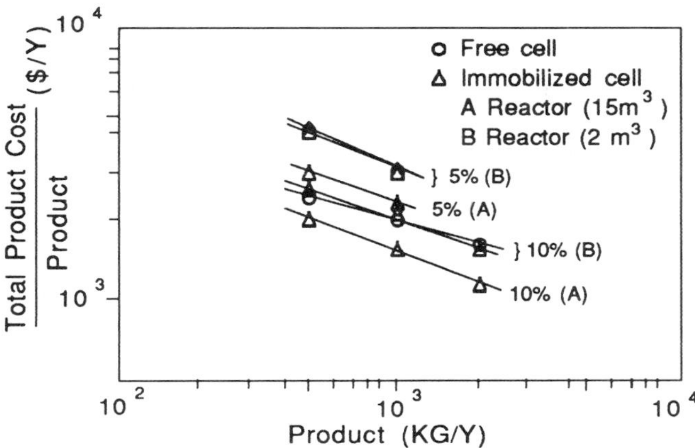

Figure 10.5 Unit total product cost as a function of product capacity, yield (5% or 10% dry weight basis) and reactor size.

An example from this latter study is shown in Table 10.4, where batch fermentation costs are projected on the basis of product yields currently obtained in plant cell cultures. The Table clearly demonstrates the advantages of high yielding immobilized systems. The case study depicted is for a 200,000 kg/yr production rate.

Table 10.4 Comparison of Production Costs in Various Plant Cell Systems (Sahai & Knuth, 1985)

	Production Cost ($/kg)	Fixed Capital Investment ($ MM)
Batch fermentation process	410	210
Batch Fermentation process	100	74
Batch fermentation process	55	45
Continuous immobilized cell process	20-25	15-20

Recently Asada and Schuler (1989) have pointed out that mass transfer within natural cellular aggregates or immobilized cell systems can be critical to product formation and organization (e.g., formation of tracheary elements). Synergistic effects of coupling cell immobilization, *in situ* product removal, elicitation, and medium optimization for ajmalicine production from *Catharanthus roseus* are possible, leading to a 60-fold increase in production.

10.1 PLANT CELL AND TISSUE CULTURE

The tools of plant cell and tissue culture are used to select for elite cell types that have application in the areas of chemicals production and improved agricultural crops. The techniques and their applications are all relatively recent. It can thus be expected that a significant amount of fundamental knowledge regarding the factors effecting plant development and variation continue to evolve. Much remains to be learned at the very basic level of cell culture in the realm of higher plants.

GAMETOCLONAL AND SOMACLONAL VARIATION

Successful growth and propagation of plant cells freely suspended in a liquid medium was realized about fifty years ago (White, 1939; Nobecourt, 1939; Gautheret, 1939). Much attention has since focused on elucidating the necessary media components that support growth and differentiation of the culture. The role of growth regulators was pointed out by Skoog and Miller (1957) and the process of somatic embryogenesis was realized by Steward et al. (1958). Work on plant protoplasts in culture (Cocking, 1960) that allows for somatic hybridization of different plant species, and work on anther cultures (Bourgin and Nitsch, 1967) which produce haploid cells and plants, opened up new highways for carrying out genetic modifications that greatly accelerated the pace at which plants could be screened for new and improved properties. Also in the late sixties, it was reported that genetic variability could be found in cell suspension cultures (Murashige and Nakano, 1967), usually after the cultures had been maintained for long periods of time. The recovery of varied plants from a somatic cell suspension culture quickly leads to the realization that this technique, referred to as somaclonal variation, could be used for crop improvement (Larkin and Scowcroft, 1981) and for the selection of cells with altered biochemical pathways (Chaleff and Parsons, 1978). Recently, the variation in plants regenerated from cell cultures of tomato have been used to ascertain the genetic basis of somaclonal variation (Evans and Sharp, 1983). The variability arises both from the explant source (intrinsic) and from spontaneous variation (extrinsic) during culture propagation. Selective pressure or simply bulk selection can and has been used to obtain regenerated plants or cells with desired characteristics.

The use of somaclonal variants in sugarcane and potato has already been successful for selection of improved varieties, particularly with respect to disease resistance (Evans et al., 1984). The applications of somaclonal or gametoclonal (where the source material is gamete derived) variation/selection in cell culture are carried out with crop species that are capable of being regenerated into plants for agricultural improvement or with species of plants that produce interesting chemicals. The traits to be selected-for are crop specific in the former case, but pertain generally to higher yield in the latter case. General strategies have been put forth (Evans et al., 1984) that are applicable in both situations.

Another method of introducing modifications into plant cell culture is by the formation of somatic hybrids via protoplast fusion. Fusion of cells between closely related species is necessary in order to obtain functional hybrids where the regenerated plants are not sterile. Successful utilization of the technique for agricultural improvement also depends on the ability to regenerate the plant from protoplasts and to direct the chromosome substitution between parental genomes in a predetermined manner. Cytoplasmic genetic traits that can be traced to the mitochondrial or chloroplast genome include herbicide resistance, disease resistance, male sterility, antibiotic resistance and pigment variegation (Evans and Sharp, 1983). Since segregation of chloroplasts and recombination of mitochondrial DNA is known to occur in many instances following protoplast fusion, novel genetic combinations (nuclear plus cytoplasmic) can take place that would otherwise never happen. The technology of protoplast fusion may also be of interest in the generation of novel cell lines for chemicals production. Cell culture of somatic hybrids is also susceptible to somaclonal variation as mentioned above, and thus further genetic modifications are possible.

AUTOTROPHIC AND AUXOTROPHIC CELL LINES

The selection of cell lines using the techniques mentioned above, and the characterization of these cell lines in interesting crop or chemicals producing plant species, is an important enabling technology. The selection is typically very specific in terms of chemicals production, or it may be very broad when applied to agriculturally interesting traits. Furthermore, the selection for chemicals production is phenotypically precise when the cells will be subsequently used in a production scheme. However, selection for agriculturally useful phenotypes in cell culture is not precise. Some important agricultural traits include cold and salt tolerance and herbicide and pathotoxin resistance.

Various methods of selection have been described (Widholm, 1980; Flick, 1983). Amino acid analogs have been used in many cases where a direct application is known. Studies on secondary metabolite biosynthesis have focused on the shikimate pathway involving phenylalanine and tryptophan amino acids. By using tryptophan analogs, such as 5-methyltryptophan, or phenylalanine analogs, such as p-fluorophenylalanine, selection of high amino acid producing (analog resistant) lines could be achieved. These lines have also demonstrated, in some cases, increased flux of precursors through the metabolic pathway, leading to higher production of some secondary metabolites (Berlin and Widholm, 1977; Sasse et al., 1983). Selection for stress tolerance in plants may also be possible by amino acid analogs, since it has been observed that proline is overproduced in some stress-resistant plants (Kueh and Bright, 1981). The amino acid analog could be hydroxyproline in this case. Also, the selection of a lysine overproducing corn plant, for example, would be useful in correcting certain nutritional deficiencies associated with the crop. Model systems include, at first, members from the family Solanaceae, such as *Nicotiana tabacum*, *Atropa belladonna* and *Petunia hybrida*, because of the large amount of background information available on cell culture and on genetic manipulations with these species.

Other methods of selection include visual screening and chemical assaying. These are indirect methods since the selection must be performed manually or semi-automatically. High yielding cell lines of *Lithospermum erythrorhizon*, *Coleus blumei* and *Catharanthus roseus* have been selected in this fashion. Selection coupled with cell growth on a production medium has been responsible for a seven-fold increase in pigment production by suspended cells, relative to entire plants. Compared also with a 2-3 year harvest period for the entire plant, and a 30 day cycle time in suspension culture, the advantages of cell culture are obvious.

Selection against disease resistance and herbicide resistance is carried out by incorporating the toxin isolated from the infectious pathogen or by incorporating the herbicide directly into the culture. Work on developing a southern leaf blight resistant corn (*Zea mays*) line was carried out in this fashion (Gengenbach and Green, 1975), although the resistance was associated with loss of male sterility. This precluded the use of the resistant line since hybrid seed cannot be efficiently produced in male fertile corn. Extensive work on herbicide resistance has been carried out with glyphosate (Monsanto's ROUNDUP herbicide). The development of tolerant carrot cell lines (Nafziger et al., 1984) clearly demonstrates the efficacy of the method. The molecular biology of glyphosate resistance is discussed below. The further development of resistant plants is of tremendous economic importance to the farming industry.

Another area where selection is used to realize a significant advantage is photoautotrophism. Such cell lines synthesize chlorophyll and express photosynthetic activity. They are thus able to use solar energy directly and grow on simple media that typically will not support the growth of otherwise competing microorganisms. The cells also have developed chloroplasts and this may be a distinct advantage for chemicals production, since the chloroplast is the site of many important enzymes involved in the secondary metabolite pathways. The number of successful photoautotrophic cultures reported is quite small, however (Yamada and Sato, 1983). Typically, the cultures show a lack of vigorous growth under the photoautotrophic conditions. Nevertheless, some work has been done on a large scale system (Yamada et al., 1981) where mixing was achieved by aeration with a CO_2 enriched air stream. A system for selecting cells or callus in culture with high photosynthetic potential is shown schematically in Fig. 10.6 (Yamada and Sato, 1983). The cell lines are also useful for investigating the genetic,

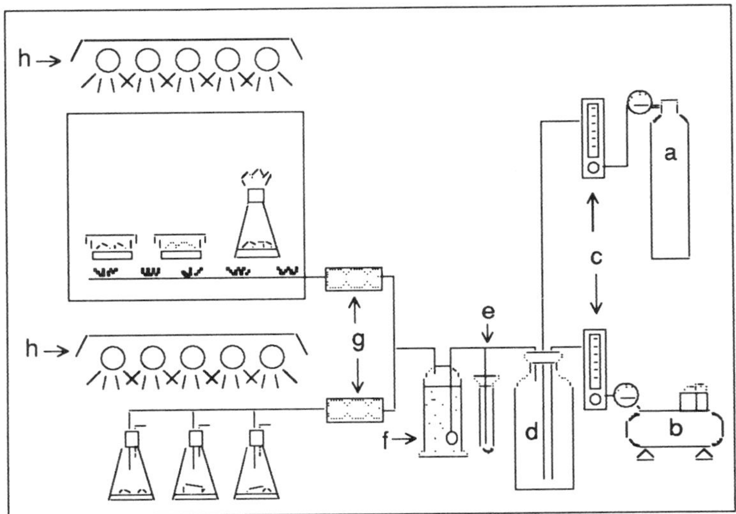

Figure 10.6 Selection system for photoautotrophic cells. (a) CO_2 gas, (b) air compressor, (c) flow control, (d) reservoir for mixed gas, (e) a safety valve, (f) distilled water (washing for gas), (g) air line filter (cotton filter, etc.), (h) illumination.

biochemical and physiological factors important in photosynthesis (Miller et al., 1983).

It is of significant value to select for auxotrophs in plant cell culture. Auxotrophs are cell lines which require special nutrient supplementation in the media in order to grow. Information from auxotrophic mutants in bacteria was and still is important in the elucidation of metabolic pathways and their regulation. A similar role for auxotrophs can be seen in higher plants. One example which has been reported (Muller and Grafe, 1981) deals with nitrate auxotrophy and nitrate reductase. Selection for auxotrophs is done by selecting for nongrowing cells or by screening vast numbers of plants. Haploids must typically be used since the auxotrophic mutant is usually recessive (Flick, 1983). To select for nongrowing cells, bromodeoxyuridine (BUdR) is effective and has been used to screen for temperature sensitive tobacco auxotrophs that have been biochemically characterized (Malmberg, 1980). The relevant goal is to gain a more thorough understanding of secondary metabolite pathways operative in higher plants. However, auxotrophs are also useful as selective markers in protoplast fusion experiments. Mutagens can be added to the media or generally exposed to the cells to increase the auxotrophic potential of cell culture.

INDUCTION SYSTEMS IN PLANT CELL CULTURE

It is well known that healthy whole plants respond to stress by production of various phytochemicals, some of which are important commercial chemicals. The selection schemes mentioned above take advantage of this process in cell culture by selecting for those cells that are best suited for responding to the stress. It is also possible, however, to use the induction of these chemicals in a nonselective scheme, i.e., to challenge the cells with an elicitor compound in order to promote the timely expression of secondary metabolites in culture. For example, microbial insult of cells and whole plants leads to production of antibiotics, and the pathways used by the plants to produce these compounds are modified by the transient expression of increased levels of key enzymes. Cultured parsley (*Petroselinum hortense*) cells have been shown to respond to a fungal elicitor preparation (a cell wall fraction of the fungus *Phytophthora megasperma* f. sp. *glycinea*) by increasing rates of gene transcription (Kuhn et al., 1984). Enzymes along the phenylpropanoid pathway are synthesized. One compound in particular, psoralen, is produced as a result of this induction of metabolic activity that has a use in the treatment of psoriasis and as an ingredient of suntan lotion (Tietjen et al., 1983; DiCosmo and Tallevi, 1985). Irradiation with UV light has a similar effect, with slightly different sets of enzymes being synthesized (Kuhn et al., 1984). It is also possible to induce lignin biosynthesis in soybean (*Glycine max*) cell cultures with the same fungal preparation (Farmer, 1985). Extensive shifts in the pattern of mRNAs synthesized in bean cultures (*Phaseolus vulgaris*) were found when the cells were challenged with a fungal preparation from *Collectotrichum lindemuthianum* (Cramer et al., 1985). Again, enzyme levels along the phenylpropanoid pathway were modified.

Brodelius et al. (1988) describe the purification and characterization of tyrosine decarboxylase from elicitor-treated plant cell suspension cultures. The authors demonstrated again that fungal elicitors have major potential to improve the productivity of plant cell cultures and to shorten fermentation times. Funk et al. (1987) show that the glucan elicitor isolated from yeast extract appears to induce enzymes of secondary metabolism in several plant species, inducing glyceollin and berberine in suspension cultures of *Glycine max* and *Thalictrum rugosum*, respectively.

Another example of induction of metabolite production in cell culture has been shown with Mexican yam (*Dioscorea deltoida*) (Rokem et al., 1984). This plant is the major source of diosgenin, a steroid precursor used in manufacture of birth control pills. It was demonstrated that the addition of autoclaved fungal mycelia could increase production in cell suspensions 72% compared to control cultures. The particular mechanism of induction and biochemical characterization was not reported in this case. However, the overall picture that is emerging from these studies suggests an effect at the level of gene transcription. The time course of the inductions is typically measured in terms of minutes to hours. This represents an extremely fine control mechanism that could greatly enhance the commercial utilization of plant cell cultures for the manufacture of specialty chemicals. Elicitors can be added to the process in a periodic or continuous manner at times when the production medium is changed to promote secondary metabolite synthesis.

The uptake of carbohydrates and oxygen by cell suspension cultures of the plant *Eschscholtzia californica* (Calfornia poppy) was studied in relation to biomass production in shake flasks, a 1-L stirred-tank bioreactor and a 1-L pneumatically agitated bioreactor, by Taticek et al. (1990; 1991). The sequence of carbohydrate uptake was similar in all cases, with sucrose hydrolysis occurring, followed by the preferential uptake of glucose. The uptake of fructose was affected by the oxygen supply rate. Carbohydrate utilization occurred at a slower rate in the bioreactors.

Returning to a previous theme, it is clear that modeling of cell growth, substrate utilization and product formation in plant cell systems is quite rare. Careful design of experiments and extensive data collection is needed for effective model identification, parameter estimation and parametric sensitivity analysis. Such work is almost routine with microbial and yeast fermentations, yet surprisingly has been avoided in discussions of plant cell suspension culture growth. Yet, any attempt at scale-up, process control and process simulation requires that such quantitative descriptions be available. Pedersen et al. (1991a; b; c) have undertaken such a task, as outlined below. A schematic diagram of the mechanisms involved appears in Fig. 10.7. Their specific application is outlined in Fig. 10.8.

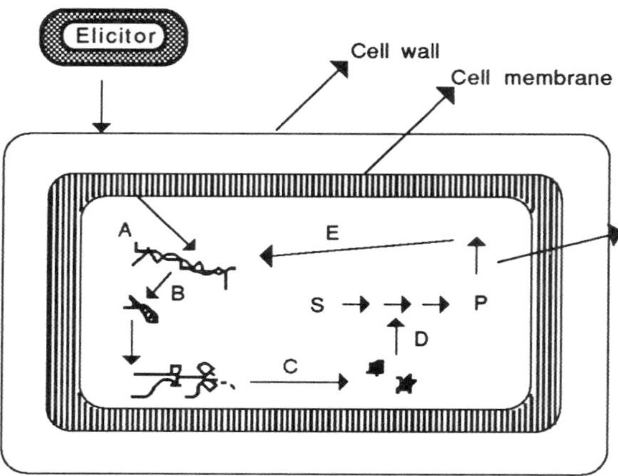

Figure 10.7 Schematic representation of secondary metabolic production in response to elicitor. A: second messenger transmission; B: transcription; C: translation; D: enzyme formation; E: product inhibition; S: substrate; P: product.

California poppy cultures (*Eschscholtzia californica*) produce benzophenanthridine alkaloids like macarpine, sanguinarine, chelerythrine and chelirubine constitutively, but under elicited conditions they can show a much higher productivity and a different relative composition profile. The elicitation process is a transient expression of signal molecules, mRNAs and proteins that all come together to make up the metabolic pathway for alkaloid synthesis. Precursors can be added and products can be extracted to greatly enhance this phenomenon. In particular, the amino acid precursor, tyrosine, can be channeled into the metabolic pathway during, but not in the absence of, elicitation. Product extraction *in situ* with polysiloxanes removes what might be a product feedback inhibition. The quantitative analysis of this system looks to develop an explanation of all these factors, and the theoretical framework found can also be applied to understanding elicited metabolism in other plant cell lines.

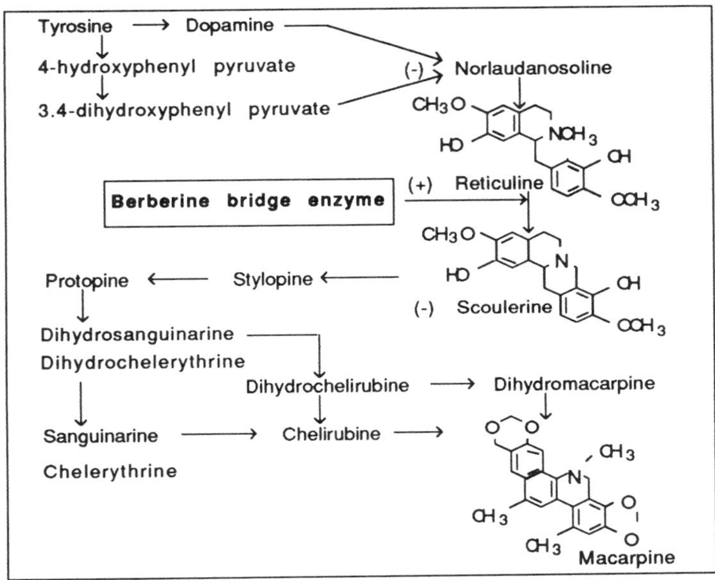

Figure 10.8 Biosynthetic sequence for macarpine production in *Eschscholtzia californica* (California poppy).

The analysis proceeds as follows:

ASSUMPTIONS

(1) The 'second messenger' evoked by elicitation induces transcription and the transcription rate is dependent on the strength of the second messenger signal.

(2) The gene concentration is constant during elicitation.

(3) The product formation rate is dependent on a limiting enzyme that is a result of mRNA translation; product accumulation regulates the transcription of mRNA.

TRANSMISSION OF SIGNAL

$$\frac{d[M]}{dt} = f(\alpha) - k_1[M] - \mu[M] \qquad [10.1]$$

TRANSCRIPTION AND TRANSLATION

$$\frac{d[mRNA]}{dt} = \eta \frac{[M]\,[G]}{g(P)} - k_2[mRNA] - \mu[mRNA] \qquad [10.2]$$

$$\frac{d[E]}{dt} = k_3[mRNA] - k_4[E] - \mu[E] \qquad [10.3]$$

PRODUCT FORMATION

$$\frac{d[P]}{dt} = k_5[E] \qquad [10.4]$$

CELL GROWTH AND SUBSTRATE CONSUMPTION

$$\frac{dX}{dt} = \mu X \qquad [10.5]$$

$$\mu = \mu_{max} \frac{S}{K_s + S} \qquad [10.6]$$

$$\frac{dS}{dt} = - k_6 \frac{dX}{dt} - k_7 \frac{dP}{dt} \qquad [10.7]$$

where:
- M = strength of second messenger signal,
- $f(\alpha)$ = generation of second messenger = constant at specific elicitor concentration,
- α = elicitor concentration,
- k_1 = decay rate constant of M,
- μ = specific growth rate,
- $[G]$ = DNA concentration,
- k_2 = decay rate constant of mRNA,
- η = transcription efficiency,
- $g(P)$ = product,
- k_s = translation rate constant,
- k_4 = decay rate constant of E,
- P = product,
- k_5 = rate constant of P formation,
- X = cell mass,
- μ_{max} = maximum specific growth rate,
- S = substrate,

K_s = Monod constant,
k_6 = yield factor by cell growth, and
k_7 = yield factor by product formation.

The model has been successfully applied, as shown in Figs. 10.9 and 10.10.

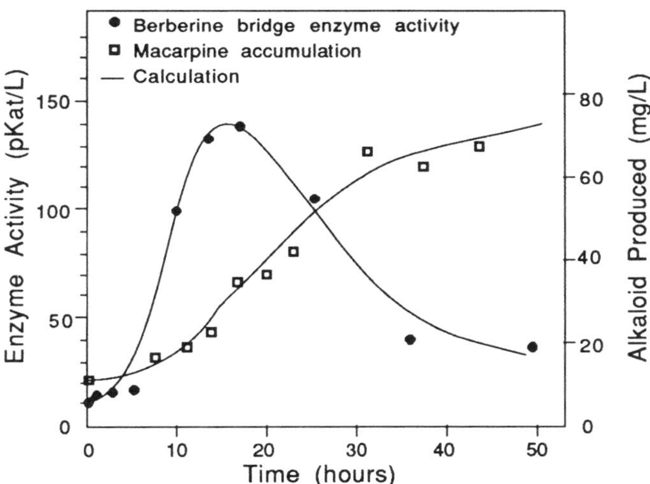

Figure 10.9 Elicitor induced changes of berberine bridge enzyme and macarpine accumulation in *Eschscholtzia californica*.

In the former case, the benzophenanthridine alkaloids produced by the California poppy plant are biochemically especially well characterized and the pathways leading to their production are known. These pathways are easily manipulated through the addition of simple oligosaccharides, i.e., elicitors, to suspension cultures of the cells. The elicitor serves as the chemical signal indicating to the plant possible adverse conditions. The time course of product formation has been followed after elicitation using oligosaccharides isolated from yeast extract. End compound benzophenanthridine alkaloids, macarpine and sanguinarine, have been identified as the major products. From this time course data and knowledge of the key regulatory points along the pathway, it was possible to quantify product formation in terms of enzyme formation rates using quasistationary approximations. This, in turn, further suggests possible genetic manipulations for modifying and controlling secondary metabolite production in plant cell cultures.

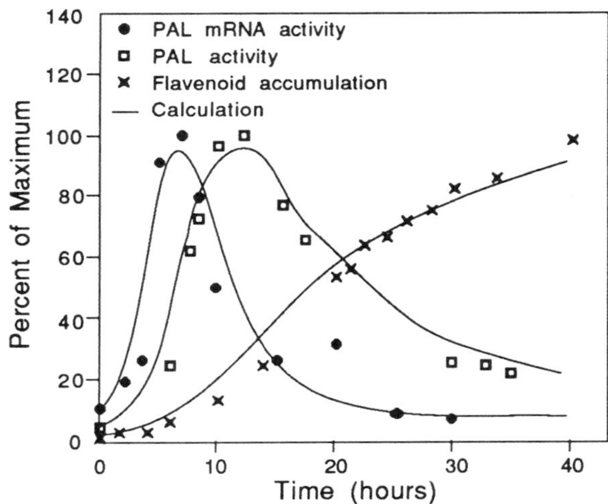

Figure 10.10 UV-light induced changes in mRNA, PAL activity and flavenoid accumulation in *Petroselinum hortense*.

PLANT MOLECULAR BIOLOGY

The techniques of molecular genetics, using recombinant DNA plasmids, have been the dominating factors in revolutionizing the industrial microbiology field. Recent results that demonstrate the ability to transform higher plants by recombinant DNA methods can also be expected to revolutionize plant biotechnology. The advantages to be gained by the use of molecular genetics are, at one level, quite similar to the advantages mentioned in the previous section on cellular genetics. The techniques, however, are much more precise and more versatile in the sense that otherwise impossible cross-kingdom genetic transformations can, in principle, be realized. Plant molecular biology is also etremely valuable in elucidating the factors that govern plant development and regulation.

Many of the traits that would be desirable to engineer into higher plants are controlled or expressed by polygenic processes. This is quite different than the current commercial emphasis in bacterial or mammalian genetic engineering, where the bacterial or mammalian cells are 'simple' vehicles for the production of specific gene products. It is unlikely that secondary metabolites found in plant cells, for instance, will be expressed in organisms other than higher plants by gene transfer techniques. Some of the desirable traits to be modified or enhanced by the techniques of plant molecular biology include:

- improved stress, herbicide and pest resistance;

- modification of undesirable characteristics such as secondary metabolites in agricultural crops;
- modification of cereal and legume seed proteins;
- increased growth rates and yields for agricultural crops;
- increased yields of secondary metabolites in cell culture;
- improved photosynthetic efficiency.

Some of the enabling genetic technologies that will allow for these traits to be realized by genetic engineering include:

- gene sequencing and cloning techniques;
- development of gene vectors or direct gene uptake mechanisms;
- control of gene stability and expression both at the level of cellular organization and at the level of plant tissue differentiation.

Bridges et al. (1990a; b) have systematically studied the *lac*-repressor/operator system as a regulatory system in plants. They found that genes for bacterial regulatory proteins expressed from plant promoters are useful in regulating heterologous genes in transgenic plants: the *lacI* gene of *Escherichia coli* was expressed from the cauliflower mosaic virus 35S promoter to very high levels in tobacco plants. In another study, pseudo-operator sequences for the repressor of bacteriophage 434 identified in the *GpAL2* gene of French beans were inserted into the -10 to -35 region of the *lac* promoter/operator. A cloned 434 repressor gene was then mutagenized and bacteria screened, using the suicide selection method to find 434 repressors acting on *lacZ* expression using this operator.

In a crossover study of the opposite variety, Misawa et al. (1990) found that the gene cluster for carotenoid biosynthesis could be cloned from *Erwinia uredovora* and expressed in *Escherichia coli*. The gene cluster comprises *ZexA*, *ZexB*, *ZexC*, *ZexD*, *ZexE* and *ZexF* and encodes enzymes for biosynthesis of zeaxanthin diglucoside from geranylgeranyl pyrophosphate. *E. coli* transformed with this gene cluster produced the same yellow pigment (1.1 mg/g dry weight) as that of the *Erwinia uredovora*, but the production with *E. coli* transformants was five-fold higher. El Hassouni et al. (1990) performed an analysis of the *Erwinia chrysanthemi arb* genes, which mediate metabolism of aromatic β-glucosides. Establishment of a functional *Arb* system in *E. coli* depended on the presence of the phosphotransferase system and on the activation by the cAMP-CRP receptor protein complex. Strains carrying mini-Mu-induced *LacZ* fusions to the *arb* genes were used to analyze *arb* genes organization and function. Three *arb* genes (*arbG*, *arbF* and *arbB*) were identified and organized in this order. Genetic and structural evidence allowed assignment of a phospho-β-glucosidase and a permease activity to the *ArbB* and *ArbF* proteins, respectively. Several Lac^+ *arb-lacZ* insertions were introduced into the *E. chrysanthemi* chromosome. Both *ArbG*$^-$ and *ArbF*$^-$ strains were unable to ferment the aromatic β-glucosides, whereas *ArbB*$^-$ strains were impaired only in salicin fermentation. None of the mutations in the *arb* genes affected cellobiose metabolism. The expression of the *arb* genes was substrate inducible and required the *ArbF* permease and, possibly, the *ArbG* protein. These results underline the resemblance between the naturally expressed *E. chrysanthemi arbGFB* and the cryptic *E. coli bglGFB* operons, yet the *arbG* gene product seemed unable to activate *E. coli bgl* operon expression.

PLANT TRANSFORMATION

Plant transformation has been demonstrated most successfully with the *Agrobacterium tumefaciens* system (Caplan et al., 1984; Bevan and Chilton, 1982). This soilbacterium contains a large tumor-inducing (Ti) plasmid. During infection of the plant, a small transferred DNA (T-DNA) segment of the plasmid is incorporated into the plant genome and expresses a tumorous phenotype. Utilization of the bacteria for transformation of plants without the tumor phenotype would obviously be necessary for practical applications in crop improvement. Considerable information has been obtained over the past few years on the genetics of the Ti-plasmid and it is now possible to modify the plasmid and the T-DNA region to allow for the transformation of plants that regenerate with a 'normal' phenotype (Bevan and Chilton, 1982). One complex yet elegant transformation system, referred to as the SEV (split end vector) system, is shown in Fig. 10.11 (Fraley et al., 1985) and possesses the desirable characteristics of being selectable (via kanamycin resistance, NPT'), scorable (via a functional nopaline synthase gene, NOS) and versatile (via a polylinker region for inserting desired DNA between the LIH and NPT' genes). DNA is inserted into pMON200 using standard techniques and transformed into *E. coli* strains that are selected for the plasmid by spectinomycin resistance (spec^r). The crossover with pTiB6S3-SE shown in Fig. 10.11 is possible because of the complementary left inside homology (LIH) regions. The integration into the plant genome takes place because of the presence of the essential left and right border sequences. The region of insertion in the plant genome is not known.

This system has been shown to be effective in transforming tobacco, petunia and tomato plants. In conjunction with a leaf disk procedure (Horsch et al., 1985), the entire process is quite simple. Transformants have so far all been members of the family Solanaceae. The stability of the integrated DNA in cell suspension cultures has not been extensively studied. In order to use the vector in a system for improved production of secondary metabolites, the transformed gene must also be constitutively expressed without effecting growth rates of the cells in culture. Usually, however, growth and secondary product formation are exclusive processes, as mentioned earlier. The important commercial benefit will result when we have identified and can focus on inserting genes that code for enzymes believed to be key factors in the expression of secondary metabolites.

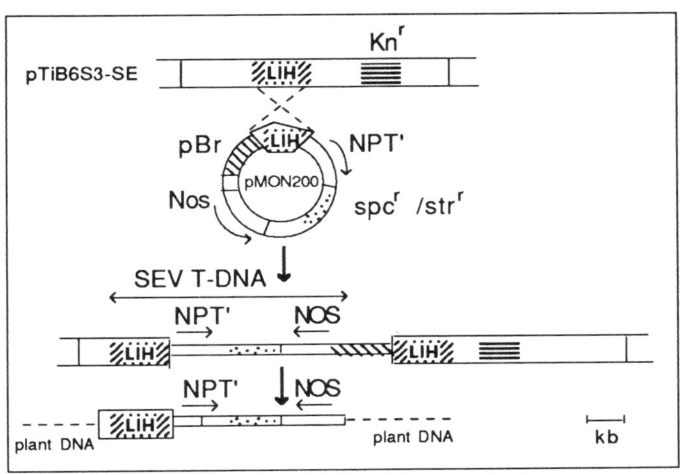

Figure 10.11 The SEV transformation system for inserting foreign DNA into plants.

The fact that the DNA is integrated directly into the plant genome may, however, be a serious problem if disruption of growth processes or other desirable traits are compromised. To avoid this potential problem, some work has been carried out with nonintegrative vectors derived from plant viruses, in particular the Caulimoviruses (Brisson et at., 1984). These vectors allow for increased copy numbers of the gene in the plant cell, as well. Other techniques, such as microinjection and protoplast fusion, appear suitable as transfer systems in place of infection with *Agrobacterium*. A rather recently described technique known as electroporation may also be a valuable gene transfer method (Messing, 1985). Electroporation is the process of inducing permeability in plant protoplasts by pulsed electric fields. Pores are formed in the membrane that facilitate the transfer of exogenous material to the cytoplasm. Having no monocot-dicot specificity, this technique allows for the transformation of all commercially important crops, including cereals. Of course, plant regeneration from protoplasts is a necessary step.

Appropriate gene vectors are still needed here, but even the *Agrobacterium* system may be used since its failure in monocots is due apparently to cell surface recognition problems.

A model gene system should possess the characteristics of being easily assayed for both gene transcription products as well as protein synthesis. One particularly useful gene in this regard is from the *lac* operon where the *lac* Z gene codes for the enzyme β-galactosidase. The size of the protein will allow for its rapid isolation, and the activity of the protein can also be measured by sensitive assays. Antibodies to the protein and nick translated DNA probes are available, or can be developed.

As an illustration, the chimeric gene can be constructed in bacteria using standard techniques; e.g., part of the gene could be from the *lac* operon and part from the 35 S promoter from cauliflower mosaic virus. In the final product the *lac* Z gene is under the control of the 35 S promoter. This chimeric *lac* Z gene is then inserted into pMON200 and mobilized as described above. Genomic DNA can be isolated and nick translated probes used to determine the presence of the model gene in transformed petunia cells. Northern blots can be used to follow the RNA synthesis (using the same probe) and Western blots can be used to follow the protein synthesis levels (using an antibody probe).

The levels of gene products and the stability of the insert can be followed in free and immobilized cell systems. Scale-up to larger reactor sizes and the effects of different media conditions on gene expression can thus be quantified to provide a rational means for designing process systems with transformed plant cells.

MODIFICATION OF PLANT SECONDARY METABOLITE PATHWAYS

The techniques mentioned above can be used to modify plant secondary metabolite pathways in higher plants by inserting genes coding for key enzymes that are identified as bottlenecks in the reaction network. The enzymes can be constitutively expressed by using appropriate promoter sequences and can be directed to the chloroplast organelles, if necessary, by fusion of transit peptides. This has been demonstrated by the incorporation of the shikimate pathway enzyme EPSP synthase into petunia plants. The enzyme is overproduced in the transformed plant cells and confers selective resistance to the herbicide glyphosate as a result. Normal plant cells are inhibited. The point of inhibition is the pathway to aromatic amino acid biosynthesis. Cloning the enzymes that are blocked back into the plant is a promising strategy for incorporating herbicide resistance and other types of stress resistance, as well.

The ability to identify differences in resistant and nonresistant cell lines at the level of enzyme action is of significant advantage in selecting a strategy for plant cell modification. The cell lines that are found to be high producers of valuable secondary metabolites have eliminated the tight regulation along the pathway from primary to secondary metabolites. The enzymes involved in amino acid synthesis and degradation are good places to look for regulation control by feedback inhibition or feedforward inhibition. Tryptophan decarboxylase and lysine decarboxylase have been implicated in the differences between high and low alkaloid producing cell lines (Berlin, 1984; Noe et al., 1984). High enzyme levels are associated with high alkaloid levels. The induction systems mentioned earlier also effect product levels by modifying enzyme levels, and it is of great interest to establish what enzymes exactly are being effected and which of those are the critical enzymes allowing for increased product synthesis.

10.2 LIGHT ENERGY COUPLING

The first step in the production of valuable products from higher plants is, of course, photosynthesis. Carbon fixation is achieved by action of the reducing power of NADPH on CO_2. This action, accompanied by oxygen evolution, is accomplished by light-induced splitting of water, or biophotolysis.

Kim et al. (1988) found that light supported better cell growth and prevented the rapid decrease in cell mass after depletion of substrate in cell suspension cultures of *Thalictrum rugosum*. Alkaloid production was enhanced with continuous illumination. In addition, light significantly suppressed the secretion of berberine, the major isoquinoline alkaloid found in the culture.

Researchers at Michigan State University have transplanted the gene for bacterial expression of polyhydroxybutyrate into members of the mustard plant family. The polymer, useful as a biodegradable plastic, appeared as a byproduct of photosynthesis, but plant growth was stunted. More effective hybrids are currently under study (Anon., 1992).

As alluded to earlier, it is clear that plants share at least some of the features which lend an elegant simplicity to membrane transport in a variety of biological systems (Uribe and Lüttge, 1984). The high energy state in the form of the electrochemical proton gradient is linked to the utilization of chemical bond energy conserved in photosynthesis and respiration through the action of the proton-pumping ATPase enzyme. Such metabolically generated energy can be coupled, in turn, to solute transport in such a way that solute composition and concentration may be maintained at levels which are optimal for the function of enzyme systems.

10.3 PHOTOSYNTHETIC ELECTRON TRANSPORT SYSTEM

INTRODUCTION

The chloroplast is the seat of photosynthetic chemistry in green plants. Shaped like pinto beans, typically there are fifty or so chloroplasts in each cell of a leaf and roughly a half million in each square millimeter of leaf surface (Guillet, 1991). An elaborate network of membranes, chiefly arranged in stacks of flat vesicles (thylakoids) enables plants to capture sunlight for photosynthesis.

Any reasonable kinetic model of the photosynthetic electron transport system (PETS) should account for the observed phenomena of (1) light saturation of the photosynthetic rate at high incident light intensities and (2) the sequential nature of the light and dark reactions involved. In formulating the following mathematical model, the intent was to achieve a balance between a detailed, realistic representation of the complexities of PETS and the need for a model of sufficient simplicity to be useful in potential biochemical process engineering applications.

Experimental evidence indicates that the electron carriers which participate in photosynthetic electron transport are vectorially arranged across the thylakoid membrane (Anderson, 1975; Trebst, 1974). It is known that the components of PETS occur in fairly fixed molar ratios relative to one another. This evidence, together with the established sequential nature of electron transport (as reflected in the Z-scheme of Fig. 10.12) suggests that the process of photosynthetic electron transport may be viewed as carried out by a collection of more or less independent electron transport 'chains' located in the thylakoid membrane. This model of the photosynthetic apparatus, including the concept of independent electron transport chains, is adopted here in developing a kinetic model of PETS.

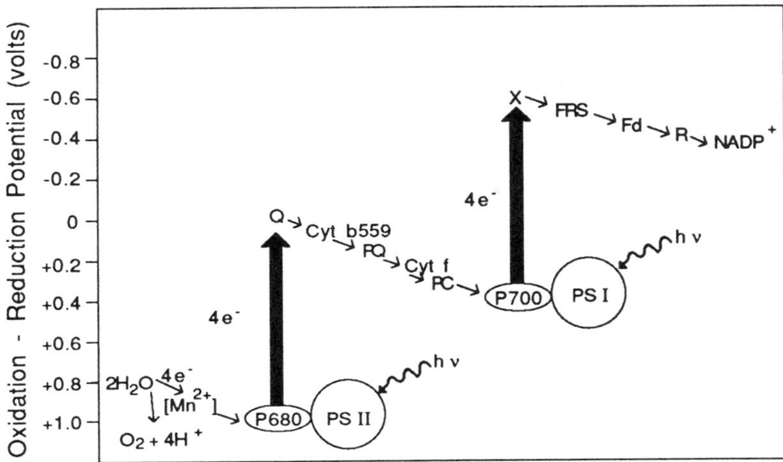

Figure 10.12 Schematic diagram of the photosynthetic electron transport system (PETS) plotted to show the approximate standard oxidation-reduction potentials (E'0) of the interacting redox components.

MODEL OF THE REACTION MECHANISM

The overall reaction of photosynthetic electron transport may be represented by a light-dependent photoexcitation step followed by a dark reaction:

$$Chl_{RC} + h\nu \underset{k_d}{\overset{k_e I}{\rightleftharpoons}} Chl^*_{RC} \qquad [10.8]$$

$$Chl^*_{RC} + \frac{1}{2}H_2O \xrightarrow{k_D} Chl_{RC} + \frac{1}{4}O_2 + H^+ + e^- \qquad [10.9]$$

Equation [10.8] represents absorption of light by the antenna chlorophyll matrix and subsequent transfer of the electronic excitation energy ($h\nu$) to a reaction-center chlorophyll molecule (Chl_{RC}). Chl^*_{RC} denotes a reaction-center chlorophyll molecule in an electronically excited state. The constants k_e and k_d are the rate constants for the excitation and de-excitation processes, respectively. (De-excitation, or relaxation, of excited Chl may occur via several pathways, two of which, fluorescence and phosphorescence, actually involve emission of a photon, as suggested by the reverse process in eqn. [10.8]).

The rate of photoexcitation of Chl_{RC} to Chl^*_{RC} is proportional to the rate of absorption of photons by antenna pigments. The latter, in turn, is directly proportional to both the incident light intensity, I, and to the concentration of absorbing pigment molecules. The light intensity (I) is shown with the excitation rate constant (k_e) in eqn. [10.8] to emphasize the fact that the incident intensity is completely independent of the photoexcitation process, and therefore independent of the reaction time.

Equation [10.9] represents the entire sequence of consecutive, coupled oxidation-reduction reactions from water to an unspecified carrier which is the terminal acceptor of the reducing equivalents (H^+ + e^-) produced. (*In vivo*, the terminal acceptor $NADP^+$ is reduced to NADPH.) This equation therefore models the entire sequence of dark redox reactions presented in Fig. 10.13. (In each photosystem there is only one truly light-dependent reaction, represented by eqn. [10.8]. The subsequent transfer of electrons along the electron transport chain, represented by eqn. [10.9], can take place in the absence of light; these oxidation-reduction reactions are referred to here as 'dark reactions'.) It is assumed that the terminal electron acceptor is always present in excess, so that electron transport through PETS is not limited by availability of the acceptor. The stoichiometric coefficients in eqn. [10.9]

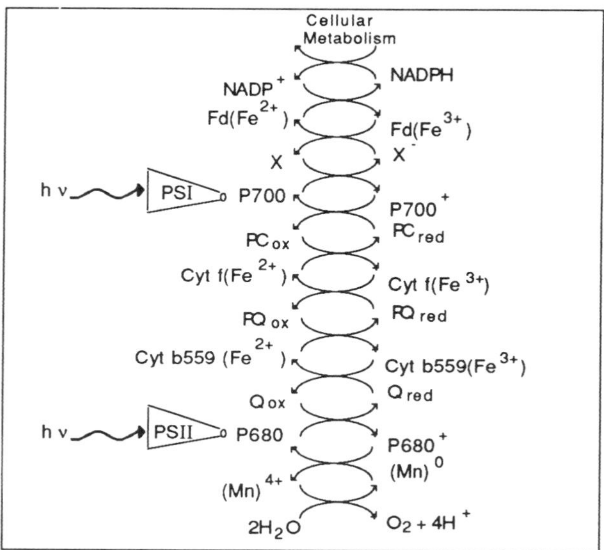

Figure 10.13 The coupled reactions in photosynthetic electron transport as currently understood.

correspond to the transfer of one electron through the electron transport chain.

Of course, in a complete biophotolytic system producing both hydrogen and oxygen, the reducing equivalents ($H^+ + e^-$) generated by PETS would be used to form H_2 by the following reaction catalyzed by hydrogenase (H_2ase):

$$2H^+ + 2e^- \underset{}{\overset{H_2ase}{\rightleftharpoons}} H_2 \qquad [10.10]$$

This reaction might be achieved by directly coupling hydrogenase to the photosynthetic electron transport chain as implied above, or by indirect coupling through some intermediate electron carrier molecule, A:

$$2H^+ + 2A^- \underset{}{\overset{H_2ase}{\rightleftharpoons}} H_2 + 2A \qquad [10.11]$$

INTERPRETATION OF THE LIGHT REACTION MODEL

For the proper interpretation of the above reaction model, it is essential to understand that the chlorophyll molecule (Chl_{RC}) represented in eqns. [10.8] and [10.9] is the *reaction-center* chlorophyll, not the bulk chlorophyll present as light-harvesting pigment molecules associated with the reaction-center. Approximately 200-300 chlorophyll molecules and other pigment molecules absorb photons and transfer their excitation energy to the 'reaction-center' (Govindjee and Govindjee, 1974; Katz et al., 1976). This reaction-center is believed to contain a single molecule of chlorophyll *a* located in a microenvironment which facilitates the primary photochemical reaction (charge separation between the primary donor-acceptor). The primary photochemical reaction (charge separation) can be described by:

$$A + Chl^{*}_{RC} \longrightarrow Chl^{+}_{RC} + A^{-} \qquad [10.12]$$

where A denotes the primary electron acceptor of either Photosystem II (i.e., Q) or Photosystem I (i.e., X). This primary act can be viewed as initiating, or 'driving' the subsequent redox reactions of the electron transport chain which result in product formation.

In the purple non-sulfur bacterium *R. capsulatus*, genes encoding structural polypeptides of the light-harvesting (LH) and reaction-center (RC) complexes incorporated into an intracytoplasmic photosynthetic membrane are induced upon lowering the oxygen tension in the media of aerobically growing cultures (Leach et al., 1991). The transcription of the polycistronic *puf* operon, which encodes pigment binding proteins of the reaction-center and light-harvesting complex I of *R. capsulatus*, is regulated by the oxygen tension in the culture (Klug, 1991; Klug et al., 1991). Sexton et al. (1990) show that the synthesis of reaction-center protein D_2 and mRNAs which encode this protein are differentially maintained at high levels in mature barley chloroplasts. These data document a novel mechanism for regulating plastid gene expression involving a light-induced switch in *psbD-psbC* promoter utilization.

Cyanobacteria harvest light energy through multimolecular structures, the phycobilisomes, regularly arrayed at the surface of the photosynthetic membranes (Houmard et al., 1990). The chromophoric protein phycocyanin is likewise the major protein in the phycobilisome rod of the cyanobacterium *Synechococcus* 6301 (formerly designated *Anacystis nidulans* (Kalla et al., 1989). In a temperature-sensitive, high CO_2-requiring mutant of *Synechococcus* sp. PCC7942, the ability to fix intracellularly accumulated inorganic carbon was severely impaired at non-permissive temperature (41°). In contrast, inorganic carbon uptake and ribulose-1.5-bisphosphate carboxylase activity in the mutant were comparable to the respective values obtained with the wild-type strain. The mutant was transformed to the wild-type phenotype (ability to form colonies at non-permissive temperature under ordinary air) with the genomic DNA of the wild-type strain. These results indicate that the cloned genomic region of *Synechococcus* sp. PCC7942 is involved in the efficient utilization of intracellular inorganic carbon for photosynthesis (Suzuki et al., 1991).

Returning now to a consideration of the spinach chloroplast *per se*, of the many electronic excitations which antenna pigment molecules continually undergo (excitations due both to photon absorption and exciton migration), only those that cause excitation of the reaction-center chlorophyll molecule can result in photochemical conversion (photo-product formation). Thus, the primary processes involving excitation of antenna pigment molecules by absorption of photons and subsequent transfer of the excitation energy to the reaction-center chlorophyll molecule may be combined and justifiably modeled by a single photoexcitation process (eqn. [10.8]). This process could also be represented conceptually as:

$$\mathrm{Chl}_{\mathrm{antenna}} + \mathrm{Chl}_{\mathrm{RC}} + h\nu \rightleftharpoons \mathrm{Chl}^{*}_{\mathrm{RC}} + \mathrm{Chl}_{\mathrm{antenna}} \qquad [10.13]$$

Since only reaction-center chlorophyll molecules participate in the primary photochemical reaction (i.e., antenna chlorophyll molecules are not photochemically active), eqn. [10.8] is a more useful representation in developing a model of the reaction kinetics.

There are, of course, two reaction-centers associated with each complete electron transport chain; one in Photosystem II (P680) and one in Photosystem I (P700). In modeling the photoexcitation process as a single photoreaction (eqn. [10.8]) it is assumed that the two photosystems excite their respective reaction-centers at equal rates. It is known that the reaction-centers of the two photosystems are present in equal concentrations in the thylakoid membrane (i.e., in a 1:1 molar ratio of P680 to P700). In order to provide optimum rates of electron transport through the linear electron transport system, absorbed quanta should be distributed equally between PS II and PS I. Available evidence indicates that the distribution of excitation energy between the two photosystems is approximately equal (Butler, 1976). Thus the implicit assumption made in modeling the two photosystems of PETS by a single photoexcitation process appears valid.

In order to simplify notation and avoid possible confusion with bulk antenna chlorophyll, the reaction model of eqns. [10.8] and [10.9] will be represented by the following scheme:

$$\begin{array}{ccl} \frac{1}{2}H_2O + P^* & \xrightarrow{k_D} & P + \frac{1}{4}O_2 + H^+ + e^- \\ k_d \Big\uparrow\Big\downarrow k_e I & & \\ P + h\nu & & \end{array} \qquad [10.14]$$

where P represents the specialized pigment (chlorophyll *a*)-protein complex of the reaction-center of either PS II or PS I (designated P680 in PS II and P700 in PS I). The vertical process represents reversible excitation of the reaction-center and the horizontal reaction represents the overall electron transport pathway from water to an unspecified electron acceptor.

The role of reaction-center chlorophyll (P) as a photocatalyst (or 'photosensitizer') is apparent from examination of the above scheme (eqn. [10.14]). The action of chlorophyll allows sequential oxidation-reduction reactions (electron transport) to proceed in response to absorbed radiation. The substrate for the reaction (water) does not itself absorb photons in the wave-length range over which the reaction is sensitive. This important distinction makes such photoassisted catalytic reactions quite distinct from non-catalyzed photochemical reactions. In the latter, the reactant (or substrate) is also the photon absorber and is not regenerated in the reaction (non-catalytic). In the case of photosynthetic electron transport, however, it is the *catalyst* that is photon-excitable, not the substrate. The absorption of incident photons is dependent on the absorption characteristics and concentrations of the light-harvesting pigment molecules within the thylakoid membrane. These light-harvesting pigment molecules are not photochemically active, but transfer the excitation energy of absorbed photons to reaction-center chlorophyll molecules. The reaction-centers undergo the primary photochemical reaction (charge separation) which initiates electron transfer. These facts, combined with the extreme rapidity of purely photophysical processes, result in a complete uncoupling of the transport of light through the photocatalyst from the kinetics of the photosynthetic reactions. The light intensity profile within the photocatalyst (thylakoid membrane) is therefore independent of the kinetics of the electron transport reactions (the converse, of course, is not true).

REACTION-CENTER EXCITATION RATE

The product of the rate constant, k_e, and the incident light intensity, I, (see eqn. [10.14]) has the units s^{-1} and represents the photoelectronic excitation rate of the reaction-center chlorophyll molecule (number of excitations per reaction-center per second). For a given light intensity, k_eI can be viewed as a first order rate constant for photoexcitation of the reaction-center. It is useful to examine some of the inherent properties of the thylakoid membrane which influence the intrinsic rate constant (k_e) of this microheterogeneous photocatalyst.

The rate of excitation (k_eI) of the reaction-center chlorophyll molecule may be expressed as:

$$k_e I = k_e' \alpha \phi_{RC} UI \qquad [10.15]$$

where α is the fraction of incident photons actually absorbed by the photosystem, ϕ_{RC} is the quantum yield, and U is the size of the photosynthetic unit. The constant k'_e is simply a modified form of the rate constant k_e defined by the relationship in eqn. [10.15].

The quantum yield for electronic excitation of the reaction-center chlorophyll with respect to photon absorption by antenna chlorophyll molecules is defined as:

$$\phi_{RC} = \frac{\text{number of excitations of the reaction center}}{\text{number of photons absorbed by } \text{Chl}_{\text{antenna}}} \quad [10.16]$$

The quantum yield defined by eqn. [10.16] simply represents the probability that a photon absorbed by one of the antenna chlorophyll molecules will result in the electronic excitation of the reaction-center chlorophyll molecule. Once a photon is absorbed, the probability of exciting the reaction-center is extremely high, so that the value of ϕ_{RC} is nearly unity (Govindjee and Govindjee, 1974).

Within the thylakoid membrane, the concentrations of the light-harvesting pigment molecules and the components of PETS are fixed. When viewed microscopically, the fraction of incident photons absorbed by the thylakoid (α) may be considered a constant for a given thylakoid membrane preparation at a given wavelength (the concentration of pigments in the thylakoid membrane and the membrane thickness determine α). One should bear in mind, however, that the chlorophyll concentration varies with plant species and with the history and treatment of the material. The factor U (for unit size) included in eqn. [10.15] represents the molar ratio of light-harvesting chlorophyll molecules to reaction-center chlorophyll molecules in the photosynthetic unit. (The reaction-center, or 'trap', and its associated light-harvesting antenna pigment molecules operate as a functional 'photosynthetic unit'. The size of the photosynthetic unit is often expressed as the number of light-harvesting pigment molecules, usually Chl, serving one reaction-center, P680 or P700). This stoichiometric factor accounts for the fact that some 200-300 antenna chlorophyll molecules absorb photons and transfer the excitation energy to a single reaction-center chlorophyll molecule where the primary photochemical reaction occurs. While the kinetic model is developed in terms of the reaction-center chlorophyll in the electron transport chain (P), it is useful to explicitly account for the relative amount of antenna chlorophyll present by incorporating the size of the photosynthetic unit (U) in the model.

Viewing the matrix of light-harvesting chlorophyll molecules in the photosynthetic unit as a photon collector and concentrator of quanta (in the sense that the excitation energies of photons absorbed by antenna Chl molecules are directed, or 'focused', at the reaction-center Chl), the unit size (U) represents the collector concentrating factor. Multiplying the electronic excitation rate of individual antenna chlorophyll molecules by the factor U gives the excitation rate of the associated reaction-center (assuming the quantum yield ϕ_{RC} is unity).

For example, calculations show that under full sunlight intensity, an individual chlorophyll molecule is electronically excited about 10 times per second (Kok, 1973). Assuming 200 antenna chlorophyll molecules per photosynthetic unit (U=200), the reaction-center would be excited at a rate of approximately 2000 times per second. In recent years, the turnover numbers of the various oxidation-reduction reactions of PETS have been determined. Kok reports that the rate-limiting step in the entire photosynthetic redox chain has a turnover number of ~100-200 per second (Kok, 1973; Zankel and Kok, 1972). This mismatch between the reaction-center excitation rate at high incident light intensities and the intrinsic turnover number of the rate-limiting dark reaction is responsible for the observed phenomenon of 'light saturation' of photosynthesis. (Under 'light-saturating' conditions it is the 'dark' oxidation-reduction reactions which are rate-limiting. Light saturation occurs when the rate of photoexcitation of the reaction-center exceeds the inherent electron transfer capabilities of the electron transport chain.) From such considerations, it is apparent that decreasing the size of the photosynthetic unit would increase the efficiency of light energy conversion by PETS at high incident light intensities.

Since efforts have been made to selectively control the size of the photosynthetic unit (Lien and San Pietro, 1975), it is worthwhile to include this potentially significant variable in the kinetic model of PETS. This can be done by using the relationship in eqn. [10.15] in all subsequent development of the mathematical model (that is, by replacing k_e with $k'_e \alpha \phi_{RC} U$).

The entire process of photon absorption and excitation of an antenna pigment molecule, migration of the excitation energy throughout the antenna pigment matrix, and final transfer of the excitation energy to the reaction-center chlorophyll molecule takes place within 10^{-9}s (Fork, 1977). That is, the photoexcitation process represented in the scheme above (eqn. [10.14]) has an intrinsic turnover number of 10^9 per second. When compared with the dark oxidation-reduction reactions of the electron transfer chain (with turnover numbers as low as 100-200 per second), it is obvious that the purely photophysical processes are never the rate-limiting factor in photosynthetic electron transport. The inherent capabilities of the photosystems to process photons (as reflected in the intrinsic excitation rate constant k_e) are never fully utilized by PETS even under conditions of full sunlight intensity. Therefore, it makes more sense to speak in terms of the rate at which the reaction-center is actually excited by the photosystem under a given light intensity (i.e., $k_e I$) rather than the intrinsic turnover rate of the process, since the latter is never approached under normal light intensities. Note that the reaction-center excitation rate is always directly proportional to the incident light intensity.

In formulating the mathematical model of PETS, it should be borne in mind that the excitation rate 'constant' represented by k_e is related to fundamental characteristics of the thylakoid membrane (α, ϕ_{RC} and U) as described above. These parameters may be considered constant for a given preparation of thylakoid membranes, but may vary from preparation to preparation.

A word about appropriate dimensional units for light intensity (I) is in order. Because physical devices (such as a photovoltaic cell) used to measure light intensities commonly measure the intensity as an energy flux ($ergs/cm^2 \cdot s$) or a photon flux ($Einsteins/cm^2 \cdot s$), (one Einstein is defined as one mole of quanta) it is convenient to express I in these units. (The light intensity could also be expressed in units of concentration such as ergs per liter, often called the energy density, or Einsteins per liter, by dividing the energy flux or photon flux, respectively, by the speed of light.) Light intensities may be interconverted from units of energy flux to units of photon flux using the relation $E = h\nu$ where h is Planck's constant and ν is the frequency of the electromagnetic radiation.

INTERPRETATION OF THE PHOTOSYNTHETIC ELECTRON TRANSPORT MODEL

Zankel and Kok (1972) have provided evidence that the entire photosynthetic electron transport process can be considered as carried out by a collection of relatively independent electron transport chains. Extensive kinetic experiments coupled with chemical analyses of chloroplast thylakoid membranes indicate the following molar ratios of the components of the photosynthetic electron transport system (Clayton, 1974; Zankel and Kok, 1972; Fork, 1977): 200 $Chl_{antenna}$:1 P680: ~7 - 10 PQ: 2 Cyt b559:1 Cyt f:1 PC; 200 $Chl_{antenna}$:1 P700:1 Fd, where the total antenna chlorophyll (~400 molecules per chain) has been partitioned rather arbitrarily into 200 molecules serving each photosystem reaction-center.

It is assumed here that the electron transfer carriers are located on, or within, the thylakoid membrane in relatively fixed positions with respect to one another, and that each component communicates with one donor and one acceptor molecule. If this is the case, the pathway of electron transport corresponds to an actual physical entity - the electron transport 'chain' - within the thylakoid membrane. Such a chain can be viewed as a tightly coupled, multicomponent catalytic complex which is photon-excitable. That is, the entire photosynthetic electron transport chain functions as a photocatalyst for the transfer of reducing equivalents from water to an electron acceptor.

The kinetic behavior of sequential oxidation-reduction reactions occurring within such a chain will be significantly different from that of conventional reactions in solution where all reactant molecules can interact. It is known that the various individual oxidation-reduction reactions in the photosynthetic electron transport chain all show first order kinetic behavior rather than second or higher order behavior (Zankel and Kok, 1972). This is apparently due to the fact that classical collision theory chemistry is not involved in the electron transfer processes. Instead, neighboring components (electron donor-acceptor pairs) in the chain are membrane-bound and are located in close proximity to one another. This natural immobilization of sequentially acting electron transport components forms a tightly coupled system in which electron transfer through the chain is not dependent on interactions between components of neighboring chains. Confinement of electrons within a tightly coupled multicomponent electron carrier sequence (the electron transport chain) results in electron transport rates and efficiencies much greater than could be achieved by electron carrier molecules interacting by random collisions in a homogeneous solution. The favorable arrangement of electron carriers in the thylakoid membrane has a profound influence on the kinetic behavior of the membrane-bound system.

Each step in the entire sequence of photosynthetic electron transport (see Fig. 10.13) consists of an oxidation-reduction reaction involving an electron donor molecule (D) and an electron acceptor molecule (A^+) and can be represented as:

$$D + A^+ \rightleftharpoons D^+ + A \qquad [10.17]$$

Viewing the overall system (collection of chains), the rate of electron transfer from D to A^+ is proportional to the concentration of chains in which the donor-acceptor pair is in the appropriate redox state (D-A^+). (The rate will *not* be proportional to the product of the two concentrations ($[D][A^+]$) if each chain operates independently of the others.) As a result, the observed kinetics of the individual oxidation-reduction reactions in PETS show first order behavior with respect to chain concentration (Zankel and Kok, 1972).

It should be realized, however, that the concept of independent electron transport chains is an idealization of the complexities of the actual photosynthetic apparatus. There is evidence of some limited cooperation between neighboring chains. The 'pool' of about 7 - 10 plastoquinone molecules per chain is thought to function as an 'electron buffer' between the two photosystems. Evidence demonstrates that several different electron transport chains can feed electrons from PS II into a single plastoquinone pool (Trebst, 1974). That is, two or more PS II reaction-centers may reduce the same plastoquinone molecule, and the chains cannot be viewed as operating completely independently of one another.

With information currently available, it is not possible to make definitive statements concerning the extent to which the concept of independent electron transport chains may be valid, or if such chains actually correspond to distinct physical entities in the thylakoid membrane. Such conclusions must await a more detailed understanding of the molecular architecture of the chloroplast thylakoid membrane and further elucidation of the relationship between membrane structure and function. Nevertheless, the concept of independent electron transport chains provides a useful model for the mathematical description of PETS.

KINETIC MODEL OF PETS

In developing a mathematical model which describes the kinetic behavior of the photosynthetic electron transport system, it is necessary to bear in mind the underlying assumptions which are made concerning the nature of the system being modeled. The assumptions discussed in preceding sections are summarized here:

1. The thylakoid membrane can be viewed as a microheterogeneous photocatalyst consisting of a collection of identical photosynthetic electron transport chains.
2. The individual electron transport chains function independently of one another.
3. There is one PS II reaction-center chlorophyll molecule (P680) and one PS I reaction-center chlorophyll molecule (P700) per electron transport chain.
4. Excitation energy from photons absorbed by the two photosystems is distributed equally between the two reaction-centers (P680 and P700). The net excitation rates of the two reaction-centers are equal and can be modeled by a single photophysical process ($P + h\nu \Leftrightarrow P^*$).
5. There is always an excess of terminal electron acceptor present; that is, the electron transport rate is zero order with respect to the concentration of the terminal electron acceptor.

From the reaction scheme presented in eqn. [10.14] and repeated below:

$$\tfrac{1}{2}H_2O + P^* \xrightarrow{k_D} P + \tfrac{1}{4}O_2 + H^+ + e^- \qquad [10.14]$$

$$k_d \upharpoonleft\downharpoonright k_e I$$

$$P + h\nu$$

The net rate of formation of P* is:

$$\frac{dP^*}{dt} = k_e\, IP - k_d\, P^* - k_D\, P^* \qquad [10.18]$$

where it is understood that P and P* represent the molar concentrations of the reaction-center chlorophyll molecule in the ground state and excited state, respectively. (When symbols representing *chemical* entities, such as P, appear in mathematical equations, it will be understood that they denote molar concentrations. The use of brackets to denote molar concentrations, [P], will be omitted in equations for the sake of simplicity.) From assumptions (3) and (4) above, P in eqn. [10.18] is the molar concentration of either P680 or P700. For a given light intensity, I, a steady state concentration of electronically excited reaction-center chlorophyll molecules (P*) is established within a very short time after illumination begins. Setting the net rate of formation of P* equal to zero in eqn. [10.18] and solving for the concentration of molecules in the excited state yields:

$$P^* = \frac{k_e\, IP}{k_d + k_D} \qquad [10.19]$$

Since all reaction-center chlorophyll molecules present are either in the excited state or the ground state, the following stoichiometric invariance can be written:

$$P_0 = P + P^* \qquad [10.20]$$

where P_0 is the total concentration of PS II or PS I reaction-center chlorophyll molecules in the system. Since there is one P680 and one

P700 per electron transport chain, P_0 in eqn. [10.20] also represents the total concentration of electron transport chains in the system.

From eqn. [10.14], the rate of photosynthetic electron transport, r_e, is given by:

$$r_e = k_D P^* \qquad [10.21]$$

where k_D denotes the overall rate constant for the dark, sequential oxidation-reduction of PETS. The electron transport rate is always directly proportional to the concentration of reaction-centers in the excited state.

Equations [10.19] and [10.20] can be used to eliminate P* and P from the formulation, leaving the following rate expression obtained from eqn. [10.21]:

$$r_e = \frac{k_D P_0 I}{\left(\frac{k_d + k_D}{k_e}\right) + I} \qquad [10.22]$$

Defining the combination of rate constants (K_I):

$$K_I = \frac{k_d + k_D}{k_e} \qquad [10.23]$$

and the maximum electron transport rate (V_m),

$$V_m = k_D P_0 \qquad [10.24]$$

Eqn. [10.22] can be expressed as:

$$r_e = \frac{V_m I}{K_I + I} \qquad [10.25]$$

The numerical value of K_I is equal to the light intensity for which the electron transport rate is half maximal (that is, when $I = K_I$, $r_e = V_m/2$).

Equation [10.25] expresses the functional relationship between the photosynthetic electron transport rate, r_e, and the light intensity, I. This reaction rate expression is in agreement with the experimentally observed behavior (Howell and Vieth, 1982). At low incident light intensities, the electron transport rate (O_2 evolution rate) is nearly proportional to the incident light intensity, I. This is often referred to as the 'light-limited' region. At high incident light intensities, the rate asymptotically approaches a maximum value, V_m (where $V_m = k_D P_0$), and is essentially zero order with respect to light intensity. When this occurs, the system is often said to be 'light-saturated.' As eqn. [10.22] indicates, the reaction rate in the light-saturated region is determined by the kinetics of the dark electron transport reactions (k_D).

The functional relationship of eqn. [10.25] (that of a rectangular hyperbola) is of the same form as the Langmuir adsorption isotherm (Langmuir, 1918), the specific growth rate relationship proposed by Monod in 1942, the heterogeneous reaction kinetics of Hougen and Watson (1948) and the standard rate equation for enzyme-catalyzed reactions involving a single substrate (Henri, 1903; Michaelis and Menten, 1913; Briggs and Haldane, 1915). The similarity of eqn. [10.25] to these other rate expressions frequently encountered in chemical and biochemical engineering practice requires a word of caution to avoid misinterpreting the kinetic model presented here. While the mathematical forms of these various rate expressions are all similar, the proper interpretation of the reaction models from which the rate expressions were derived, and the simplifying assumptions made in the derivations, are quite different. Assumptions of equilibrium (Henri, 1903; Michaelis and Menten, 1913) or use of the quasi-steady state approximation (Briggs and Haldane, 1915) are not necessary in developing eqn. [10.25]. Therefore, the recognized limitations and restrictions which such simplifying assumptions and approximations place on the kinetic model (Bailey and Ollis, 1977)) do not pertain to the model of eqn. [10.25].

In particular, it should be emphasized that the quasi-steady state approximation commonly invoked in deriving enzyme kinetic models is *not* implied in the kinetic model of photosynthetic electron transport developed here. (In enzyme kinetics, the concentration of enzyme-substrate complex, E-S, necessarily decreases with time as substrate is consumed; hence the need for a quasi-steady state approximation. Such an approximation is not justified under conditions of high enzyme concentration or low substrate concentration.) Equation [10.18] was derived by assuming that the concentration of electronically excited reaction-center chlorophyll molecules (P*) actually reaches a steady state level within a short time interval after the onset of illumination. As long as the incident light intensity remains constant (time invariant), the steady state assumption is rigorously valid. Even when the incident light intensity is very low, the assumption of steady state conditions with respect to P* is justified.

The unique feature of the reaction scheme presented in eqn. [10.14] which distinguishes it from other frequently encountered reaction schemes having similar rate expressions is, of course, the photoexcitation process. As discussed, the light intensity during the course of the reaction is completely independent of the kinetics of the photosynthetic electron transport reactions. This uncoupling of light intensity (I) from the reaction kinetics must be kept in mind when interpreting the rate expression in eqn. [10.25]. Parameter estimation is described in Appendix B, together with the light transport model which must also be incorporated for this purpose.

Thomas and his group (Thomasset et al., 1988; 1986; Cocquempot et al., 1986; Jeanfils et al., 1986; Breton et al., 1985; Cocquempot and Thomas, 1984) have incorporated the photosynthetic electron transfer rate model into an analysis of photoinactivation of immobilized chloroplast thylakoid membranes.

Cocquempot and Thomas (1984) immobilized chloroplast membranes in a serum albumin-glutaraldehyde matrix. The effect of incident light intensity on the rate of potassium ferricyanide reduction by free and immobilized chloroplasts was studied in a batch stirred reactor. The observed behavior was in agreement with the light transport-kinetic model described by Howell and Vieth (1982). Likewise, Carpentier et al. (1989; 1988) apply a similar model to interpret their photoacoustic data describing photosynthetic energy storage in photosystem II submembrane fractions.

10.4 REFERENCES

Al-Abta, S., I. J. Galpin and H. A. Collin, *Plant Sci. Lett.*, **16**, 129-134 (1979).

Alfermann, A. W., I. Schuller and E. Reinhard, *Planta Med.*, **40**, 218-223 (1980).

Ammirato, P. V., *Bio/Technology*, **1**, 68-74 (1983).

Ammirato, P. V., D. E. Evans, C. E. Flick, R. J. Whitaker and W. R. Sharp, *Trends Biotechnol.*, **2**, 1-6 (1984).

Anderson, J. M., *Biochim. Biophys. Acta*, **416**, 191 (1975).

Anonymous, *Time*, May 4 (1992).

Archambault, J., B. Volesky and W. G. Kurz, *Biotechnol. Bioeng.*, **35**, 702-711 (1990).

Asada, M. and M. L. Shuler, *Appl. Microbiol. Biotechnol.*, **30**, 475-481 (1989).

Bailey, J. E. and D. F. Ollis, "Biochemical Engineering Fundamentals," pp. 90-105. McGraw Hill: New York (1977).

Balandrin, M. F., J. A. Klocke, E. S. Wurtele and W. H. Bollinger, *Science*, **228**, 1154-1160 (1985).
Barnabas, N. J. and S. B. David, *Biotechnol. Lett.*, **10**, 593-596 (1988).
Beaumont, M. D. and D. Knorr, *Biotechnol. Lett.*, **9**, 377-382 (1987).
Berlin, J., *Endeavour*, **8**, 5-8 (1984).
Berlin, J. and J. M. Widholm, *Plant Physiol.*, **59**, 550-553 (1977).
Bevan, M. and M. D. Chilton, *Annu. Rev. Genet.*, **16**, 357-384 (1982).
Bisson, W., R. Beiderbeck and J. Reichling, *Planta Med.*, **47**, 164-168 (1983).
Bornman, C. H. and A. Zachrisson, In "Methods in Enzymology," K. Mosbach, Ed. Academic Press: San Diego, CA (1987).
Bourgin, J. P. and J. P. Nitsch, *Ann. Physiol. Veget.*, **9**, 377-382 (1967).
Bramble, J. L., D. L. Graves and P. Brodelius, *Biotechnol. Prog.* **6**, 452-457 (1990).
Breton, J., D. Thomas and J. F. Hervagault, *Eur. J. Biochem.*, **152**, 509-514 (1985).
Bridges, I. G., S. W. Bright, A. J. Greenland, W. W. Schuch, D. Pioli and A. Merryweather, Patent WO 9008827 A1, Use of Microbial Repressors and Operators to Regulate Expression in Plants (1990).
Bridges, I. G., S. W. Bright, A. J. Greenland and W. W. Schuch, Patent WO 9008829 A1, The *lac*-Repressor/Operator System as a Regulatory System in Plants (1990).
Briggs, G. E. and J. B. S. Haldane, *Biochem. J.*, **19**, 338 (1915).
Brisson, N., J. Paszkowski, J. R. Penswich, P. Gronenborn, J. Potrykus and T. Hohn, *Nature*, **310**, 511-514 (1984).
Brodelius, P., In "Immobilized Cells and Organelles," B. Mattiasson, Ed. CRC Press: Boca Raton, FL (1983).
Brodelius, P., I. Marques, K. Gügler and C. Funk In "Enzyme Engineering 9," H. W. Blanch and A. M. Klibanov, Eds. *Ann. N. Y. Acad. Sci.*, **542**, 159-165 (1988).
Buech, G. W., L. Enders, R. G. Berger and F. Drawert, In "Biochem. Eng." - Stuttgart (Proc. Int. Symp.), 2nd, Mtg. Date 1990, 232-235, M. Reuss, Ed. Fischer: Stuttgart (1991).
Buitelaar, R. M., A. C. Hulst and J. Tramper, *J. Biotechnol. Tech.*, **2**, 109-114 (1988).
Buitelaar, R. M., A. A. M. Langenhoff, R. Heidstra and J. Tramper, *Enzyme Microb. Technol.*, **13**, 487-494 (1991).
Butler, W. L., In "Chlorophyll - Proteins, Reaction Centers, and Photosynthetic Membranes," J. M. Olson and G. Hind, Eds. Brookhaven Symposia in Biology No. 28, pp. 338-346 (1976).
Cabral, J. M. S. and J. F. Kennedy, In "Methods in Enzymology," K. Mosbach, Ed. Academic Press: San Diego,CA (1987).
Caplan, A., L. Herrera-Estrella, D. Inze, E. Van Haute, M. Van Montague, J. Schell and P. Zambryski, *Science*, **222**, 815-821 (1984).
Carew, D. P. and E. J. Staba, *Lloydia*, **28**, 1-12 (1965).
Carpentier, R., R. M. Leblanc and M. Mimeault, *Biotechnol Bioeng.*, **32**, 64-67 (1988).
Carpentier, R., R. M. Leblanc and M. Mimeault, *Biochim. Biophys. Acta*, **975**, 370-376 (1989).

Chaleff, R. S. and M. F. Parsons, *Proc. Nat. Acad. Sci.*, USA, **75**, 5104-5107 (1978).
Cho, G. H., D.-I. Kim, H. Pedersen and C-K. Chin, *Biotechnol. Prog.* **4**, 184-188 (1988).
Chotani, G. K., Ph. D. Thesis, Dept. of Chemical and Biochemical Engineering, Rutgers University (1984).
Clayton, R. K., "Photosynthesis: How Light is Converted to Chemical Energy," Addison-Wesley Module in Biology No. 13, Addison-Wesley Publ. Co., Inc.: Reading, MA (1974).
Cocking, E. C., *Nature*, **187**, 927-929 (1960).
Cocquempot, M. F. and D. Thomas, *Enzyme & Microbial Technol.*, **6**, 321-324 (1984).
Cocquempot, M. F., J. F. Hervagault and D. Thomas, *Enzyme & Microbial Technol.*, **8**, 533-536 (1986).
Cramer, C. L., T. B. Ryder, J. N. Bell and C. J. Lamb, *Science*, **227**, 1240-1243 (1985).
Curtin, M. E., *Bio/Technology*, **1**, 649-457 (1983).
Denis, S. and P. Boyaval, *Appl. Microbiol. Biotechnol.*, **34**, 608-612 (1991).
DiCosmo, F. and S. G. Tallevi, *Trends Biotechnol.*, **3**, 110-111 (1985).
DiCosmo, F., P. J. Facchini and A. W. Neumann, *Trends in Biotechnol.*, **6**, 137-140 (1988).
El Hassouni, M., M. Chippaux and F. Barras, *J. Bacteriol.*, **172**, 6261-6267 (1990).
Esquivel, M. G., M. M. R. Fonseca, J. M. Novais, J. M. S. Cabral and M. S. S. Pais, In "Plant Cell Biotechnology," M. S. S. Pais et al., Eds. NATO ASI Series, Vol. H18. Springer-Verlag: Berlin (1988).
Evans, D. A. and W. R. Sharp, *Science*, **221**, 104-107 (1983).
Evans, D. A., W. R. Sharp and H. P. Medina-Filho, *Amer. J. Bot.*, **71**, 759-774 (1984).
Facchini, P. J., F. DiCosmo, L. G. Radvanyi and Y. Giguere, *Biotechnol. Bioeng.*, **32**, 935-958 (1988).
Facchini, P. J., A. W. Neumann and F. DiCosmo, *Appl. Microbiol. Biotechnol.*, **29**, 346-355 (1988).
Facchini, P. J., A. W. Neumann and F. DiCosmo, *Biotechnol. & Appl. Biochem.*, **11**, 74-82 (1989).
Facchini, P. J. and F. DiCosmo, *Biotechnol. Bioeng.*, **37**, 397-402 (1991).
Farmer, E. E., *Plant Physiol.*, **78** 338-342 (1985).
Felix, H. R., P. Brodelius and K. Mosbach, *Anal. Biochem.*, **116**, 462-470 (1981).
Flick, C. E., In "Handbook of Plant Cell Culture," D. A. Evand, W. R. Sharp, P. V. Ammirato and Y. Yamada, Eds. Macmillan: New York (1983).
Flores, H. E., Abstract, Biochemical Engineering VII, Engineering Foundation Conf., Santa Barbara, CA, March 3-8 (1991).
Fork, D. C., In "The Science of Photobiology," K. C. Smith, Ed., pp. 329-369. Plenum Press: New York (1977).
Fowler, M. W. and A. H. Scragg, In "Plant Cell Biotechnology," M. S. S. Pais, F. Mavituna and J. M. Novais, Eds. NATO ASI Series Vol. H18, p. 165. Springer-Verlag: Berlin (1988).

Fraley, R. T., S. G. Rogers, R. B. Horsch, D. A. Eichholtz, J. S. Flick, C. L. Fink, N. L. Hoffman and P. R. Sanders, *Bio/Technology*, **3**, 629-635 (1985).

Fujita, Y., Y. Hara, T. Ogino and C. Suga, *Plant Cell Rep.*, **1**, 59-60 (1981).

Fujita, Y., Y. Hara, C. Suga and M. Morimoto, *Plant Cell Rep.*, **1**, 61-63 (1981).

Fujita, Y., T. Yoshioka and Y. Hara, Paper presented at 1989 Int. Chem. Cong. of Pacific Basin Soc., Honolulu, Hawaii, Dec. 17-22 (1989).

Fuller, K. W., *Chemistry and Industry*, **3 Dec.**, 825-833 (1984).

Funk, C., K. Mosbach and P. Brodelius, In "Enzyme Engineering 8," A. I. Laskin, K. Mosbach, D. Thomas and L. B. Wingard, Eds. *Ann. N. Y. Acad. Sci.*, **501**, 347-350 (1987).

Furusaki, S., T. Nozawa, T. Isohara and T. Furuya, *Appl. Microbiol. Biotechnol.*, **29**, 437-441 (1988).

Furuya, T., T. Yoshikawa and M. Taira, *Phytochem.*, **23**, 999-1001 (1984).

Gautheret, R. J., *Comp. Rend. Hebd. Sci. Acad. Sci.*, **208**, 118-121 (1939).

Gengenbach, B. G. and C. E. Green, *Crop Sci.*, **15**, 645-649 (1975).

Govindjee and R. Govindjee, *Scientific American*, **231**, 68 (1974).

Guillet, J., *The Sciences*, **Nov./Dec.**, 23-28 (1991).

Haldimann, D. and P. Brodelius, *Phytochem.*, **26**, 1431-1434 (1987).

Hamada, H., N. Umeda, N. Otsuka and S. Kawabe, *Plant Cell Reports*, **7**, 493-494 (1988).

Hamilton, R., H. Pedersen and C-K. Chin, *Biotechnol. Bioeng.*, Symp. No. 14, 383-396 (1984).

Hamilton, R., H. Pedersen and C-K. Chin, *Biotechniques.*, **Mar./Apr.**, 96 (1985). See also *Agricell Report*, **4**, 4 (1985).

Henri, V., "Lois Generales de l'Action des Diastases," Herman: Paris (1903).

Hirasuna, T., Ph. D. Thesis, Cornell University, Ithaca, New York (1990).

Horsch, R. B., J. E. Fry, N. L. Hoffman, D. Eichholtz, S. G. Rogers and R. T. Fraley, *Science*, **227**, 1229-1231 (1985).

Hougen, O. A. and K. M. Watson, "Chemical Process Principles," Vol. III, Kinetics and Catalysis. Wiley & Sons: New York (1948).

Houmard, J., V. Capuano, M. V. Colombano, T. Coursin and N. Tandeau de Marsac, *Proc. Natl. Acad. Sci.* USA., **87**, 2152-2156 (1990).

Howell, J. M. and W. R. Vieth, *J. Mol. Catal.*, **16**, 245-298 (1982).

Ishida, B. K., *Plant Cell Reports*, **7**, 270-273 (1988).

Jeanfils, J., M. F. Cocquempot and D. Thomas, *Enzyme & Microbial Technol.*, **8**, 157-160 (1986).

Jolicoeur, M., C. Chavarie, P. J. Carreau and J. Archambault, *Biotechnol. Bioeng.*, **39**, 511-521 (1992).

Kalla, R., L. K. Lind and P. Gustafsson, *Mol. Microbiol.*, **3**, 339-347 (1989).

Kargi, F., *Biotech. Lett.*, **10**, 181-186 (1988).

Kargi, F. and I. Freidel, *Biotech. Lett.*, **10**, 409-414 (1988).

Katz, J. J., J. R. Norris and L. L. Shipman, In "Chlorophyll - Proteins, Reaction Centers, and Photosynthetic Membranes," J. M. Olson and G. Hind, Eds. Brookhaven Symposia in Biology No. 28, pp. 16-55 (1976).

Kessel, R. H. J. and A. H. Carr, *J. Exp. Bot.*, **23**, 996-1007 (1972).

Ketel, D. H., A. C. Hulst, H. Gruppem, H. Breteler and J. Tramper, *Enzyme Microb. Technol.*, **9**, 303-307 (1987).
Kim, D.-I., Master's Thesis, Dept. of Chemical and Biochemical Engineering, Rutgers University (1987).
Kim, D.-I., H. Pedersen and C.-K. Chin, *Biotechnol. Lett.*, **10**, 709-712 (1988).
Kim, D.-I., G. H. Cho., H. Pedersen and C. -K. Chin, *Appl. Microbiol. Biotechnol.*, **34**, 726-729 (1991).
Kim, D. J., H. N. Jang and J. R. Liu, *Biotechnol. Tech.*, **3**, 139-144 (1989).
Klug, G., *Mol. Gen. Genet.*, **226**, 167-176 (1991).
Klug, G., N. Gad'on, S. Jock and M. L. Narro, *Mol. Microbiol.*, **5**, 1235-1239 (1991).
Knoop, B. and R. Beiderbeck, Z. *Naturforsh*, **38**, 484-486 (1983).
Knorr, D. and J. Berlin, *J. Food Sci.*, **5**, 1397-1400 (1987).
Kobayashi, Y., H. Fukui and M. Takata, *Plant Cell Reports*, **6**, 185-186 (1987).
Kobayashi, Y., H. Fukui and M. Takata, *Plant Cell Reports*, **7**, 249-252 (1988).
Kok, B. In "Proceedings of the Workshop on Bio-Solar Conversion," M. G. Gibbs, A. Hollaender, B. Kok, L. O. Krampitz and A. San Pietro, Eds. Bethesda, MD, supported by the NSF/RANN, Grant GI-40253, pp. 22-30 (1973).
Kondo, O., H. Honda, M. Taya and T. Kobayashi, *Appl. Microbiol. Biotechnol.*, **32**, 291-294 (1989).
Kreis, W. and E. Reinhard, *J. Biotechnol.*, **16**, 123-135 (1990).
Kudo, I., Masters Thesis, Dept. of Chemical and Biochemical Engineering, Rutgers University (1985).
Kueh, J. S. H. and S. W. J. Bright, *Planta*, **153**, 166-171 (1981).
Kuhn, D. N., J. Chappell, A. Boudet and K. Hahlbrock, *Proc. Nat. Acad. Sci.*, USA, **81**, 1102-1106 (1984).
Langmuir, I., *J. Am. Chem. Soc.*, **40**, 1361 (1918).
Larkin, P. J. and W. R. Scowcroft, *Theoret. App. Genet.*, **60**, 197-214 (1981).
Leach, F., G. A. Armstrong and J. E. Hearst, *J. Gen. Microbiol.*, **137**, 1551-1556 (1991).
Leckie, F., A. H. Scragg and K. C. Cliffe, *Enzyme Microb. Technol.*, **13**, 801-810 (1991).
Leckie, F., A. H. Scragg and K. C. Cliffe, *Enzyme Microb. Technol.*, **13**, 296-305 (1991).
Levy, L. W., *Environ. Exp. Bot.*, **21**, 389-395 (1981).
Lien, S. and A. San Pietro, "An Inquiry into Biophotolysis of Water to Produce Hydrogen," NSF/RANN Report sponsored under Grant GI-40253 to Indiana Univ. (1975).
Malmberg, R. L., *Cell*, **22**, 603-609 (1980).
Matsushita, T., K. Ishibashi, M. Kizu and K. Funatsu, *Kagaku Kogaku Ronbunshu*, **17**, 649-654 (1991).
Mavituna, M., A. K. Wilkinson and P. D. Williams, In "Separations for Biotechnology," M. S. Verrall and M. T. Hudson, Eds. Ellis Horwood Ltd.: Chichester, England (1987).
McDonald, K. A. and A. P. Jackman, *Plant Cell Rep.*, **8**, 455-458 (1989).
Messing, R. A., Personal Communication (1985).

Michaelis, L. and M. L. Menten, *Biochem. Z.*, **49**, 333 (1913).
Miller, M. E., J. E. Jurgenson, E. M. Reardon and C. A. Price, *J. Biol. Chem.*, **258**, 14478-14484 (1983).
Misawa, N., K. Kobayashi and K. Nakamura, Eur. Patent, EP 393690 A1, Cloning and Expression of Genes for *Erwinia* carotenoid Biosynthesis Enzymes (1990).
Muller, A. J. and R. Grafe, *Molec. Gen. Genet.*, **161**, 67-76 (1981).
Murashige, T. and R. Nakano, *Amer. J. Bot.*, **54**, 963-970 (1967).
Nafziger, E. D., J. M. Widholm, H. C. Steinbrucken and J. L. Killmer, *Plant Physiol.*, **76**, 571-574 (1984).
Nobecourt, P., *Comp. Rend. Sci Soc. Biol. Fil.*, **130**, 1270-1271 (1939).
Noe, W., C. Mollenschott and J. Berlin, *Plant Mol. Biol.*, **3**, 281-288 (1984).
Noguchi, M., T. Matsumato, Y. Hirata, H. Yamamato, A. Katsuyama, A. Kato, S. Azechi and K. Kato, In "Plant Tissue Culture and Its Biotechnological Application," W. Barz, E. Reinhard and M. H. Zenk, Eds. Springer Verlag: Berlin (1977).
Park, J. M., W. S. Hu and E. John Staba, *Biotechnol. Bioeng.*, **34**, 1209-1213 (1989).
Park, J. M., K. L. Giles and D. D. Songstad, Proc. Asia-Pacific Biochem. Eng. Conf. '90, Kyungju, Korea, April 22-25 (1990).
Payne, G. F., V. Bringi, C. L. Prince and M. L. Shuler, "Plant Cell and Tissue Culture in Liquid Systems." Hanser Publishers: Munich; Dist. in U. S. by Oxford University Press: New York (1991).
Payne, G. F., N. N. Payne and M. L. Shuler, *Biotechnol. Bioeng.*, **31**, 905-912 (1988).
Pedersen, H., Paper Presented at 13th Ann. Spring Symp., AIChE, May (1990).
Pedersen, H., D.-I. Kim, G. H. Cho and C.-K. Chin, *Appl. Microb. Biotech.*, **34**, 726-729 (1991).
Pedersen, H., D.-I. Kim and C.-K. Chin, *Biotechnol. Lett.*, **13**, 213-216 (1991).
Pedersen, H., S. Y. Byun and C.-K. Chin, "Bioprocessing of Higher Plant Cell Cultures," *ACS Symp. Series XXX*, S. Sikdar, P. Todd and M. Bier Eds. (1991).
Petersen, M., A. W. Alfermann and H. U. Seitz, In "Plant Cell Biotechnology," M. S. S. Pais et al., Eds. NATO ASI Series, Vol. H18. Springer-Verlag: Berlin (1988).
Prenosil, J. E. and H. Pedersen, *Enzyme Microb. Technol.*, **5**, 323-331 (1983).
Prenosil, J. E., M. Hegglin, J. R. Bourne and R. Hamilton, In "Enzyme Engineering 8," A. I. Laskin, K. Mosbach, D. Thomas and L. B. Wingard, Eds. *Ann. N. Y. Acad. Sci.*, **501**, 390-395 (1987).
Prenosil, J. E. and M. Hegglin, In "Enzyme Engineering," H. Okada, A. Tanaka and H. W. Blanch, Eds. *Ann. N. Y. Acad. Sci.*, **613**, 234-247 (1990).
Pu, H. T. and R. Y. K. Yang, *Biotechnol. Bioeng.*, **32**, 891-896 (1988).
Renaudin, J. P., *Plant Sci. Lett.*, **22**, 59-69 (1981).
Rhodes, M. J. C., J. I. Smith and R. J. Robins, *Appl. Microbiol. Biotechnol.*, **26**, 28-35 (1987).
Rodriguez-Mendiola, M. A., A. Stafford, R. Cresswell and C. Arias-Castro, *Enzyme Microb. Technol.*, **13**, 697-702 (1991).
Rokem, J. S., J. Schwartzberg and I. Goldberg, *Plant Cell Rep.*, **3**, 159-160 (1984).

Sahai, O. and M. Knuth, *Biotechnol. Prog.*, **1**, 1-9 (1985).
Sasse, F., M. Buchholz and J. Berlin, *Z. Naturforsch.*, **38**, 910-915 (1983).
Schmidt, A. J., J. M. Lee and G. An, *Biotechnol. Bioeng.*, **33**, 1437-1444 (1989).
Sexton, T. B., D. A. Christopher and J. E. Mullet, *EMBO J.*, **9**, 4485-4494 (1990).
Sharp, W. R., D. R. Evans and P. V. Ammirato, *Food Technol. (Overview)*, **Feb.**, 112-119 (1984).
Shuler, M. L., *Ann. N. Y. Acad. Sci.*, **369**, 65-79 (1981).
Shuler, M. L., G. A. Hallsby, J. W. Pyne and T. Cho, *Ann. N. Y. Acad. Sci.*, **469**, 271-278 (1986).
Skoog, F. and C. O. Miller, "Chemical Regulation of Growth and Organ Formation in Plant Tissues Cultured in Vitro," llth Symposium Society Experimental Biology, pp. 118-131 (1957).
Sonomoto, K., H. Nakajima, F. Sato, K. Ichimura, Y. Yamada and A. Tanaka, In "Enzyme Engineering," H. Okada, A. Tanaka and H. W. Blanch, Eds. *Ann. N. Y. Acad. Sci.*, **613**, 542-546 (1990).
Steward, F. C., M. O. Mapes and K. Mears, *Am. J. Bot.*, **45**, 704-708 (1958).
Su, W. W. and A. E. Humphrey, *Biotechnol. Lett.*, **13**, 889-892 (1991).
Suzuki, E., H. Fukuzawa and S. Miyachi, *Mol. Gen. Genet.*, **226**, 401-408 (1991).
Tabata, M. and Y. Fujita, In "Biotechnology in Plant Science," M. Zatlin, P. Day and A. Hollaender, Ed., p. 207-218. Academic Press: Orlando, FL (1985).
Taticek, R. A., M. Moo-Young and R. L. Legge, *Appl. Microbiol. Biotechnol.*, **33**, 280-286 (1990).
Taticek, R. A., M. Moo-Young and R. L. Legge, *Appl. Microbiol. Biotechnol.*, **35**, 558 (1991).
Thomasset, B., A. Friboulet, J. N. Barbotin and D. Thomas, *Biotechnol. Bioeng.*, **28**, 1200-1205 (1986).
Thomasset, B., D. Thomas and R. Lortie, *Biotechnol.* Bioeng., **32**, 764-770 (1988).
Tietjen, K. G., D. Hunkler and U. Matern, *Eur. J. Biochem.*, **131**, 401-407 (1983).
Trebst, A., *Annu. Rev. Plant Physiol.*, **25**, 423 (1974).
Tulecke, W. and L. G. Nickell, *Science*, **130**, 863-864 (1959).
Ulbrich, B., W. Wiesner and H. Arens, In "Primary and Secondary Metabolism of Plant Cell Cultures," K. H. Neumann, W. Barz and E. Reinhard, Eds., p. 293-303. Springer-Verlag: Berlin (1985).
Ulbrich, B., H. Osthoff and W. Wiesner, In "Plant Cell Biotechnology," M. S. S. Pais, F. Mavituna and J. M. Novais, Eds. NATO ASI Series Vol. H18, p. 461-474. Springer-Verlag: Berlin (1988).
Uribe, E. G. and U. Lüttge, *American Scientist*, **72**, 567-573 (1984).
Vanek, T., V. V. Urmantseva, Z. Wimmer and T. Macek, *Biotech. Lett.*, **11**, 243-248 (1989).
Vankova, R. and C. H. Bornman, *Physiol. Plantarum*, **70**, 1-7 (1987).
Vasil, I. K., In "Tissue Culture in Forestry and Agriculture," R. H. Henke, K. W. Hughes, M. P. Consyantin and A. Hollaender, Eds. Plenum: New York (1985).
Wagner, F. and H. Vogelmann, In "Plant Tissue Culture and Its Biotechnological Application," W. Barz, P. Reinhard and M. H. Zenk, eds. Springer Verlag: Berlin (1977).
Wang, C. J. and E. J. Staba, *J. Pharm. Sci.*, **52**, 1058-1062 (1963).

White, P. R., *Am. J. Bot.*, **26**, 59-64 (1939).
Whitney, P. J., *Enzyme Microb. Technol.*, **14**, 13-17 (1992).
Widholm, J. M., in "Plant Culture as a Source of Biochemicals," E. J. Staba, Ed. CRC Press: Boca Raton, FL (1980).
Yamada, Y., K. Imaizumi, F. Sato and T. Yasuda, *Plant Cell Physiol.*, **22**, 917-922 (1981).
Yamada, Y. and Y. Fujita, In "Handbook of Plant Cell Culture," D. E. Evans, W. R. Sharp, P. V. Ammirato and Y. Yamada, Eds. Macmillan: New York (1983).
Yamada, Y. and F. Sato, In "Handbook of Plant Cell Culture, D. E. Evans, W. R. Sharp, P. V. Ammirato and Y. Yamada, Eds. Macmillan: New York (1983).
Yamakawa, T., S. Kato, K. Ishida, T. Kodama and Y. Minoda, *Agric. Biol. Chem.*, **47**, 2185-2191 (1983).
Yu, A. H. C. and K. Hosono, *Biotechnol. Lett.*, **13**, 411-416 (1991).
Zankel, K. L. and B. Kok, In "Methods in Enzymology," A. San Pietro, Ed., **24**, pp. 218-238. Academic Press: New York (1972).
Zenk, M. H., H. El-Shagi, H. Arens, J. Stockigt, E. W. Weller and B. Deus, In "Plant Tissue Culture and Its Biotechnological Application," W. Barz, P. Reinhard and M. H. Zenk, Eds. Springer Verlag: Berlin (1977).

APPENDIX A

A.0 BIOREACTOR TRANSIENT MODEL: LAC OPERON EXPRESSION

Many biochemical reactions and events which occur in the growth of a recombinant cell need to be considered before one can formulate a working model, capable of predicting the bioreactor dynamics of a genetically-engineered microorganism. Production of protein from cloned genes depends primarily on cell growth rate, plasmid copy number and the kinetics of transcription and translation. The molecular mechanisms involved in protein biosynthesis can be used in formulation of the rate equations necessary to yield a cohesive mathematical model. Obvious simplifications must be made in the development of any working model, due to the complexity of secondary interactions between cell growth, plasmid replication and gene expression.

The basis of any recombinant cell growth model is a set of appropriate equations describing the growth of wild-type cells. Superimposed on this must be a structured model describing the kinetics of plasmid replication, segregation and gene expression. For example, a structured kinetic single-cell model of wild-type *E. coli* growth was developed (Shuler et al., 1979) and modified to incorporate the effects of plasmid replication and gene expression kinetics for recombinant cells (Ataai and Shuler, 1985).

IMRC REACTOR BALANCE EQUATIONS, E. COLI

Recombinant and plasmid-free cell concentrations may be obtained from *component* balances for each species in the following reactor model. The cell concentration in the immobilized phase is assumed to remain constant with respect to time and dilution rate. The idea is that live cells grow to saturate a thin layer near the particle surface, accessible to substrate. Any additional replication is then released to the bulk. The liquid (free) phase concentrations of recombinant and wild-type cells may be calculated for a given set of operating conditions. The net rate of free-cell production of each species is the sum of contributions from the liquid and immobilized phases, minus the rate at which cell removal occurs from the bioreactor. The following assumptions are made in the development of the subsequent *component* balance equations for recombinant and plasmid-free cells.

1. All cells reproduce by binary fission.
2. All recombinant cells contain the same number of plasmids corresponding to the mean plasmid copy number.
3. Plasmid is divided randomly between daughter cells at the time of cell division.
4. No plasmid-free cell or its progeny can regain the plasmid.

Recombinant cell balance:

$$\frac{dX_F^+}{dt} = (1 - \theta_I)\, \mu_I^+\, X_I\, \phi_I\, \frac{1 - \varepsilon}{\varepsilon} + (1 - \theta_F)\, \mu_F^+\, X_F^+ - \frac{D}{\varepsilon}\, X_F^+ \qquad [A.1]$$

The first term on the right hand side of eqn. [A.1] represents release of plasmid-bearing cells from the immobilized phase to the bulk phase; the other two terms are the customary ones for chemostats, expressing plasmid-bearing cell growth and removal in the reactor effluent.

Plasmid-free cell balance:

$$\frac{dX_F^-}{dt} = [\theta_I\, \mu_I^+\, X_I\, \phi_I + \mu_I^-\, X_I\, (1 - \phi_I)]\, \frac{1 - \varepsilon}{\varepsilon} + \theta_F\, \mu_F^+\, X_F^+ + \mu_F^- X_F^- - \frac{D}{\varepsilon}\, X_F^- \qquad [A.2]$$

The composite term appearing first on the right hand side of eqn. [A.2] represents release of plasmid-free cells, while the second term represents cell reversion in the free suspension culture.

Substrate balance:

$$[\varepsilon + \xi_I\, (1 - \varepsilon)\,]\, \frac{dS_F}{dt} = D\, (S_0 - S_F) - \frac{1}{Y_X}\, r\,(X) \qquad [A.3]$$

where:

$$r\,(X) = [\phi_I\, \mu_I^+ + (1 - \phi_I\,)\, \mu_I^-]\, X_I\, (1 - \varepsilon) + \mu_F^+\, X_F^+\, \varepsilon + \mu_F^-\, X_F^-\, \varepsilon \qquad [A.4]$$

The immobilized cell growth rate of each species can be related to the corresponding free cell growth rate as follows:

$$\mu_I^+ = \eta_I \cdot \mu_F^+ \qquad \text{and} \qquad \mu_I^- = \eta_I \cdot \mu_F^- \tag{A.5}$$

The terms η_I and ξ_I are bioreaction and substrate distribution effectiveness factors, respectively. These effectiveness factors may be evaluated using the following equations:

$$\eta_I = \frac{\int_{r^*}^{R} r(S) \cdot 4\pi a^2 da}{\left(\int_{r^*}^{R} 4\pi a^2 da\right) \cdot r(S_F)} \tag{A.6}$$

and
$$\xi_I = \frac{\int_{r^*}^{R} S \cdot 4\pi a^2 da}{\left(\int_{r^*}^{R} 4\pi a^2 da\right) \cdot S_F} \tag{A.7}$$

Equations [A.6] and [A.7] are included here for the sake of completeness. In the evaluation of steady state bioreactor kinetics, the term containing ξ_I drops out of the analysis. In the IMRC bioreactor simulations, already shown together with the experimental data in Chapter 8, all of the experiments were essentially steady state with respect to substrate concentration, and η_I was treated as a parameter and evaluated by simulation to match the predicted cell concentration to that observed experimentally.

A structured model for *lac* promoter-operator function in wild-type and high copy number plasmids has been developed (Lee and Bailey, 1984a, b). The effect of plasmid copy number on protein production was modeled with results showing that the rate of transcription with respect to plasmid copy number deviates from linearity at high copy number, as one would expect. The fraction of plasmid-containing cells in the immobilized phase, ϕ_I, may be evaluated using the following relation developed by Bailey et al. (1983). The fraction of plasmid-containing cells in the immobilized phase is given by:

$$\frac{1}{\phi_I(t)} = 1 + \frac{\theta(1-\omega)}{(1-\theta)(2-\omega)}\left[1 - \left(\frac{2^{\alpha}}{2-\theta}\right)^{g}\right] \tag{A.8}$$

where: $\omega = (2 - \theta)^{1/\alpha}$, $\theta = 2^{1-N_p}$, $\alpha = \dfrac{\mu_I^-}{\mu_I^+}$, $g = \dfrac{t}{G_I^+}$

The recombinant cell fraction in the bioreactor effluent is the same as the fraction of recombinant cells in the liquid phase. It may be calculated directly from the integrated values of X_F^+ and X_F^-. The fraction of plasmid-containing cells in the liquid phase is given by:

$$\phi_F = \frac{X_F^+}{X_F^+ + X_F^-} \qquad \text{[A.9]}$$

GENE EXPRESSION KINETICS

Control of mRNA synthesis for β-galactosidase takes place at two independent sites (Fig. A.1):

1. The repressor gene (i) constitutively produces a protein (R) which, in the absence of inducer (lactose), binds to the operator gene (o), blocking transcription from the promoter;
2. In the presence of high levels of cyclic adenosine monophosphate (cAMP), a complex forms between cAMP and the catabolite receptor protein (CRP) which then binds to the P_I region of the promotor gene, facilitating RNA polymerase binding at the P_{II} site (de Crombrugghe et al., 1984).

Thus negative regulation takes place at the operator gene, while positive regulation occurs at the promoter P_I site. A high lactose concentration is sufficient to prevent the repressor protein from binding to the operator gene, but in the presence of glucose, low cAMP levels prevent the CRP-cAMP complex from forming and binding to the promoter. As a result, extremely low levels of β-galactosidase are synthesized by the cell. There is evidence that the cAMP-CRP complex is not the sole mediator of catabolite repression (Wanner et al., 1978). This is confirmed by the fact that catabolite repression can take place in *crp* mutants, in the absence of a functional cAMP-CRP complex (Guidi-Rontani et al., 1980). An intermediate in the metabolism of glucose is thought to bind free RNA polymerase (RNAP), inhibiting the binding of RNAP to the P_{II} site of the promoter gene (Dessein et al., 1978). This intermediate has been given the name "catabolite modulating factor."

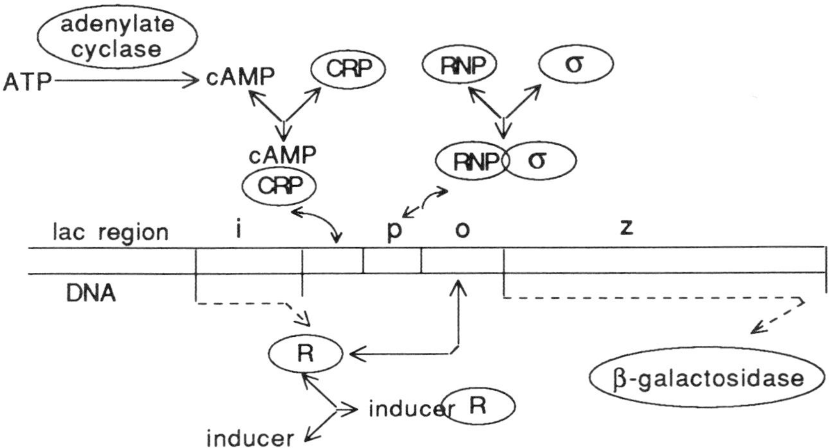

Figure A.1 Schematic of lac operon transcription control system.

The effects of catabolite modulating factor have been incorporated into a rigorous mathematical model of β-galactosidase biosynthesis in wild-type *Escherichia coli* (Ray et al., 1987). The model provides a quantitative description of active inducer transport (Kaushik et al., 1985) and the role of PTS (phosphoenolpyruvate dependent sugar transport system) enzymes in the regulation of cAMP levels and in inducer exclusion. It is important to note that the *lac* promoter is widely used as a genetic regulatory unit for initiation of transcription for cloned genes in a variety of multicopy recombinant plasmids, so the modeling effort to be presented has a certain generality.

Once the recombinant cell concentration in the bulk phase is known, *lac* operon kinetics are used to evaluate the concentration of β-galactosidase in the effluent stream (Ray et al., 1986). The concentration of β-galactosidase is dependent upon the concentration of mRNA which codes for the enzyme. The β-galactosidase mRNA concentration is dependent upon the rate of initiation of transcription.

β-galactosidase mRNA balance:

$$\frac{d[\mathrm{mRNA} \cdot X_F^+]}{dt} = K_{+M} \cdot N_p \cdot F \cdot Q \cdot \mu_F^+ \cdot X_F^+ - \left(K_{-M} + \frac{D}{\varepsilon}\right)[\mathrm{mRNA} \cdot X_F^+] \qquad [A.10]$$

where the *gene copy number*, N_p, appears explicitly. The concentration of β-galactosidase can readily be calculated from the mRNA concentration, based on the translation kinetics.

β-galactosidase enzyme balance:

$$\frac{d[E \cdot X_F^+]}{dt} = K_{+E}\,[mRNA \cdot X_F^+] - K_{-E}\,[E \cdot X_F^+] - \frac{D}{\varepsilon}\,[E \cdot X_F^+] \qquad [A.11]$$

By now familiar equations are used to evaluate the fraction of unbound operator gene, Q, and the binding efficiency, F, of RNA polymerase at the promoter site (Ray et al., 1987).

Operator index:

$$Q = \frac{1 + b_1 \cdot S_{in}}{1 + b_1 \cdot S_{in} + b_2\,[R]_t} = \frac{1 + b_1 \cdot S_{in}}{1 + b_1 \cdot S_{in} + b_2} \qquad [A.12]$$

The intracellular lactose concentration, S_{in}, is typically a multiple of the lactose level in the extracellular environment. It may be evaluated with the aid of the following equation.

Intracellular lactose concentration:

$$S_{in} = \frac{A' \cdot S_F}{B + S_F} \qquad [A.13]$$

Catabolic repression index:

$$F = \frac{K_1}{K_2 + [M]} \qquad [A.14]$$

The catabolite modulating factor concentration ([M]) has a negative effect on the RNAP binding efficiency. It it responsible for the existence of an optimal dilution rate at which the specific enzyme concentration reaches a maximum. Reiterating briefly, this has been modeled by assuming that a critical growth rate exists such that the concentration of catabolite modulating factor is negligible at growth rates below the critical value, and increases hyperbolically at growth rates above the critical value (Ray, 1985).

Catabolite modulating factor:

$$[M] = \frac{k_m \cdot (\mu_F^+ - \mu_{crit})}{Y_X \cdot [k_{-m} + (\mu_F^+ - \mu_{crit})]} \quad , \quad \mu_F^+ \geq \mu_{crit} \qquad [A.15]$$

Lastly, the growth rate of cells in the bulk phase may be evaluated using the following growth model.

Cell growth rate:

$$\mu = \frac{\mu_m \cdot S_F}{S_F + K_s\,(1 + K_i\,X)} = \frac{\mu_m \cdot S_F}{S_F + K_s\,[1 + K_p\,(S_0 - S_F)]} \qquad [A.16]$$

A.1 α-AMYLASE STUDIES

EXPERIMENTAL RESULTS

The growth of the recombinant strains, 1A289, pC194 and the production of α-amylase was investigated in chemostat culture. The results indicated the level of production could be controlled through manipulating the dilution rate. At steady state the growth rate is equal to dilution rate. As expected, higher levels of α-amylase activity were seen at lower growth rates. The span of a sub-phase of the experimental program is illustrated in Figs. A.2 and A.3.

Splitting out an individual set of data, consider the chemostat experiment carried out with 10 g/L maltose in the feed medium. The maximum cell concentration, 5.2 g/L dcw, and the highest specific amylase activity, 8200 units/g dcw, were attained at a dilution rate of 0.045 hr^{-1} (Fig. A.4). As the dilution rate ranged from 0.11 to 0.50 hr^{-1}, cell density decreased to the lower limit of 1.5. Also the specific α-amylase activity decreased sharply from 7,200 units to a negligible level at the highest dilution rates.

Next, a continuous fermentation was carried out with the maltose concentration in the feed medium equal to 2 g/L (Fig. A.5). The lower dilution rate of 0.045 hr^{-1} resulted in both the highest cell concentration and enzyme activity that were observed over the range of the dilution rates studied. The maximum cell concentration attained was 4.0 g dcw/L and decreased as the dilution rate was increased. The highest specific amylase activity attained was 12.5 units/g dcw. The enzyme concentration also decreased sharply from 10,274 to 10 units/g dcw.

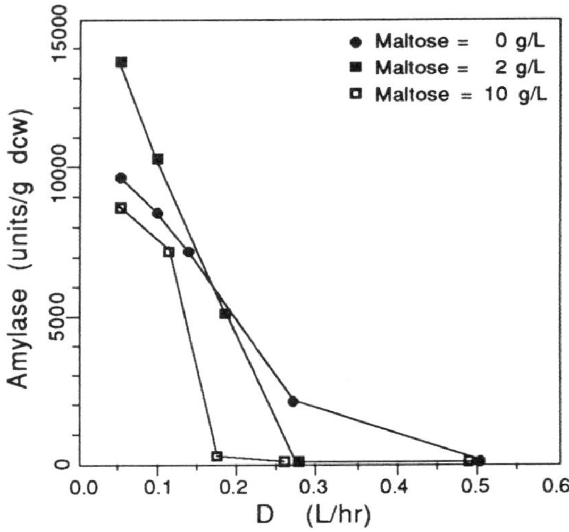

Figure A.2 Comparison of amylase activity vs dilution rate for a 1A289, pC194 grown in chemostat with 0 g/L, 2 g/L and 10 g/L maltose as the carbon source.

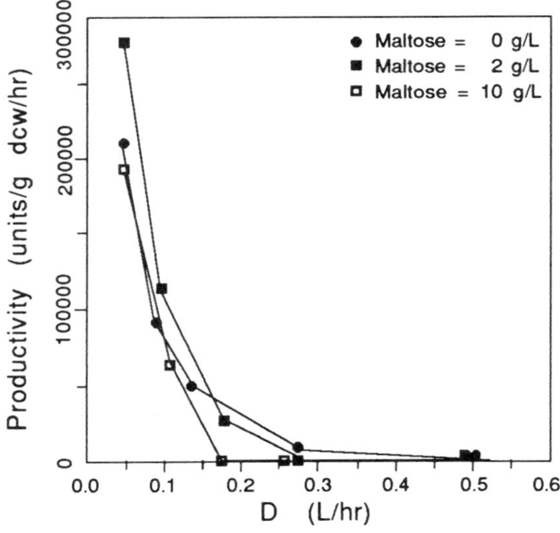

Figure A.3 Productivity vs Dilution Rate for 1A289, pC194 grown in a complex medium with 0 g/L, 2 g/L and 10 g/L maltose as the carbon source.

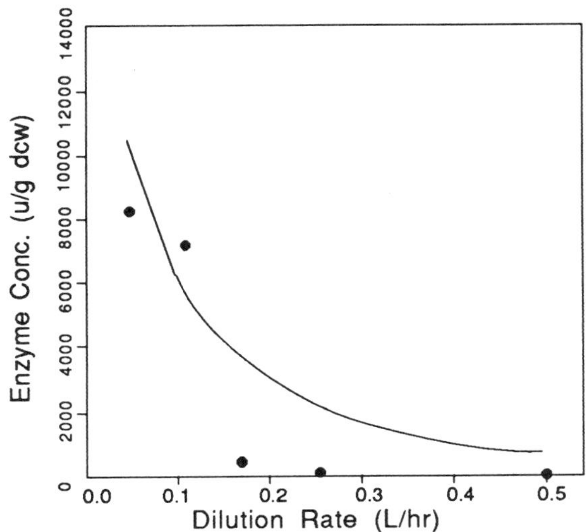

Figure A.4 α-Amylase concentration versus dilution rate for free cell chemostat with 10 g/L maltose in medium, model and data results.

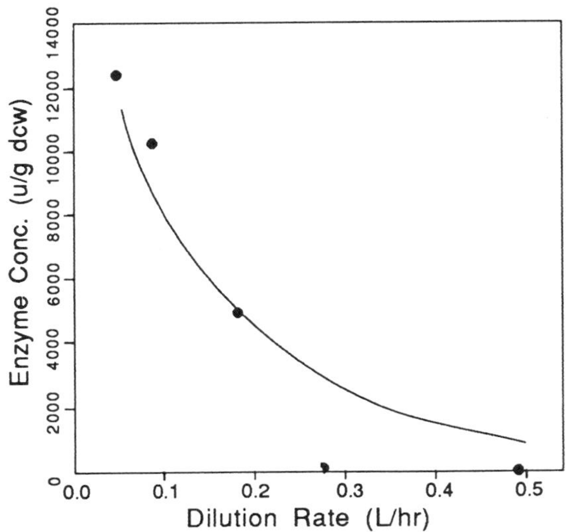

Figure A.5 α-Amylase concentration versus dilution rate for free cell chemostat with 2 g/L maltose in medium, model and data results.

Figs. A.6 and A.7 illustrate the bulk substrate-dilution rate profiles corresponding to the enzyme profiles shown in Figs. A.4 and A.5, respectively. As a control, a continuous fermentation was run with no maltose or any other carbohydrate in the feed. The maximum cell concentration was 2.8 g dcw/L. The highest specific enzyme concentration realized was 9000 units/g dcw; although this is slightly higher than was observed in the chemostat with 10 g/L maltose in the feed, it is not significantly higher. Thus while any repression due to maltose has been removed, it appears that the lack of a carbohydrate poses too extensive a carbon deficiency for the highest concentrations of enzyme to be produced. This opposes what was observed in batch when the amylase activity for the fermentations with maltose and with no carbohydrate resulted in the same specific activity.

As the dilution rate is increased, the same trend as seen in the previous experiments is observed: there is a decrease in dcw and enzyme activity. However, the enzyme concentration did not drop off to zero at a dilution rate of .27 hr^{-1}; there was still a production rate of 2000 units/g dcw. This suggests that the repression effects due to the presence of maltose have been alleviated, resulting in a higher rate of transcription.

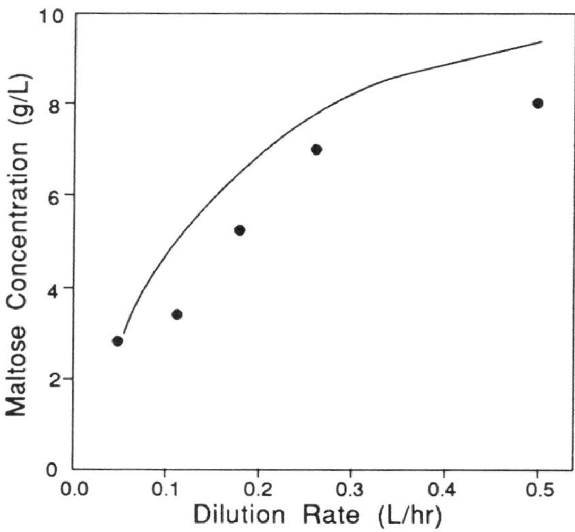

Figure A.6 Steady state maltose concentration versus dilution rate for free cell chemostat with 10 g/L maltose in medium, model and data results.

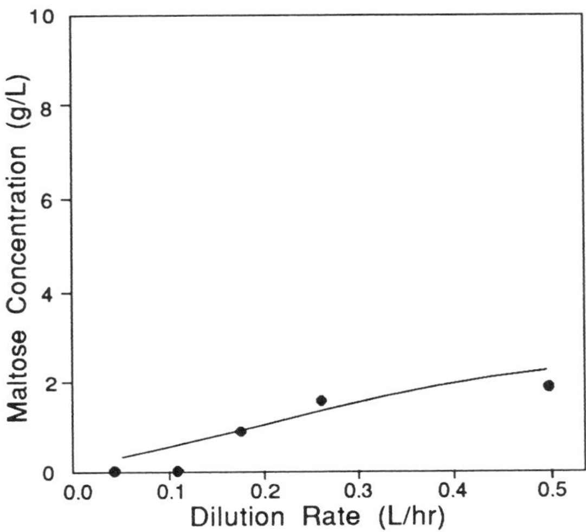

Figure A.7 Steady state maltose concentration versus dilution rate for free cell chemostat with 2 g/L maltose in medium, model and data results.

GROWTH RATE AND SUBSTRATE UTILIZATION

The change in the cell concentration, X, and substrate concentration, S, in continuous culture can be expressed as:

$$\frac{dX}{dt} = \mu X - DX \qquad [A.17]$$

and,

$$-\frac{dS}{dt} = \frac{1}{Y_{X/S}}\frac{dX}{dt} - D(S_0 - S) - mX \qquad [A.18]$$

D is the dilution rate, μ is the specific growth rate, and $Y_{X/S}$ is the yield coefficient. The maintenance coefficient is represented by m.

The specific growth rate can be described using Monod kinetics with product inhibition (Ponzo et al., 1986).

$$\mu = \mu_m \frac{S}{S + K_S\left(1 + \frac{P_i}{k_i}\right)} \qquad [A.19]$$

where μ_m is the maximum specific growth rate, S is the substrate concentration within the chemostat, K_S is the Monod saturation constant, P_i is the concentration of growth inhibiting product and k_i represents the coefficient of inhibition. The concentration of inhibitory product in a chemostat at steady state is:

$$P_i = Y_{P_i}(S_0 - \tilde{S}) \qquad [A.20]$$

S_0 is the concentration of the substrate in the feed and Y_{P_i} is the yield coefficient for inhibiting product and is equal to the product of K_p, the product inhibition coefficient, and k_i. Therefore the growth rate at steady state can be expressed as:

$$\mu = \frac{\mu_m \tilde{S}}{\tilde{S} + K_S\,[1 + K_p(S_0 - \tilde{S})]} \qquad [A.21]$$

and the substrate and cell yields at steady state are:

$$\tilde{S} = \frac{DK_S(1 + K_p S_0)}{\mu_m + D(K_S K_p - 1)} \qquad [A.22]$$

and,

$$\tilde{X} = \frac{D(S_0 - \tilde{S})Y_{x/s}}{D + mY_{x/s}} \qquad [A.23]$$

ENZYME PRODUCTION

Total RNA is a function of the specific growth rate. The amount of specific RNA (abbreviated hence as r) is dependent on the transcription efficiency, the amount of repressor-free operators and the efficiency of transcription initiation with RNA polymerase. The mRNA balance equation can be expressed as:

$$\frac{dr}{dt} = q_{MAX} N_p f(\mu) q - k_{rd}\, r - \mu r \qquad [A.24]$$

q_{MAX} is the specific rate of mRNA synthesis and is multiplied by N_p, the number of plasmids per cell at a specified dilution rate. $f(\mu)$ and q are two efficiency factors. Ponzo et al. (1986) multiplied the maximum specific growth rate by the two effectiveness factors to account for the effects due to induction and repression. $f(\mu)$ embodies the notion that the transcriptional capacity varies with specific growth rate. k_{rd} is the first order decay constant; Ponzo determined through independent experiments that k_{rd} was equal to 2.55 hr^{-1}.

Jeong et al. (1990) developed a detailed mechanistic model for *B. subtilis*. The hypothesis that GTP may be a signal for differentiation leads to the incorporation of purine nucleotide synthesis in the model. The present model does predict the onset of sporulation; further, the authors feel that the ability to predict cellular differentiation will eventually lead to more accurate predictions of exoenzyme synthesis.

Reiterating briefly, Van Dedem and Moo Young (1973) assumed the interactions between corepressor, such as glucose, and a cytoplasmic repressor, such as some DNA binding protein are at equilibrium. q is the number of operators which are bound to the corepressor-repressor complex divided by the total number of operators. They obtained the following expression for q:

$$q = \frac{1 + K_1 s^n}{1 + K_1 s^n + K_1 K_2 P_t s^n} \qquad \text{[A.25]}$$

where s is the intracellular concentration of the corepressor. Maltose is converted to phosphorylated glucose, the corepressor. The amount of corepressor is assumed to be proportional to the maltose concentration in the media and the cellular growth rate. At higher growth rates and higher maltose concentrations the amount of maltose the cell transports and converts to phosphorylated glucose is higher; thus, the repression level is higher.

P_t is the cytoplasmic repressor, n is the number of binding sites; the number of binding sites was set equal to one. As discussed earlier, Laoide et al. (1989) determined the site for catabolite repression was a cis-acting site downstream of the promoter. K_1 and K_2 are equilibrium constants for the interactions of repressor and corepressor and repressor/corepressor and operator, respectively.

MALTOSE UPTAKE

Maltose is transported across the cytoplasmic membrane via chemiosmotically coupled active transport. Transport across the cytoplasmic membrane normally alters most sugars. However it appears that maltose is not altered until it is within the cell, where it is converted to glucose-6-phosphate or glucose-1-phosphate, which is subsequently converted to glucose-6-phosphate (Schartz, 1987). Utilization of maltose by the cells proceeds via hydrolysis of maltose to glucose by α-glucosidase (maltase).

Yoon et al. (1989) found during their investigation of α-amylase production, that an increased uptake rate of maltose and the subsequent fast conversion to glucose could cause repression of the extracellular enzymes. The maltose regulon cannot be induced in the presence of glucose, due to inducer exclusion and catabolite repression.

Maltose is transported by active transport as is lactose; thus, a similar equation can be developed for intracellular maltose concentration (s) based on the bulk concentration, (S), of maltose.

$$\frac{ds}{dt} = \frac{AS}{B + S} - \mu s \qquad [A.26]$$

at steady state,

$$s = \frac{A\tilde{S}}{B + \tilde{S}}\mu^{-1} \qquad [A.27]$$

Wei et al. (1990) describe a mathematical model for continuous and fed-batch fermentations of *Bacillus subtilis*, where the specific rates for key processes, such as cell growth, substrate consumption and metabolite production are dependent solely on the concentration of the limiting substrate. The effects of dynamic interactions are illustrated by numerical simulations and experimental results.

EFFICIENCY FACTOR

The efficiency factor, f(μ), takes into account the increase in transcription with respect to a decrease in growth rate. Freese et al. proposed that GTP may suppress sporulation directly through a binding protein (1979). The mechanism developed by Ray (1985) for the catabolite modulating factor is similar to the repression of transcription initiation with respect to a repressor molecule or the GTP molecule, which is referred to in the model as R.

$$P + RNAP \overset{k_1}{\Leftrightarrow} OPN \qquad [A.28]$$

$$R + RNAP \overset{k_2}{\Leftrightarrow} RNAP.R \qquad [A.29]$$

P is the promoter, RNAP is RNA polymerase, R is the molecule responsible for repression and OPN is the open promoter complex where transcription is initiated. $f(\mu)$ is equal to the fraction of promoters where transcription is initiated divided by the maximum fraction of promoters that can be initiated.

$$k_1 = \frac{[OPN]}{[P]\,[RNAP]} \qquad [A.30]$$

$$k_2 = \frac{[RNAP.R]}{[R]\,[RNAP]} \qquad [A.31]$$

At steady state the total number of promoters and the total number of RNA polymerase molecules is constant.

$$P_t = P + OPN \qquad [A.32]$$

$$RNAP_t = RNAP + RNAP.R \qquad [A.33]$$

$$f(\mu) = \frac{f}{f_{max}} \qquad [A.34]$$

$$f = \frac{OPN}{P_t} \qquad [A.35]$$

$$[P] = \frac{[P]_t(k_2[R] + 1)}{k_1[RNAP]_t + k_2[R] + 1} \qquad [A.36]$$

$$[OPN] = \frac{[P]_t k_1 [RNAP]_t}{k_1[RNAP]_t + k_2[R] + 1} \qquad [A.37]$$

$$[RNAP] = \frac{[RNAP]_t}{k_2[R] + 1} \qquad [A.38]$$

$$[RNAP.R] = \frac{[RNAP]_t k_2[R]}{k_2[R] + 1} \qquad [A.39]$$

The eqns. [A.36] through [A.39] can be substituted into the equation for f so that f is represented with constants and R.

$$f = \frac{P_t k_1 RNAP_t}{P_t(k_1 RNAP_t + k_2 R + 1.0)} \qquad [A.40]$$

$$K_A = k_1 RNAP_t \qquad [A.41]$$

$$K_B = k_2 \qquad [A.42]$$

If R is equal to zero at $f = f_{max}$, then:

$$f(\mu) = \frac{1 + K_A}{K_A + K_B R + 1} \qquad [A.43]$$

The concentration GTP is proportional to the growth rate, (Jeong et al., 1990). Jeong measured the changes in concentration of GTP as a function of growth rate in *B. subtilis*; as growth rate decreased, the level of GTP decreased. Considering this hypothesis and a first order decay of GTP the following equation can be used to express the rate of synthesis:

$$\frac{dR}{dt} = k_R \mu - k_{-R} R - \mu R \qquad [A.44]$$

and at steady state,

$$R = \frac{k_R D}{k_{-R} + D} \qquad [A.45]$$

At steady state the concentration of mRNA is:

$$r = \frac{q_{MAX} N_p f(\mu) q}{D + k_{rd}} \qquad [A.46]$$

Equations [A.43] and [A.45] are combined, with the result:

$$f(\mu) = \frac{K'_A(k_{-R} + D)}{K'_A(k_{-R} + D) + K'_B D} \qquad [A.47]$$

It follows that the expression for amylase is:

$$\frac{de}{dt} = k_e r - r_e - \mu e \qquad [A.48]$$

where e is the intracellular α-amylase concentration. r_e is the rate of transport of the enzyme through the cell membrane. However, as was discussed earlier, the amount of intracellular amylase is negligible. Therefore the extracellular enzyme concentration, E, is described using the following equation:

$$\frac{dE}{dt} = k_e r - \mu E = r_e \qquad [A.49]$$

And at steady state one can derive the extracellular concentration of amylase, E, expressed as:

$$E = \frac{k_e \cdot r}{D} \qquad [A.50]$$

where k_e is a translational coefficient.

PLASMID COPY NUMBER

Ataai and Shuler determined through experimentation and model comparison that plasmid copy number is predicted best by the random mode of replication (1987), for a plasmid with a ColE1 origin of replication. Further, the model incorporates equal partitioning of the plasmid at replication. In their nomenclature, RNAI is the binding protein which inhibits RNAII from initiating replication. Based on independent experiments for the transcription and binding of RNAI and RNAII, the values for the parameters were estimated.

Similarly, an equation for plasmid pC194 might be developed based on the concentration of Rep which initiates plasmid replication. However, the amount of information is incomplete and parameter estimation on that basis would be impossible at this time. Alternatively, an empirical equation is used to represent plasmid copy number variations. Through experimentation it was observed that the plasmid copy number increases monotonically with increasing growth rate. As seen in Fig. A.8, the following equation adequately describes the plasmid copy number:

$$N_P = N_{P\,max} \cdot \left(\frac{1}{1 - e^{k_{p1} D + k_{p2}}} \right) \quad [A.51]$$

where $N_{P\,max}$ is the maximum plasmid copy number, which is equal to 40. k_{p1} and k_{p2} are constants.

Table A.1 Parameter Values (Feed Maltose Concentration of 2 g/L)

Parameters	Values	Parameters	Values
K_s	0.996 g/L	K'_2	46.9 g/L
K_p	1.25 L/g	K'_A	1.18
K_e	2129	K'_B	16.75
A	84.6 g/L	K_{-R}	251
B	0.43 g/L	K_{p1}	19.24
K_1	3.14 L/g	K_{p2}	-1.662

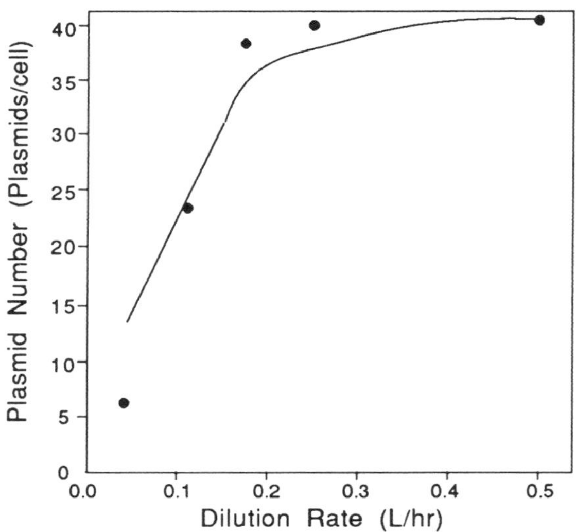

Figure A.8 Plasmid number versus dilution rate for free cell chemostat, model and data results.

Values of individual parameters are listed above in Table A.1 for the case of lower feed-maltose concentration. The model simulates the data in Figs. A.4 to A.8, already described, indicating that the molecular mechanisms accurately represent the kinetics. A crucial element is the inclusion of the inducer transport equation, as signified above by the parameters A and B. Without this element, the simulations fail.

A.2 NOMENCLATURE

a	particle radius, cm.
D	dilution rate, hr^{-1}.
e	intracellular α-amylase concentration, g/L.
$[E]$	β-galactosidase specific activity, units/g cells.
$[E \cdot X_F^+]$	β-galactosidase concentration, units/L
$f(\mu)$	transcriptional effectiveness factor, α-amylase.
g	number of generations.
G_I	immobilized cell doubling time, hr.
k_i	inhibition constant, L/g.
k_{rd}	decay constant for mRNA, hr^{-1}.
K_A, K_B	dimensionless constants.
K_1, K_2	promoter binding constants.
m	maintenance coefficient.
$[mRNA \cdot X_F^+]$	mRNA concentration, mg/L
N_p	plasmid copy number.
P_i	concentration of inhibitory product, g/L.
q	fraction of repressor-free operators, α-amylase.
q_{max}	specific rate of mRNA synthesis, mg/L/hr, α-amylase.
r	specific mRNA concentration, mg/L.
s	intracellular maltose concentration, g/L.
S	substrate concentration, g/L.
$\tilde{S}$	steady state bulk substrate concentration, g/L.
S_F	liquid phase substrate concentration in the microenvironment of a cell, g/L.
S_{in}	intracellular lactose concentration, g/L.
S_o	inlet substrate concentration, g/L.
X	cell concentration, g/L.
Y_{pi}	product yield coefficient.
Y_x	biomass yield coefficient.

SUBSCRIPTS

F	free (liquid) phase.
I	immobilized phase.

SUPERSCRIPTS

+	recombinant (plasmid-containing).
-	wild-type (plasmid-free).

GREEK SYMBOLS

α	ratio of wild-type to recombinant cell growth rates.
ε	liquid fraction.
η_I	ratio of IMC to free cell growth rate $[\mu_I / \mu_F]$.
θ	probability of obtaining plasmid-free segregant.
μ	specific growth rate, hr^{-1}.
μ_m	maximum specific growth rate, hr^{-1}.
σ	standard deviation.
ξ_I	substrate diffusion effectiveness factor.
ϕ	recombinant cell fraction.

A.3 REFERENCES

Ataai, M. M. and M. L. Shuler, Paper, AIChE Ann. Mtg., Chicago, November (1985).

Ataai, M. M. and M. L. Shuler, *Biotechnol. Bioeng.*, **30**, 389-397 (1987).

Bailey, J. E., M. Hjortso and F. Srienc, *Ann. N. Y. Acad. Sci.*, Biochem. Eng. III, **413**, 71 (1983).

De Crombrugghe, B., S. Busby and H. Bus, *Science*, **224**, 831 (1984).

Dessein, A., F. Tillier and A. Ullmann, *Molec. Gen. Genet.*, **162**, 89 (1978).

Freese, E., J. E. Heinze and E. M. Galliers, *J. Gen. Microbiol.*, **115**, 193-205 (1979).

Guidi-Rontani, C., A. Danchin and A. Ullmann, *Proc. Natl. Acad. Sci. USA,* **77**, 5799 (1980).

Jeong, J. W., J. Snay and M. M. Ataai, *Biotechnol. Bioeng.*, **35**, 160-184 (1990).

Kaushik, K. R., W. R. Vieth and Y. S. Jiang, *J. Memb. Sci.*, **30**, 39-56 (1985).

Laoide, B. M., G. H. Chambliss and D. McConnell, *J. Bacteriol.*, **171**, 2435-2442 (1989).

Lee, S.B. and J.E. Bailey, *Biotechnol. Bioeng.*, **26**, 1372 (1984).

Lee, S.B. and J.E. Bailey, *Biotechnol. Bioeng.*, **26**, 1383 (1984).

Ponzo, J., F. Keller, S. J. Parulekar and W. A. Wiegand, *Ann. N. Y. Acad. Sci.*, **469**, 617-625 (1986).

Ray, N. G., Ph.D. Thesis in Chemical and Biochemical Engineering, Rutgers Univ. (1985).
Ray, N. G., W. R. Vieth and K. Venkatasubramanian, *Ann. N.Y. Acad. Sci., Biochem. Eng. IV*, **469**, 212 (1986).
Ray, N. G., W. R. Vieth and K. Venkatasubramanian, *Biotechnol. Bioeng.*, **29**, 1003 (1987).
Schartz, M., Escherichia coli and Salmonella typhimurium, *ASM*, 1482-1498 (1987).
Shuler, M. L., S. Leung and C. C. Dick, *Ann. N. Y. Acad. Sci.*, **326**, 35-55 (1979).
Van Dedem, G. and M. Moo Young, *Biotechnol. Bioeng.*, **15**, 419-439 (1973).
Wanner, B. L., R. Kodaira and F. C. Neidhardt, *J. Bacteriol.*, **136**, 947 (1978).
Wei, D., S. J. Parulekar and W. A. Weigand, *Ann. N. Y. Acad. Sci.*, Biochem. Eng. VI, **589**, 508-258 (1990).
Yoon, M. Y., Y. J. Yoo and T. W. Cadman, *Biotech. Lett.*, **11**, 57-60 (1989).

APPENDIX B

B.0 LIGHT TRANSPORT MODEL

In Chapter 10, eqn. [10.25] expresses the photosynthetic electron transport rate as a function of local light intensity and therefore represents the microscopic rate. When thylakoid membranes are immobilized in a film of reconstituted collagen, it is necessary to account for the variation in local light intensity within the film in order to model the overall, or macroscopic, electron transport rate observed for the entire film. The local light intensity within the photosynthetically useful wavelength range (400 nm $< \lambda >$ 700nm) will decrease with increasing distance (depth) from the film surface due to the progressive absorption of photons as the latter pass through the photocatalytically active film.

Due to the nature of photoassisted catalysis, the process of light transport through the film is uncoupled from the kinetics of the photosynthetic reaction. Therefore, a mathematical description of the light intensity profile within the photocatalytic film can be obtained without consideration of reaction kinetics.

Light impinging upon the surface of a film of immobilized thylakoid membranes can be absorbed, reflected, or transmitted. Due to the 'rough' nature of the collagen film surface, reflection of radiation from the surface is expected to be nearly isotropic or diffuse; that is, the reflected radiation is distributed approximately uniformly in all directions (as opposed to specular, or regular, reflection from a polished surface). Similarly, internal reflection, or scattering, within the collagen-thylakoid film results in scattering of the incident light in all directions. It should be noted that scattering of light within the film does not change the wavelength of the incident photons. Scattered photons are therefore just as likely to be absorbed by light-harvesting pigment molecules as are photons which have undergone no scattering. By increasing the effective path length of the light through the photocatalytic film, scattering actually increases the probability that a photon is absorbed.

An exact mathematical description of the light intensity profile for diffuse (non-collimated) polychromatic light impinging upon a heterogeneous catalytic film would be a very complex function of: (1) the concentrations of the absorbing and scattering species present in the film; (2) the effective absorption and scattering cross sections of these species; (3) the thickness of the film; and (4) the wavelengths of the incident photons (the absorption and scattering cross sections both vary with wavelength).

It will be assumed that the fraction of the incident light which is absorbed is proportional to the thickness of the film traversed and the concentration of light-absorbing pigment molecules, c. If I(x) is the light intensity at some depth, x, in the film, the decrease in intensity in traversing a thickness dx is:

$$-dI(x) = aI(x)\,cdx \qquad [B.1]$$

where a is the effective absorption coefficient (or effective absorption cross section) for the catalytic film. Upon integration, eqn. [10.26] yields:

$$I(x) = I_i\,e^{-acx} \qquad [B.2]$$

where I_i is the incident light intensity at the film surface (x = 0).

This expression is formally similar to the Bouguer-Lambert-Beer Law (Smith, 1977) for the absorption of light by solutions containing light-absorbing solutes. The latter law is rigorously valid only under the following conditions: (1) the incident light is monochromatic and collimated; (2) the absorbing medium is homogeneous (i.e., no light scattering); and (3) the absorbing species act independently of one another. For diffuse, polychromatic light impinging upon a heterogeneous collagen-thylakoid film, the light intensity profile described by eqn. [B.2] is obviously an approximation. Such an equation could certainly not be used in this system for quantitative chemical analysis by absorption spectrophotometry. However, as a model describing light attenuation in the collagen-thylakoid film, eqn. [B.2] incorporates the significant system parameters and is a very useful approximation.

It is worth re-emphasizing that the light intensity profile within the collagen-thylakoid film is uniquely determined by the physical properties of the film itself and is not influenced by the kinetics of the photocatalyzed reactions. By using otherwise identical films of varying thickness, it would be possible to experimentally determine the actual light transmission as a function of film thickness and wavelength using a spectrophotometer. For the determination of *total* light transmitted, the detector (photomultiplier tube) would have to be provided with a large window close to the sample in order to collect as much of the scattered light as possible. Such experimental data could be used either to justify the validity of the functional form of eqn. [B.2] or to develop an alternative expression to describe light attenuation in the film.

COMBINED LIGHT TRANSPORT-KINETIC MODEL

Substitution of the expression for the light intensity profile within the film (eqn. [B.2]) into the rate equation for photosynthetic electron transport (eqn. [10.25]) yields the following:

$$r_e(x) = \frac{V_m I_i e^{-acx}}{K_I + I_i e^{-acx}} \qquad [B.3]$$

Equation [B.3] gives the microscopic electron transport rate as a function of position in the film, where x is the distance measured from the film surface (x=0) in a direction perpendicular to the plane of the film and I_i is the incident light intensity at the surface of the film. As previously discussed, the light transport model and the kinetic model can be combined in this manner due to the fact that light transport through the film is independent of the kinetics of the photosynthetic electron transport reactions. This uncoupling of light transport from the reaction kinetics greatly simplifies the mathematical description of the system.

Consider a collagen-thylakoid film of thickness L exposed to a uniform incident light intensity, I_i, at the film surface (Fig. B.1). The macroscopic reaction rate for the photocatalytic film can be obtained by integrating the microscopic rate (eqn. [B.3]) with respect to x between the limits x=0 and x=L:

$$R_m = \int_0^L r_e(x)\,dx = \int_0^L \frac{V_m I_i}{K_I \exp(acx) + I_i}\,dx \qquad [B.4]$$

where R_m is the macroscopic photosynthetic electron transport rate. Evaluation of the definite integral in eqn. [B.4] yields the following expression for R_m:

$$R_m = \frac{V_m}{ac} \ln\left[\frac{I_i + K_I}{I_i \exp(-acL) + K_I}\right] \qquad [B.5]$$

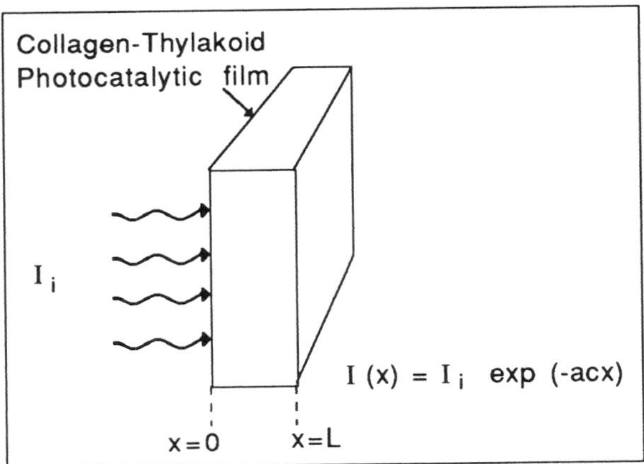

Figure B.1 Schematic representation of collagen-thylakoid photocatalytic film. L = film thickness; I_i = incident light intensity.

Equation [B.5] represents the integrated light transport-kinetic model for the collagen-thylakoid film and can be used for the design and analysis of systems in which this photocatalyst is used. The macroscopic electron transport rate, R_m, is expressed in units of moles of electrons per unit area of illuminated film surface per unit time (one mole of electrons is defined as one faraday). From eqn. [10.14] it is seen that one mole of oxygen is evolved for every four moles of electrons transferred through PETS. Thus the macroscopic O_2 evolution rate is numerically equal to one-fourth the macroscopic electron transport rate.

It should be noted that increasing the concentration of thylakoid membranes in the photocatalytic film changes two of the model parameters: V_m and c. As defined in eqn. [10.24], $V_m = k_D P_0$. Therefore, the macroscopic rate (eqn. B.5]) can be written:

$$R_m = \frac{k_D P_0}{ac} \ln\left[\frac{I_i + K_I}{I_i \exp(-acL) + K_I}\right] \qquad [B.6]$$

The factor P_0/c in eqn. [B.6] represents the ratio of the molar concentration of PS II (or PS I) reaction-center chlorophyll molecules to the total molar concentration of light-absorbing pigment molecules in the collagen-thylakoid film. For a given preparation of thylakoid membranes, the molar ratio of reaction-centers to total light-harvesting pigment molecules is fixed. Therefore the ratio P_0/c appearing in eqn. [B.6] is determined by the inherent characteristics of the thylakoid membrane itself and cannot be changed by increasing or decreasing the concentration of thylakoid membranes in the collagen-thylakoid film.

The molar ratio P_0/c is related to the photosynthetic unit size (U) already discussed. The quantity U was defined as the molar ratio of antenna chlorophyll molecules to reaction-center chlorophyll molecules in the photosynthetic unit. In terms of molar concentrations,

$$U = \frac{Chl_{antenna}}{2P_0} \qquad [B.7]$$

where the factor of two in the denominator accounts for the fact that the total reaction-center chlorophyll concentration is twice the concentration of PSII (or PS I) reaction-centers.

If the total concentration of light-absorbing pigment molecules (c) is proportional to the concentration of antenna chlorophyll molecules ($Chl_{antenna}$), then, from eqn. [B.7], the ratio c/P_0 is proportional to the photosynthetic unit size U. Therefore, the factor P_0/c in eqn. [B.6] is inversely proportional to U. This relationship between the two variables P_0 and c should be noted when using the mathematical model of eqn. [B.6]. Unless efforts to selectively control the size of the photosynthetic unit (U) are successful, the ratio P_0/c can be considered a constant which is characteristic of a given thylakoid membrane preparation, and is not subject to control in designing the photocatalytic system.

While increasing the concentration of thylakoid membranes in the collagen-thylakoid film (i.e., the catalyst loading) does not alter the P_0/c ratio, it does increase the concentration of light-absorbing pigment molecules (c) in the film. As eqn. [B.6] indicates, increasing c increases the value of the expression in brackets, thus increasing the macroscopic rate, R_m.

The macroscopic rate of photosynthetic electron transport, R_m, has units of moles $e^-/cm^2 \cdot s$ and is therefore a measure of the areal productivity of the photocatalytic film. The thickness (L) of the photocatalytic film and the concentration of light-absorbing pigment molecules (c) in the film are not as fundamentally significant as the areal productivity (R_m). The same macroscopic rate, R_m, can be obtained with essentially an infinite number of combinations of film thickness and pigment concentration.

An important requirement for an applied system is that the product of film thickness (L) and pigment concentration (c) be great enough to insure essentially complete absorption of photosynthetically active radiation within the film, so that none of the incident light is wasted. For large values of cL, the macroscopic electron transport rate (eqn. [B.6]) reduces to the following asymptotic expression:

$$R_m = \frac{k_D P_0}{ac} \ln\left[\frac{I_i}{K_I} + 1\right] \qquad [B.8]$$

Under these conditions, R depends only on the incident light intensity (I_i) and the inherent properties of the thylakoid membranes immobilized in the collagen film (k_D, P_0/c, a and K_I).

OPTICAL PROPERTIES OF RECONSTITUTED COLLAGEN FILM

From the standpoint of biochemical process engineering, the properties of the support material itself often play a critical role in determining the usefulness of the immobilized biomaterial for a particular application. Such carrier properties as durability, permeability to substrate and product, dimensional stability, resistance to microbial degradation, and capacity of the carrier for the biomaterial to be immobilized are all very important. In addition to these considerations affecting carrier suitability, the optical properties of the support matrix are of concern in the present application. When immobilizing photosynthetic membranes isolated from chloroplasts, the ability of the support material to transmit radiation in the photosynthetically active wavelength region (400-700 nm) is an important consideration.

When radiation in the visible portion of the spectrum falls on thylakoid membranes immobilized in a support matrix, the incident light may be reflected, absorbed or transmitted. Only incident radiation which is actually absorbed by the light-harvesting pigment molecules will result in photochemical reaction and product formation. Any absorption of light by the photocatalytically inactive support material does not lead to product formation and only reduces the energy conversion efficiency of the system. A support matrix used to immobilize photosynthetic membranes should therefore be non-absorbing in the wavelength range from 400 to 700 nm.

Figure B.2 shows the percent transmission of incident radiation through a reconstituted collagen film as a function of wavelength over the photosynthetically active spectral range. The film is cast from a 1.5% (w/v) suspension of dialyzed collagen and dried by evaporation in air at room temperature (using the procedures already described, but omitting chloroplast thylakoids). The thickness of the air-dried film is

51 μm (0.002 in.). The film is swollen in distilled water and placed against the inner face of a 1-cm glass cuvette filled with distilled water. The thickness of the hydrated film is approximately 100 μm (0.004 in). Transmission of the incident radiation varys from slightly less than 60% at a wavelength of 400 nm to over 70% at 700 nm (Fig. B.2).

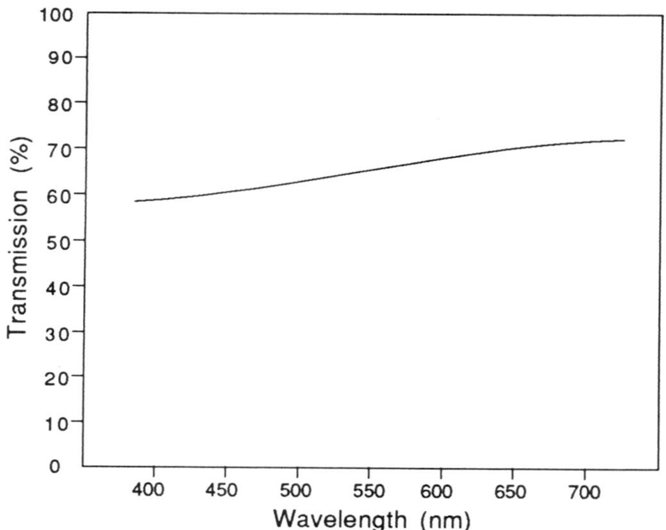

Figure B.2 Transmission of visible radiation through a collagen film as a function of wavelength.

Less than 100% light transmission indicates that some of the incident light is either reflected or absorbed by the collagen film. The fact that there are no peaks or valleys in the visible transmission spectrum indicates the absence of any strongly absorbing species in the film. The dried collagen film is colorless and appears fairly transparent to the eye. After swelling in distilled water, the film became translucent. These observations indicate that the 30-40% loss in measured light transmission is due to scattering of incident light by swollen collagen fibrils in the film and not to absorption by the film. (When light is scattered so widely that it is not intercepted and measured by the photomultiplier tube of the spectrophotometer, a decrease in transmitted light is observed.) The increase in percent transmission with increasing wavelength (Fig. B.2) can be attributed to the strong inverse dependence of light scattering on wavelength.

It should be noted that scattered radiation is as effective in promoting the photosensitized catalytic reactions of PETS as is direct radiation. The effect of the collagen support matrix is to cause scattering of photons with a change in direction only (i.e., with no loss in energy). For photosynthetic membranes immobilized in such a light scattering matrix, the result is to increase the effective path length of photons as they traverse the film, thereby increasing the probability of their absorption by light-harvesting pigment molecules. However, scattering of light from the film surface back in the direction from which the incident light originated is undesirable. Such backscattering represents a reflection loss and may reduce the attainable energy conversion efficiency of the system. The photocatalytic film should be designed so that as much of the incident light as possible is absorbed by the pigments present in the thylakoids.

The ability of collagen to transmit visible light is well established. The organic material of the transparent cornea of the vertebrate eye is almost pure collagen (Lehninger, 1975), while the vitreous humor of the eye contains a collagen matrix. The use of collagen membranes in corneal implant surgery (Dunn et al., 1967) and the replacement of diseased vitreous humor in the eye with collagen gels (Hamed and Rodriquez, 1975) further demonstrate the excellent light transmission capabilities of collagen. The transparency of gelatin (prepared by thermal denaturation of collagen) permits its widespread use in photographic film emulsions.

The optical properties of reconstituted collagen films are influenced by the purity of the collagen used, the size and arrangement of collagen fibrils in the film, the film thickness, and the texture of the film surface. Unlike the hyaloid collagen membrane found in the cornea of the vertebrate eye, collagen films prepared by the methods described are translucent when swollen in water. In their native state, the collagen fibrils of bovine hide collagen form an interlacing network laid down in sheets (Lehninger, 1975). As the reconstituted films used in this work are prepared from comminuted bovine hide collagen, it is likely the fibrils are randomly oriented in the films, resulting in the observed light scattering. In the transparent cornea, the collagen fibrils have very narrow cross sections and lie parallel to one another (White et al., 1973). This suggests that improvements in the light transmission properties of reconstituted collagen films might be made by using suspensions of smaller diameter collagen fibrils and developing film-forming methods capable of orienting the fibrils in a parallel arrangement.

EFFECT OF PHOTOCATALYST CONCENTRATION ON OBSERVED ACTIVITY

The concentration of the biocatalyst (thylakoid membranes) in the assay vessel cannot be varied without altering the light intensity distribution within the reaction mixture. A linear dependence of reaction rate on chlorophyll concentration is predicted by the kinetic model of photosynthetic electron transport:

$$r_e = \frac{V_m I}{K_I + I} \tag{10.25}$$

where $V_m = k_D P_0$.

For a given preparation of thylakoids, the concentration of light-absorbing pigment molecules, c, will be related to the total concentration of PS II or PS I reaction-center chlorophyll molecules, P_0, by some constant of proportionality. Therefore, the photochemical electron transport rate, r_e, can be expressed as:

$$r_e = \frac{k_D' c I}{K_I + I} \tag{B.9}$$

where $V_m = k'_D\, c$.

The influence of light intensity gradients within the assay vessel is reduced to a minimum by vigorous mixing of the reaction mixture during the initial-rate measurements. The reaction vessel in which the polarographic oxygen electrode is inserted is specifically designed for intense agitation. The 5-ml reaction volume is agitated with a built-in magnetic stirring system at a stirrer speed of 480 rpm. For such a well-stirred suspension, the effective light intensity actually 'seen' by the thylakoid membranes can be approximated by a mean light intensity, $\bar{I}$, defined by the following equation (Rabe and Benoit, 1962):

$$\bar{I} = \frac{\int_0^d I(x)\, dx}{\int_0^d dx} \tag{B.10}$$

where I(x) is the light intensity at some point within the assay vessel and d is the characteristic path length of the light passing through the thylakoid suspension. Assuming the validity of eqn. [B.2] as a model describing one-dimensional light attenuation in the assay vessel:

$$I(x) = I_i\, e^{-acx} \qquad \text{[B.2]}$$

The mean light intensity in the reaction vessel is given by:

$$\bar{I} = \frac{\int_0^d I_i\, e^{-acx}\, dx}{\int_0^d dx} \qquad \text{[B.11]}$$

$$\bar{I} = \frac{I_i\,(1 - e^{-acd})}{acd} \qquad \text{[B.12]}$$

The mean light intensity depends not only on I_i and c, but also upon the geometry of the assay vessel. For a rectangular assay vessel, the distance d would represent the width of the vessel. For the cylindrical assay vessel used in the present work, d represents the average effective light path.

It is realized that individual thylakoid membranes in the stirred suspension are actually exposed to a time-varying light intensity depending on their position within the assay vessel. A more realistic mathematical description of the system would require incorporation of a stochastic, kinematic model of thylakoid motion in the light transport-kinetic model. Such a refinement has been proposed by Sheth et al. (1977) for the description of algal photosynthesis in turbulent channel flow.

$$r_e = \frac{k_D' c \bar{I}}{K_I + \bar{I}} \qquad \text{[B.13]}$$

$$r_e = \frac{k_D' c I_i}{K_I \left[\dfrac{acd}{1 - e^{-acd}} \right] + I_i} \qquad \text{[B.14]}$$

Eqn. [B.14] represents the combined light transport-kinetic model applicable to the assay of stirred suspensions of thylakoid membranes.

It expresses the effect of incident light intensity on the observed reaction rate. For small values of cd, eqn. [B.14] reduces to the following expression:

$$r_e = \frac{k_D' c I_i}{K_I + I_i} \qquad [B.15]$$

For large values of cd, the rate (eqn. [B.14]) approaches:

$$r_e = \frac{k_D' c I_i}{K_I\, acd + I_i} \qquad [B.16]$$

From consideration of eqn. [B.14] and its two asymptotic forms at low and high chlorophyll concentrations (eqns. [B.15] and [B.16], respectively) it is possible to predict the conditions under which a linear dependence of reaction rate on chlorophyll concentration is to be expected. The observed reaction rate (μmol O_2/ml·h) will be proportional to the chlorophyll concentration (μg Chl/ml) if either or both of the following conditions are met: (1) the entire sample is light-saturated (i.e., the rate is zero order with respect to incident light); (2) the entire sample is uniformly exposed to approximately equal light intensities as the chlorophyll concentration is varied. The second condition is met in an optically thin suspension (i.e., low values of cd) in which only a small fraction of the incident light is absorbed by the sample. The non-linear rate-concentration behavior shown in Fig. B.3 indicates that under the assay conditions employed, neither of the above two situations prevailed.

The variation in the observed reaction rate with chlorophyll concentration (Fig. B.3) seems qualitatively to fit that of a rectangular hyperbola, as predicted by eqn. [B.16]. In view of the complexities of the optical arrangement used in the assay system (i.e., an externally illuminated cylindrical assay vessel) and the approximations used in deriving the above mathematical model (eqn. [B.14]), the qualitative agreement between the observed behavior and the model is noteworthy.

These results clearly demonstrate the importance of accounting for the influence of the photocatalyst concentration on the microenvironmental light intensity when interpreting the observed kinetics of a photoassisted catalytic reaction.

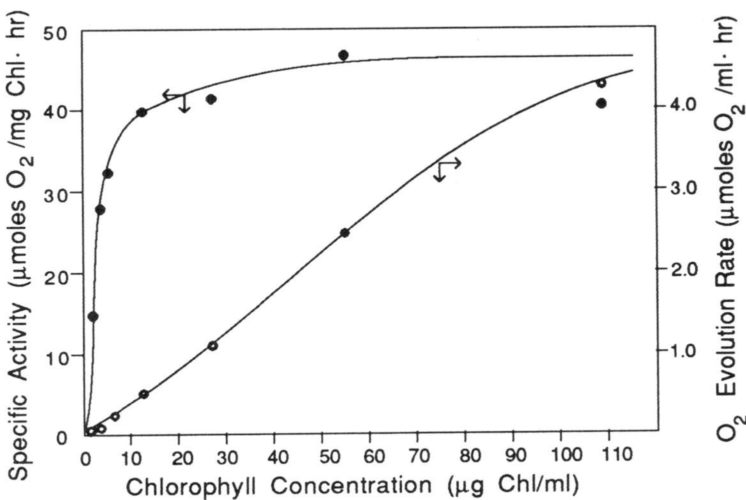

Figure B.3 Effect of chlorophyll concentration on the observed electron transport activity of suspensions of chloroplast thylakoid membranes. Assay temperature: 25°C. Incident light intensity: $1x10^6$ ergs/$cm^2 \cdot$ s (100 mW/cm^2).

EFFECT OF PHOTOCATALYST CONCENTRATION ON OBSERVED ACTIVITY OF ISOLATED THYLAKOID MEMBRANES

As previously discussed, the dependence of the observed electron transport rate on the photocatalyst concentration can be explained in terms of a combined light transport-kinetic model (eqn. [B.14]) for the stirred batch reactor assay system used in these studies. The effect of photocatalyst concentration (measured as total chlorophyll) on the electron transport activity of suspensions of isolated chloroplast thylakoid membranes is then analyzed. Reaction rates are measured as oxygen evolution accompanying electron transport from water to potassium ferricyanide.

As shown in Fig. B.3, the oxygen evolution rate is nearly proportional to the chlorophyll concentration within the range 13-60 μg Chl/ml. Consequently, the specific activity remains relatively constant within this region. Combined with information on the effect of incident light intensity on the electron transport rate, these results indicate that the assay system is light-saturated for chlorophyll concentrations below 60 μg Chl/ml.

The rapid decline in specific activity observed at low chlorophyll concentrations (less than 13 μg Chl/ml) is likely due to photoinhibition of electron transport under conditions in which the thylakoids are exposed to high light intensities. As eqn. [B.12] indicates, the mean light intensity in the reaction vessel increases with decreasing chlorophyll concentration and approaches the incident light intensity in the case of very dilute thylakoid suspensions. Under high illumination, isolated thylakoid membranes are unable to dissipate all of the excitation energy of absorbed quanta by normal photosynthetic electron flow. In such instances, some of the excitation energy is thought to be dissipated via pathways which cause irreversible photochemical damage to the electron transport apparatus (Kok, 1956; Lien and San Pietro, 1975). The data of Fig. B.3 are consistent with the interpretation that photoinhibition occurs at low chlorophyll concentrations.

MASS TRANSFER CONSIDERATIONS

Water serves as the substrate for all electron transport reactions studied in this work. As all reactions were carried out in aqueous solutions, it can be concluded that the mass transport of substrate had no influence on the observed reaction kinetics.

In developing the mathematical model describing the kinetic behavior of the photosynthetic electron transport system, it was assumed that there is always an excess of the terminal electron acceptor present. That is, the electron transport rate is zero order with respect to the concentration of the electron acceptor. It was not the intent of this work to conduct a detailed investigation of the possible influence of external and internal mass transport of the electron acceptor on the kinetic behavior of immobilized thylakoid membranes. However, a number of considerations support the conclusion that the assay system employed was free from complications due to diffusional effects.

Initially, there was some concern that the electron acceptor (potassium ferricyanide) used to assay the electron transport activity of immobilized thylakoid membranes might bind to the collagen support matrix, thereby reducing its concentration in the reaction mixture to less than saturating levels. To test this possibility, complete reaction mixtures containing 3 mM $K_3Fe(CN)_6$ were incubated for 30 minutes with 50 mg of blank collagen film containing no thylakoids. After removal of the collagen film, the reaction mixture was used to assay the Hill activity of a chloroplast thylakoid suspension. The activities were the same as those measured in assays using reaction mixtures which had not been pre-incubated with a blank collagen film. Such experiments demonstrated that any binding of ferricyanide to collagen which may have occurred was not sufficient to reduce the free ferricyanide concentration to non-saturating levels.

Under the vigorous stirring conditions employed in the assay system, it was assumed that external diffusional resistances to ferricyanide transport had no significant influence on the observed reaction kinetics. The possibility still exists that diffusion of ferricyanide within the collagen-thylakoid film may limit the observed electron transport rate. In experiments in which the initial ferricyanide concentration was varied over a ten-fold range, from 1.5 mM to 15 mM $K_3Fe(CN)_6$, no significant variation in electron transport activity was observed for collagen-thylakoid films prepared by the freeze-drying technique. In published studies of the effect of ferricyanide concentration on the Hill activity of broken chloroplasts (Lumry et al., 1954; Katoh and San Pietro, 1966), the rate of photoreduction was found to be independent of ferricyanide concentration within the range of 0.005 to 1.0 mM $K_3Fe(CN)_6$. Films of collagen-immobilized chloroplast thylakoids prepared by freeze-drying have a very open sponge-like fibrous structure. The bulk density of a typical film prepared by this method was measured to be 0.05 g/cm^3 film in the dry state. For such a macroporous support, it is unlikely that internal diffusional resistances would reduce the concentration of ferricyanide in the vicinity of the immobilized thylakoids to such low levels that the observed electron transport rate is affected. To do so would require more than a 600-fold reduction of the microenvironmental ferricyanide concentration in the film, relative to the bulk concentration used in these studies (3 mM $K_3Fe(CN)_6$).

The manner in which the photochemical assays were conducted also helps to eliminate internal mass transport influences. The complete reaction mixture is pre-equilibrated in the dark for approximately 5 minutes prior to initiating the reaction by turning on the light. During this dark incubation period, no photoreduction of ferricyanide occurs. It is therefore assumed that a uniform concentration of the electron acceptor is established throughout the collagen-thylakoid film during this period. For a bulk ferricyanide concentration of 3 mM $K_3Fe(CN)_6$, it is reasonable to assume that the equilibrated concentration within the film is well in excess of that required to saturate PETS. Subsequent initiation of the reaction by turning on the light allows initial reaction rates to be obtained free of significant internal mass transfer effects. A light intensity gradient within the photocatalytic film is established almost instantaneously when illumination begins. In the absence of any electron acceptor concentration gradient, initial variation in rate with position in the film is due to a non-uniform light intensity and not to diffusional limitations. In fact, as long as the intrafilm ferricyanide concentration remains rate-saturating throughout the film, the observed reaction kinetics will not be influenced by internal mass transport even in the presence of concentration gradients. That is, the rate of mass transport of the electron acceptor to the photocatalytic site within the collagen-thylakoid film does not disguise the intrinsic kinetics of the reaction during the period that initial rates are measured. The fact that oxygen evolution remains linear with time during the first 2 to 3 minutes of illumination provides a good indication that any changes in the microenvironmental ferricyanide concentration during this time interval did not affect initial rates. Mass transport of the oxygen evolved will result in a slight time delay in the product concentration measured by the oxygen electrode, but this does not affect the initial rate determination.

EFFECT OF LIGHT INTENSITY ON OBSERVED ACTIVITY OF FREE AND IMMOBILIZED THYLAKOIDS

In evaluating a photocatalyst for potential use in an applied solar energy conversion system, the behavior of the catalyst over a wide range of incident light intensities is of obvious interest. Experiments were also conducted to determine the effect of incident light intensity on the observed activity of free and immobilized chloroplast thylakoid membranes.

Incident light intensities were varied from 0 to 117 mW/cm^2 (0-1.17 x 106 $ergs/cm^2 \cdot s$). It is noted that the solar irradiance varies from 0 to 93 mW/cm^2, measured at sea-level for an air-mass value of 1 and a solar angle of 90°C (Hollaender, 1956). The spectral distribution of the radiation from the incandescent lamp used to illuminate the reaction vessel differs, of course, from that of natural sunlight. Therefore, the observed behavior under artificial illumination cannot be equated to expected results under the same intensities of solar radiation.

The photovoltaic cell used to measure light intensities was sensitive to photons in the wavelength range 400-1000 nm. Thus the incident intensities were measured over a wavelength region considerably broader than the photosynthetically active region (400-700 nm). Solar radiation covers a very broad spectrum (Jagger, 1977). The use of a broad-spectrum light source was considered desirable in order to subject the photocatalytic preparations to light conditions more closely approximating those which would be encountered in an applied biophotolytic system. Accordingly, the spectral output of the incandescent lamp was not narrowed using filters (other than the water shield, or 'heat filter,' used to limit heating by infrared radiation, as discussed).

A film of chloroplast thylakoid membranes immobilized in collagen was prepared using the freeze-drying technique described earlier. Seven ml of a chloroplast thylakoid suspension containing 3.54 mg Chl/ml in G-R (II) medium were added to 80 ml of a dialyzed collagen suspension containing 1.5% (w/v) collagen. After thorough mixing with a glass stirring rod at 0°C, 0.174 ml of 5% (w/v) glutaraldehyde in 50 mM potassium phosphate buffer, pH 6.8, was slowly added to the collagen-thylakoid suspension with stirring. The concentration of glutaraldehyde in the final mixture was 0.01% (w/v). The mixture was maintained at 0°C for 20 minutes prior to casting and freeze-drying.

The electron transport activity of the collagen-thylakoid film was assayed as a function of incident light intensity. Results presented in Figure B.4 were obtained after storage of the collagen-thylakoid film at -20°C for 24 days.

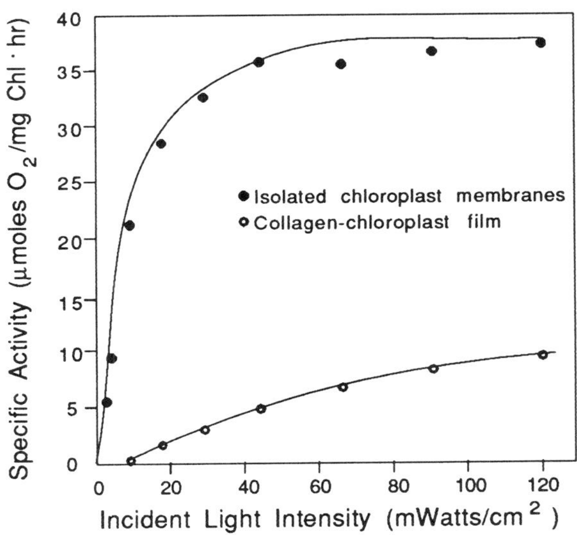

Figure B.4 Specific activity of free and immobilized chloroplast thylakoid membranes as a function of incident light intensity. Initial rates of O_2 evolution were measured, using ferricyanide as the electron acceptor. Reaction mixtures (4 ml) for assay of collagen-thylakoid film samples had the same composition, with the exception that 50mg of collagen-thylakoid film were added in place of free thylakoid membranes.

The effect of light intensity on the Hill activity of a suspension of free (non-immobilized) thylakoid membranes is shown in Fig. B.4 for comparison. The observed behavior is in agreement with the light transport-kinetic model (eqn. [B.14]) developed to describe the kinetics of a stirred suspension of thylakoid membranes:

$$r_e = \frac{k_D P_0 I_i}{K_I\left[\dfrac{acd}{1-e^{-acd}}\right] + I_i} \qquad [B.17]$$

where $V_m = k_D P_0 = k'_D c$.

From eqn. [B.7], the total concentration of Photosystem II (or Photosystem I) reaction-center chlorophyll molecules (P_0) can be related to the concentration of antenna chlorophyll molecules (C_{chl}) and the photosynthetic unit size (U) by the following expression:

$$P_0 = \frac{C_{chl}}{2U} \qquad [B.18]$$

Since there is one PS II reaction-center (P680) and one PS I reaction-center (P700) per electron transport chain, P_0 in eqn. [B.17] also represents the total molar concentration of photosynthetic electron transport chains in the system.

As previously discussed, the size of the photosynthetic unit, U, is defined as the molar ratio of light-harvesting chlorophyll molecules to reaction-center chlorophyll molecules. Based on the classical experiment of Emerson and Arnold (1932), the photosynthetic unit size for oxygen evolution in the green alga *Chlorella* is 300 chlorophyll molecules per reaction-center. Subsequent measurements (Zankel and Kok, 1972; Clayton, 1977) indicate that each reaction-center is served by about 200 antenna chlorophyll molecules (U=200). The interpretation is that there is one complete electron transport chain for every 400 chlorophyll molecules in the thylakoid membrane.

Utilizing this information, an expression can be obtained for the observed specific oxygen evolution rate based on chlorophyll content (μmol O_2/mg Chl·h). Substituting eqn. [B.18] into eqn. [B.17] and re-arranging yields the following expression for the observed specific activity as a function of incident light intensity:

$$\frac{r_e}{C_{chl}} = \frac{\left(\dfrac{k_D}{2U}\right) I_i}{K_I\left[\dfrac{acd}{1 - e^{-acd}}\right] + I_i} \qquad [B.19]$$

The maximum reaction rate, V_m, can now be expressed as:

$$V_m = k_D P_0 = \left(\frac{k_D}{2U} \right) C_{chl} \qquad [B.20]$$

The quantity ($k_D/2U$) represents the maximum specific activity observed under light-saturating conditions.

As shown by the data in Fig. B.4, at low light intensities the specific activity is nearly proportional to the incident light intensity; the rate is light-limited (pseudo-first order region). At high intensities, the specific activity asymptotically approaches a maximum value ($k_D/2U$); the rate is light-saturated (pseudo-zero order region).

It was observed that below some low threshold value of incident light intensity (ca. 8 mW/cm^2), the immobilized thylakoids did not evolve oxygen. A possible explanation for this behavior is that thylakoid membranes immobilized near the film surface are inactive, perhaps due to more adverse conditions in this region of the film during the immobilization process. Because inactive thylakoids at the film surface would continue to absorb incident light, a threshold light intensity would be required for light penetration into the active interior of the film.

B.1 EVALUATION OF KINETIC PARAMETERS

Free thylakoids:

Estimates of the apparent kinetic parameters, (K_I(app), V_m(app) and k_D(app)/2U), for suspensions of non-immobilized thylakoid membranes were obtained by using the combined light transport-kinetic model of eqn. [B.19] to describe the reaction system.

To minimize the recognized limitations of parameter estimation using traditional graphical methods (e.g., Lineweaver-Burk plots, Eadie or Hanes plots, and Hofstee plots), the direct linear plotting technique of Eisenthal and Cornish-Bowden (1974) was used to analyze the kinetic data of Fig. B.4. Apparent parameter values determined by this method were:

$$\frac{k_D\,(\text{app})}{2U} = 39.2\ \frac{\mu\text{mol } O_2}{\text{mg Chl} \cdot \text{h}}$$

$$V_m\,(\text{app}) = 1.68\ \frac{\mu\text{mol } O_2}{\text{ml} \cdot \text{h}}$$

$$K_I\,(app) = 6.9\ mW/cm^2$$

A double-reciprocal plot of the data (Fig. B.4) is shown in Fig. B.5 with the predicted behavior using the above parameter estimates. Knowing the value of the maximum specific activity (k_D(app)/2U), the apparent reaction rate constant for the dark oxidation-reduction reactions, k_D(app), can be calculated. The ratio of chlorophyll *a* (M.W.=893.48) to chlorophyll *b* (M.W.=907.46) was determined by assay to be 2.5:1 for the thylakoid membrane preparation used in this experiment. Based on a weighted average molecular weight for chlorophyll of 897.47 and a photosynthetic unit size U=200 (i.e., 400 Chl:1 chain) the apparent reaction rate constant is calculated to be k_D(app) = 15.6 s^{-1}:

$$k_D = \left(39.2\frac{\mu mol\ O_2}{mg\ Chl \cdot h}\right)\left(\frac{897.47 \times 10^{-3} mg\ Chl}{1\ \mu mol\ Chl}\right)\left(\frac{400\ Chl}{1\ chain}\right)\left(\frac{4e^-}{1\ O_2}\right)\left(\frac{1\ h}{3600\ s}\right) =$$

$$(39.2)\ (0.399)\ s^{-1} = 15.6\ s^{-1}$$

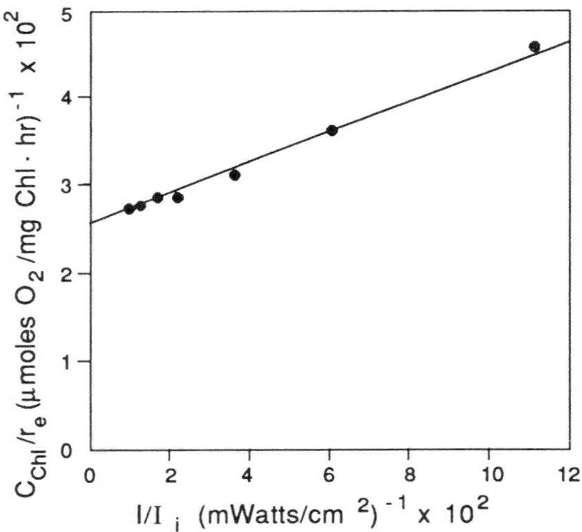

Figure B.5 Determination of apparent kinetic parameters for photosynthetic electron transport catalyzed by free thylakoid membranes. Data of Fig. B.4 are shown as a double-reciprocal plot. Values of apparent kinetic parameters: k_D(app)/2U=39.2 μmol O_2/mg Chl·h; V_m(app)=1.68 μmol O_2/ml·h; K_I(app)=6.9 mW/cm^2.

(Multiplication of the specific activity (μmol O_2/mg Chl·h) by 0.399 gives the turnover rate in s^{-1} for the conditions noted in the text, U=200.) The apparent turnover rate of the photosynthetic electron transport system in this thylakoid membrane preparation is therefore 15.6 electrons per electron transport chain per second.

The maximum specific activity observed under light-saturating conditions (39.2 μmol O_2/mg Chl·h) is based on the *total* chlorophyll content of the sample. Because some unknown fraction of the total chlorophyll content may be associated with inactive electron transport systems, the above estimate of the kinetic rate constant k_D (that is, the turnover rate) must be regarded as an observed or apparent value. To determine the intrinsic turnover rate would require knowledge of the actual concentration of active, functional electron transport chains present in the thylakoid membranes.

It is clear from eqn. [B.19] that the value of the lumped kinetic parameter (K_I) estimated by this procedure represents an apparent kinetic constant for the free thylakoid membranes. This apparent constant, K_I(app), is related to the intrinsic constant, K_I, by the expression:

$$K_I\,(\text{app}) = K_I\left[\frac{acd}{1 - e^{-acd}}\right] \qquad \text{[B.21]}$$

The bracketed quantity in the above expression (whose value is ≥ 1) accounts for the effect of light absorption by the reaction mixture on the observed kinetic behavior of the system. Only in the limit as acd approaches zero will K_I(app) equal K_I. The mathematical form of the bracketed expression is a consequence of the exponential attenuation law earlier assumed to model the light intensity profile in the reaction system. Alternative models to describe light attenuation would of course lead to different expressions relating K_I(app) and K_I.

For the chlorophyll concentration used in the experiment reported in Fig. B.4 (i.e., 43 μg Chl/ml) and the size of the reaction vessel (2 cm diameter), the absorption of incident light by the reaction mixture is expected to be significant. Schwartz (1972) has measured the percentage absorption of incident light in the wavelength range 640-730 nm for suspensions of spinach chloroplasts. Measurements were made with a spectrophotometer equipped with an integrating sphere accessory to avoid complications due to light scattering. For a chloroplast suspension containing 40 μg Chl/ml (comparable to the concentration used here) and a 1-cm light path, the percent absorption varied from 71% at

640 nm to 3.9% at 730 nm, with a maximal absorption of 96.6% at 680 nm. Due to the complexities in the lamp-reactor configuration used in these studies, a valid estimate of the fraction of incident light actually absorbed by the reaction mixture ($1 - e^{-acd}$) could not be made. If the value of acd is known, the intrinsic kinetic parameter K_I can be determined. The essential point of eqn. [B.21] is that the apparent kinetic parameter estimated from observed reaction kinetics is systematically larger than the intrinsic value by a factor which depends on the optical properties of the reaction system. If, for example, 50% of the incident light had been absorbed by the reaction mixture ($1 - e^{-acd}) = 0.5$; acd = 0.693), the bracketed expression in eqn. [B.21] would have a value of 1.39. In this instance, a value of 5.0 mW/cm^2 would be calculated for the intrinsic kinetic constant (K_I) based on the apparent value of 6.9 mW/cm^2 estimated above.

Immobilized thylakoids:

Estimates of the apparent kinetic parameters of immobilized thylakoid membranes were obtained using the same procedure described in the preceding section for free thylakoids. Apparent parameter values determined by the direct linear plotting technique of Eisenthal and Cornish Bowden (1974) were:

$$\frac{k_D\,(\text{app})}{2U} = 42.1\ \frac{\mu\text{mol } O_2}{\text{mg Chl} \cdot \text{h}}$$

$$V_m\,(\text{app}) = 6.79\ \frac{\mu\text{mol } O_2}{\text{ml} \cdot \text{h}}$$

$$K_I\,(\text{app}) = 389\ \text{mW/cm}^2$$

A double-reciprocal plot of the data (Fig. B.4) is shown in Fig. B.6 with the behavior predicted using the above parameter estimates. Assuming 400 chlorophyll molecules per electron transport chain (U=200), the apparent reaction rate constant is:

$$k_D(\text{app}) = 16.8\ \text{s}^{-1}$$

Table B.1 presents a comparison of the apparent kinetic parameters for free and immobilized thylakoid membranes. The maximum reaction rates (V_m) are reported per ml of fluid volume in the

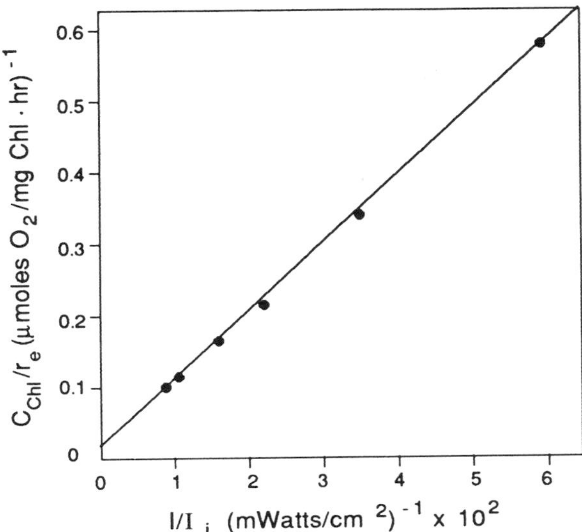

Figure B.6 Determination of apparent kinetic parameters for photosynthetic electron transport catalyzed by a film of immobilized thylakoids. Data of Fig. B.4 are shown as a double-reciprocal plot. Values of apparent kinetic parameters: k_D(app)/2U=42.1 µmol O_2/mg Chl·h; V_m(app)=6.79 µmol O_2/ml·h; K_I(app)=389 mW/cm^2.

batch reaction mixture. The four-fold increase in V_m(app) of the immobilized thylakoids compared to free thylakoids is the result of the difference in chlorophyll concentrations used in the activity assays (Fig. B.4). The apparent maximum turnover rates under light-saturating conditions (k_D(app)) are nearly the same for the 12-hour old suspension of thylakoid membranes and the 24-day old collagen-thylakoid film. This demonstrates that the immobilization procedure developed in this work does not, in itself, cause a drastic decrease in the ability of the thylakoid membranes to catalyze dark oxidation-reduction reactions of the photosynthetic electron transport chain. During the course of these investigations, the most active preparation of free chloroplast thylakoid membranes had a specific activity of 61 µmol O_2/mg Chl·h measured 2 h after isolation from spinach, using potassium ferricyanide as the electron acceptor and NH_4Cl as an uncoupler of photophosphorylation. This specific activity corresponds to an apparent turnover rate of 24.3 s^{-1}. The maximum turnover rate of the immobilized thylakoids reported above (16.8 s^{-1}) is equal to 69% of this value.

Table B.1 Comparison of Apparent Kinetic Parameters for Free and Immobilized Chloroplast Thylakoid Membranes

Parameter	Units	Free thylakoids	Immobilized thylakoids
$\frac{k_D(app)}{2U}$	$\frac{\mu mol\ O_2}{mg\ Chl \cdot h}$	39.2	42.1
$k_D(app)$	s^{-1}	15.6	16.8
$V_m(app)$	$\frac{\mu mol\ O_2}{ml \cdot h}$	1.68	6.79
$K_I(app)$	mW/cm^2	6.9	389

The 56-fold increase in K_I(app) for immobilized thylakoids compared to free thylakoids can be attributed to the significantly different light intensity profiles which prevailed in the two reaction systems. The kinetic model upon which the parameter estimates are based (eqn. [B.19]) assumes that all the thylakoid membranes are exposed to the same space-averaged light intensity (eqn. [B.12]). This is equivalent to assuming complete mixing of the batch-tank reactor in which the activity assays were conducted. While this assumption may be justified for vigorously stirred suspensions of free thylakoid membranes, it is not valid for the situation in which collagen-thylakoid film chips are suspended in the reaction mixture. In the latter case, individual chloroplast membranes remain fixed in position relative to one another within the collagen film. Even though the film chips are agitated during assay, the light intensity necessarily varies with position in the film. Neither the light transport-kinetic model derived for a static film of immobilized thylakoids (eqn. [B.6]) nor the model developed for a completely mixed suspension of thylakoid membranes (eqn. [B.19]) is entirely adequate for describing the behavior of a stirred suspension of collagen-thylakoid film chips. The large increase in the value of K_I(app) for immobilized thylakoids indicates the expected existence of significant light intensity gradients in the collagen-thylakoid film. The light attenuation properties of the collagen-thylakoid film would have to be fully characterized in order to quantitatively account for the effect of the non-uniform light distribution on the observed reaction kinetics.

The significant increase in K_I(app) of immobilized thylakoids compared with free thylakoids may be due to factors in addition to the influence of light attenuation within the photocatalytic film. The intrinsic kinetic constants which characterize the immobilized thylakoid membranes may, in fact, significantly differ from those for non-immobilized thylakoids. Alterations in the thylakoid membrane structure during the immobilization procedure may result in light-harvesting pigment molecules which have become energetically uncoupled, either partially or totally, from an active reaction-center. An increase in the de-excitation rate constant (k_d) or a decrease in the excitation rate constant (k_e) would result in increased values for the lumped kinetic constant K_I (eqn. [B.7]).

As Fig. B.4 indicates, the observed specific activity of collagen-immobilized thylakoid membranes is significantly less than the specific activity of free thylakoids in suspension under identical light intensities. From the preceding analysis, it is clear that the observed decrease in specific activity is not due to inactivation of dark redox reactions by the immobilization procedure, but rather to significant light attenuation within the collagen-thylakoid film.

From an applications point of view, the light intensity profile across the photocatalytic film will have a profound effect on the solar energy conversion efficiency of the system. The photosynthetic apparatus of green plants saturates at light intensities corresponding to approximately 10-20% of full sunlight intensity. This rate saturation at low intensities precludes the use of solar concentrators to directly increase reaction rates by increasing the irradiance incident on the photocatalytic film surface. To utilize light at maximum efficiency, the entire photocatalytic film should operate at intensities well below light-saturating levels (i.e., in the pseudo-first order region).

Efficient use of incident solar radiation, rather than efficient use of the immobilized photocatalyst, would be of major concern in an applied biophotolysis system. Instead of maximizing specific electron transport activities based on chlorophyll content (i.e., μmol O_2/mg Chl·h), an applied system should seek to maximize the productivity based on a unit area of illuminated collector surface.

The data of Fig. B.4 are plotted in Fig. B.7 in units which express the photosynthetic electron transport rate per unit liquid volume in the stirred batch reactor (μmol O_2/ml·h). If the illuminated surface area is taken as the longitudinal cross-sectional area of the cylindrical reaction vessel (liquid height x diameter), then the volumetric production rates can be converted to areal productivities by multiplying the former by a constant factor which represents the ratio of the reactor volume to the cross-sectional area. For the reaction vessel employed here (inside diameter 2.0 cm), this ratio ($\pi D/4$) has a constant value of 1.57 cm. An applied system would be designed to insure complete absorption of all photosynthetically active radiation. In comparing the areal productivities (Fig B.7) of free and immobilized thylakoids, it must be noted that the total chlorophyll concentration present in the assay of the free thylakoids (43 μg Chl/ml) was approximately one-fourth of that present in the assay of immobilized thylakoids (161 μg Chl/ml). Higher areal productivities could be achieved by increasing the photocatalyst loading to the point of complete light extinction. This was attained in a study by Delachapelle et al. (1991a; b), who studied hydrogen production by a photosynthetic bacterium, *Rhodobacter capsulatus*. Maximum hydrogen yield, 85 mL/l-L for 2000 h, was observed at a bacterial dilution rate of 0.04/h with 5mM glutamate. But higher bacterial concentrations also caused a decrease in hydrogen production due to self-shielding from

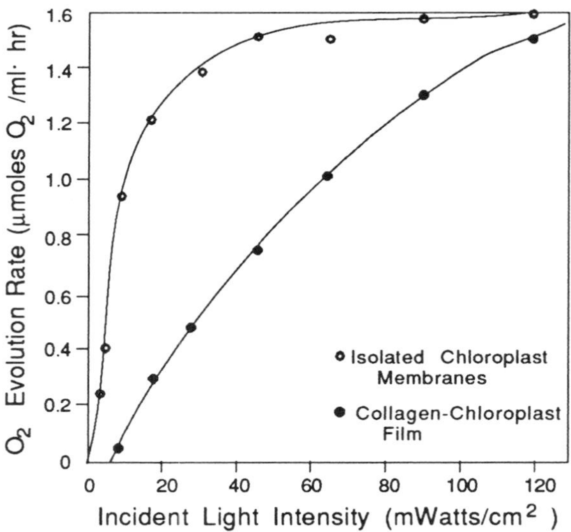

Figure B.7 Oxygen evolution rates of free and immobilized chloroplast thylakoid membranes as a function of incident light intensity. See Fig. B.4 legend for assay conditions. For the assay system used, an estimate of the areal production rate (μmol $O_2/cm^2 \cdot h$) is obtained by multiplying the oxygen evolution rate (μmol O_2/ml·h) by the factor 1.57 cm, which takes into account the ratio of reactor volume to illuminated cross-section.

light, and led to a change in metabolism from photosynthetic to fermentative. Under fermentative conditions, lactate was converted to formate, acetate, etc., and no hydrogen was produced.

The continuous production of hydrogen in a nozzle loop reactor was investigated using calcium alginate-immobilized *Rhodospirillum rubrum* KS-301, with glucose as the growth-limiting substrate (Lee et al., 1989). In the continuous reactor, the hydrogen production rate and residual glucose concentration were increased with increase of input glucose concentration, dilution rate and recycle rate. The maximum production rate under partial recycle conditions was 91 mL/L-h at a dilution rate 0.4/h, with input glucose concentration set at 5.4 g/L. Treat et al. (1990) studied the growth and biomass characteristics of photobioreactor culture of photosynthetic soybean cells. They found that a second growth phase, prolonging cell viability, could be initiated by increasing the CO_2 level from 2% to 5%, just before the onset of stationary phase. The biomass contained approximately 44% cellulosics and 14.5% protein. Sasaki and Hashimoto (1991) showed that odoriferous components, especially hydrogen sulfide, can be efficiently removed from the aqueous phase in a closed type aquarium by treatment with immobilized photosynthetic bacteria, *Rhodopseudomonas palustris* and *Rhodobacter spaeroides denitrificans*.

PHOTOCATALYTIC FILM

The technique developed in the present study to immobilize active thylakoid membranes within a reconstituted collagen support matrix represents an enabling technology which has several potential advantages. Immobilization provides the means for retaining the photocatalyst within a continuous flow reactor (Howell, 1981). The collagen-thylakoid film can be continuously contacted with substrate (water) and any other substance (such as an intermediate redox carrier) which might be needed to couple the photocatalytic component to a catalyst system capable of evolving hydrogen. Immobilization permits a degree of flexibility in reactor design not attainable with suspensions of free (non-immobilized) thylakoid membranes. Thus, thylakoid membranes immobilized in thin films or sheets can be arranged in a variety of physical configurations, thereby facilitating the development of practical reactor systems employing these photocatalytic surfaces.

Using immobilized thylakoid membranes, it is possible to design a two-stage biophotolytic reactor system in which the sites of oxygen and hydrogen production can be physically separated from one another (Egan and Scott, 1978). This is a very important consideration in view of the known oxygen-sensitivity of hydrogenase enzymes and the need to separate H_2 from O_2 prior to storage of the hydrogen fuel.

The ability to prepare immobilized thylakoid membranes in the form of thin sheets or films (as opposed to pellets or beads, for example) makes this immobilization method particularly well-suited for potential solar energy applications. The thylakoid loading of the film and the film thickness are easily controlled variables, permitting fabrication of photocatalytic films having desired light intensity profiles.

Although glutaraldehyde has been demonstrated to have a stabilizing effect on the electron transport capabilities of thylakoid membranes, it is also inhibitory (Zilinskas and Govindjee, 1976; Hardt and Kok, 1976). An advantage of the immobilization method developed here is that the thylakoid membranes are exposed to a glutaraldehyde concentration of only 0.01% (w/v). Even with this low concentration of glutaraldehyde, the final cross-linked collagen-thylakoid matrix which is formed upon lyophilization possesses very good mechanical properties. Fragile thylakoid membranes can thus be irreversibly immobilized under fairly mild conditions. It was found that omitting the glutaraldehyde resulted in freeze-dried films which quickly disintegrated when immersed in substrate solution and subjected to shear stresses in the magnetically stirred reactor vessel.

Selection of a glutaradehyde concentration of 0.01% (w/v) was made, following experiments in which the concentration of glutaraldehyde was varied from 0.01% to 1.0%. Results of these experiments demonstrated a continuous decline in specific activity (μmol O_2/mg Chl·h) of the collagen-thylakoid films as the glutaraldehyde concentration was increased. A concentration of glutaraldehyde as low as 0.01% (w/v) produced films having good mechanical strength. This concentration was therefore used in all subsequent work in which isolated thylakoids were immobilized in collagen using the freeze-drying procedure. Due to the simultaneous effects of inactivation and stabilization of thylakoid membranes by glutaraldehyde (Zalinskas and Govindjee, 1976; Hardt and Kok, 1976; Cocquempot et al., 1981), the immobilization procedure could likely be optimized with respect to the glutaraldehyde concentration employed.

A particularly noteworthy feature of collagen-thylakoid films prepared using the freeze-drying technique is their very open, sponge-like fibrous structure. The bulk density of the film preparation used in the present reactor study was estimated to be 0.049 g/cm^3 film based on the measured dimensions of a dry film sample and the film weight. It was assumed that the density of the solid phase in the film was approximately equal to that of dry collagen. Using a dry collagen density of 0.35 g/cm^3, reported by Fels (1964), the void volume of the collagen-thylakoid film is calculated to be 20 cm^3/g film and the void fraction of the film is 0.96 (96% of the dry film consists of void space and 4% is occupied by the solid phase). Properties of the collagen-thylakoid film are summarized in Table B.2.

Table B.2 Properties of Collagen-Thylakoid Film[a]

Weight:	0.0521 g
Thickness:	1.0 mm
Chl content:	12.8 mg Chl/g film
Bulk density:	0.049 g/cm^3
Void volume:	20 cm^3/g film
Void fraction:	0.96
Size of film chips:	*ca,* 0.4 cm square

[a]All values are for a dry film sample.

The very open, macroporous nature of the film is a direct consequence of the fabrication method used. The film was prepared by casting a collagen-thylakoid suspension containing less than 2% (w/v) solids, followed by freezing and lyophilization. Very large pores were observed to penetrate from one surface of the dried film to the other. These continuous macropores in the freeze-dried film correspond to regions occupied by bulk water when the aqueous collagen-thylakoid suspension was cast and then frozen.

The open-pored structure of the composite insured good contact of substrate solution with the immobilized thylakoid membranes. The macroporous matrix of the film was observed to offer little hydraulic resistance to fluid flow. As a result, the major mass transport mechanism in film chips suspended in substrate solution in a vigorously stirred reactor is likely to be convective mass transfer by bulk fluid flow through the network of macropores in the film.

When a dried film sample was immersed in substrate solution (pH 6.8) for 1 h, the film thickness increased from 1.0 mm (dry) to 1.8 mm (wet) with no significant change in the lateral dimensions (length x width). This 80% increase in bulk film volume upon wetting is the result of a swelling of the cross-linked collagen-thylakoid matrix. The void fraction for the swollen film will, of course, differ from that estimated for the dry film. The observation that the collagen-thylakoid phase constitutes only 4% by volume of the dry film, however, makes it reasonable to assume that only a small protion of the 80% increase in bulk volume seen upon immersing the film in substrate solution can be attributed to sorption of water by the collagen-thylakoid phase itself. Most of the volume increase is due to the absorption of bulk water by the sponge-like matrix. Therefore, a very large fraction of the swollen film consists of void space occupied by substrate solution. The macroporous structure of films formed by the freeze-drying technique thus provides excellent contact between substrate solution and the immobilized biocatalyst.

B.2 REFERENCES

Clayton, R. K., In "Chlorophyll - Proteins, Reaction Centers, and Photosynthetic Membranes," J. M. Olson and G. Hinds, Eds. Brookhaven Symposia in Biology No. 28, pp. 1-15 (1977).

Cocquempot, M. F., B. Thomasset, J. N. Barbotin, G. Gellf and D. Thomas, *Eur. J. Appl. Microbiol. Biotechnol.*, **11**, 193 (1981).

Delachapelle, S., M. Renaud and P. M. Vignais, *Rev. Sci. Eau*, **4**, 83-99 (1991).

Delachapelle, S., M. Renaud and P. M. Vignais, *Rev. Sci. Eau*, **4**, 101-20 (1991).

Dunn, M. W., T. Nishihara, K. H. Stenzel, A. W. Branwood and A. L. Rubin, *Science*, **157**, 1329 (1967).

Egan, B. Z. and C. D. Scott, In "Biotechnology in Energy Production and Conservation," C. D. Scott, Ed., pp. 489-500, Biotechnol. Bioeng. Symp. No. 8. Wiley: New York (1978).

Eisenthal, R. and A. Cornish-Bowden, *Biochem. J.*, **139**, 715 (1974).

Emerson, A. and W. Arnold, *J. Gen. Physiol.*, **15**, 391 (1932).

Fels, I. G., *J. Appl. Polymer Sci.*, **8**, 1813 (1964).

Hamed, G. and F. Rodriquez, *J. Appl. Polymer Sci.*, **19**, 3299 (1975).

Hardt, H. and B. Kok, *Biochim. Biophys. Acta*, **449**, 125 (1976).

Hollaender, A., Ed. "Radiation Biology," Vol. III, p. 155. McGraw-Hill: New York (1956).

Howell, J. M., Ph. D. Thesis in Chemical and Biochemical Engineering, Rutgers University (1981).

Jagger, J., In "The Science of Photobiology," K. C. Smith, Ed., pp. 1-26. Plenum Press: New York (1977).

Katoh, S. and A. San Pietro, *J. Biol. Chem.*, **24**, 3575 (1966).

Kok, B., *Biochim. Biophys. Acta*, **21**, 234 (1956).

Lee, C. G., Y. H. Seon, J. W. Han, H. S. Lee and Y. I. Joe, *Sanop Misaengmul Hakhoechi*, **17**, 629-633 (1989).

Lehninger, A. L., "Biochemistry," 2nd Ed., p. 135; pp. 603-605. Worth: New York (1975).

Lumry, R., J. D. Spikes and H. Eyring, *Ann. Rev. Plant Physiol.*, **5**, 271 (1954).

Rabe, A. E. and R. J. Benoit, *Biotechnol. Bioeng.*, **4**, 377 (1962).

Sasaki, T. and G. Hashimoto, Jpn. Patent 03147726 A2, Treatment of aquarium water using immobilized photosynthetic bacteria (1991).

Schwartz, M., In "Methods in Enzymology," A. San Pietro, Ed. Vol. 24, pp. 139-146 (1972).
Sheth, M., D. Ramkrishna and A. G. Fredrickson, *A. I. Ch. E. J.*, **23**, 794 (1977).
Smith, K. C., "The Science of Photobiology," p. 346. Plenum Press: New York (1977).
Treat, W. J., J. Castillon and E. J. Soltes, *Appl. Microbiol. Biotechnol.*, **24-25**, 497-510 (1990).
White, A., P. Handler and E. L. Smith, "Principles of Biochemistry," p. 978; pp. 1007-1009. McGraw-Hill: New York (1973).
Zilinskas, B. A. and Z. Govindjee, *Pflanzenphysiol.*, **77**, 302 (1976).

POSTSCRIPT

One spring weekend, fishing on the Outer Banks of Cape Hatteras, trying for migrant bluefish, I happened to work my way down to the vicinity of a rather remarkable chap: a surfcaster who had but one leg, sitting just above the wash in a lawn chair, no one else in the neighborhood. "It's a great life!" he said as I approached. Enjoying the warm air and the clear skies which had taken the harsh edge off winter, I had to agree. Over the next few hours, fishing nearby, I saw him struggle with the ordinary techniques that are the surfman's stock in trade; for example, unable to stride into a cast, he was having difficulty reaching the slough beyond the wash where passing fish would be swimming in search of prey. "It's a great life," he said again, but privately, the other side of that old expression began to surface in my thoughts.

Next day, I retraced my path out near Hatteras Point, looking forward to seeing a fellow fisherman. There he was, just as before, flailing away at the suds. As I walked up I expected to go through our little ritual of the day before. Instead, as I arrived, his words of greeting were, "Damned if I'll weaken!" That's good enough for me!

AUTHOR INDEX

SUBJECT INDEX

RETURN TO → CHEMISTRY LIBRARY 8210
100 Hildebrand Hall 642-3753

LOAN PERIOD 1	2	3
7 DAYS	1 MONTH	
4	5	6

ALL BOOKS MAY BE RECALLED AFTER 7 DAYS
Renewable by telephone

DUE AS STAMPED BELOW

NON-CIRCULATING
UNTIL: JUL 21 1994

JAN 03 1995
APR 24 1995
SEP 01 1995
NOV 08 REC'D
DEC 10 1995
AUG 18 1997
APR 04 1999
DEC 16 1999
MAY 25 2002
MAY 24 2003
JUL 12 '04

JUN 14 1994

UNIVERSITY OF CALIFORNIA, BERKELEY
FORM NO. DD5, 3m, 12/80 BERKELEY, CA 94720 ®s